Collins

Cambridge IGCSE™

Biology

TEACHER'S GUIDE

Sue Kearsey, Mike Smith

William Collins' dream of knowledge for all began with the publication of his first book in 1819.

A self-educated mill worker, he not only enriched millions of lives, but also founded a flourishing publishing house. Today, staying true to this spirit, Collins books are packed with inspiration, innovation and practical expertise. They place you at the centre of a world of possibility and give you exactly what you need to explore it.

Collins. Freedom to teach.

Published by Collins
An imprint of HarperCollins*Publishers*
The News Building
1 London Bridge Street
London
SE1 9GF

HarperCollins*Publishers*
1st Floor
Watermarque Building
Ringsend Road
Dublin 4
Ireland

Browse the complete Collins catalogue at
www.collins.co.uk

ISBN 978-0-00-843087-0

British Library Cataloguing-in-Publication Data. A catalogue record for this publication is available from the British Library.

Updating author: **Mike Smith**
Author of previous edition: **Sue Kearsey**
In-house editor: **Letitia Luff**
Project manager: **Nivedhitha Souriyan**
Copyeditor and proofreader: **Sarah Binns**
Safety checker: **Joe Jefferies**
Internal designer, typesetter and illustrator:
  Jouve India Private Limited
Cover designer: **Gordon MacGilp**
Cover artwork: **Maria Herbert-Liew**
Production controller: **Lyndsey Rogers**
Printed and Bound in the UK by:
Ashford Colour Press Ltd

MIX
Paper from
responsible sources
FSC™ C007454

Contents

S indicates activities that cover the extended syllabus only and so may be omitted if students are not studying the extended syllabus. The core syllabus only includes core content. The extended syllabus includes both core and supplement content.

Introduction

Welcome to the Collins Cambridge IGCSE™ Biology Teacher's Guide, which has been written by experienced authors and teachers to help you deliver an effective and successful Cambridge IGCSE Biology course.

Overview

This **printed book** follows the Student Book topics and includes:

- **learning episode plans** with resource lists, learning objectives, and detailed guidance
- **practical activities** including teacher demonstrations and student experiments
- **worksheets** for students with instructions to follow and questions to answer
- **technician's notes** so **practicals and demonstrations** can be safely planned and executed
- **answers** to all the questions in the Student Book
- **a content overview** that maps all practical activities to the syllabus topics and learning objectives.

Downloadable files

Everything in the printed book is available in PDF and in MS Word format so that you can tailor lessons and activities. It also includes:

- **technician's notes** with lists of equipment, resources and set-up information for each section collated in one document that you can hand to your technician
- printable **end of topic checklists** that students can use to map their own progress
- one document with all the **Student Book answers** to in-text, end of topic and exam-style questions.

Learning episode plans

The Teacher's Guide has been matched section by section to the Student Book to allow easy cross-referencing between the books. Each section is introduced with a brief introduction, links to other topics and an overview of the learning episodes in an easy-to-navigate table. Learning episode plans give guidance on delivering learning episodes that can be combined for longer lessons to give you complete flexibility and control. Your school may wish to join an organisation such as CLEAPSS for health and safety support www.cleapss.org.uk.

Teachers should make sure that they do not contravene any school, education authority or government regulations. Responsibility for safety matters rests with schools. In addition, the following general advice should be given to students:

'Be careful with chemicals. Never ingest them and always wash your hands after handling them.'

Learning episode plans cover the following:

- **learning objectives** from the syllabus linked to the activities and the Student Book
- **learning aims** which reflect the intended outcomes of the consolidation and summary sessions.
- **common misconceptions** held by students, which may need addressing
- **resources** you will need for activities
- a detailed pick-up-and-teach approach for **differentiated**, varied tasks
- **demonstration** and **practical activity** technician's notes so you can easily check that you have everything you need
- **answers** to worksheets and all the questions in the Student Book.
- **career links** are included and can be used to enrich teaching by providing context and links to careers in different fields of biology. These highlight scientific careers in the area of biology being studied. Careers in many other fields use the knowledge and skills gained studying science.

Worksheets

Worksheets provide engaging activities for students. Those that deal with practical work give step-by-step guidance for students to ensure that they carry out their investigations safely and successfully.

All worksheets are available in editable format as well as in one printable document containing all the worksheets in each section.

Technician's notes

Detailed **technician's notes** are provided for every lesson, with lists of resources needed for practical activities and/or demonstrations, set-up instructions and safety notes.

For up-to-date safety information, which you will need to check before you set up any practical activity or demonstration, please refer to your employer's guidelines or visit: www.cleapss.org.uk

Technician's notes appear within the lesson plans for your reference, and as separate printable files to download.

Answers

Every learning episode plan in this book includes the **answers** to the associated questions in the Student Book and any worksheets. Answers to the **end of topic** and **exam-style questions** are provided. A document with all the answers collated for easy printing is available to download.

Exam-style questions and sample answers have been written by the authors. In examinations, the way marks are awarded may be different. References to assessment and/or assessment preparation are the publisher's interpretation of the syllabus requirements and may not fully reflect the approach of Cambridge Assessment International Education.

Checklists

Editable checklists that students can use to record their progress are provided to download. These have the same content as those at the end of each topic in the Student's Book.

Content overview

The detailed **content overview** at the back of this book provides a comprehensive overview of our Cambridge IGCSE course, matching each learning episode to the learning objectives in the syllabus. It also shows at a glance where the practical activities and demonstrations occur so that you can plan ahead. An editable version of the content overview is available to download.

Teachers should always refer back to the current syllabus when devising a scheme of work for students.

Science in context boxes

Please note that the **science in context** boxes in the Student Book put the ideas that students are learning into real-life context. It is not necessary for students to learn the content of these boxes as it is beyond the requirements of the syllabus. However, they do provide interesting examples of scientific applications that are designed to enhance their understanding. Some science in context boxes contain one or more questions which provide the opportunity to explore the content more deeply.

Teaching sequence

The course can be taught in syllabus order and it will work well to do so. If you want to change the teaching sequence for the course, you should keep in mind the considerations below. An example alternative teaching sequence has been provided.

- There should be a logical progression in the development of concepts and knowledge in order to aid students' understanding.
- The course should have a 'storyline', so that wherever possible the sequence can be easily explained to students. You can achieve this in different ways for this syllabus, as some sections on body systems are covered in relation to animals (generally humans) and others focus on flowering plants. You might approach this from the view of the systems, or from the view of the organisms by covering all systems relating to humans and then covering plant systems at a different time.
- You should consider carefully how to introduce the more difficult concepts. For example, you should ensure that they are not introduced too early in the course, or put in a sequence where they all come one after another. Either of these approaches could result in demotivation. There should be time to consolidate more difficult concepts in the planned sequence.
- Sections 19–21 depend on a wide range of earlier learning and are best left to the end of the course to make sure students can apply knowledge they have gained earlier, so that they can develop a better understanding of the new content.
- Maintain the emphasis on practical and investigative work throughout the course. Note that the investigative work in Section 19 involves fieldwork to collect organisms. This may only be practical at certain times of the year, and so should be planned accordingly. However, you should not consider this any earlier than the third term of the first year, in order to ensure that students have sufficient background knowledge to make best use of the investigations.
- You will need to allocate different amounts of teaching time to different sections, depending on levels of difficulty or extent of the content. The syllabus does not recommend teaching times, but an idea of the amount of learning covered in each activity can be gained from looking at the Approach section for each activity in the Teacher Pack.
- It is important to retain some flexibility so that difficult ideas can be revisited, or new ideas applied in different contexts.
- Some topics may need consideration of seasons and geographical location of teaching, for example growing seedlings may be better in the spring/summer rather than the winter.
- You will need time for revision at the end of the course. It is also recommended that you encourage students to revisit material for revision throughout the course. This could be done using the flashcards created at the end of each topic.

Alternative teaching sequence

This example alternative teaching sequence is based on a two-year course taught in six terms. It is mapped over five terms on the assumption that much of the sixth term is spent on revision. For different numbers of terms or semesters, where the course is started early, or where the time allocated each week is either generous or limited, you will need to adjust this sequence.

First year, 1st term

B1 Characteristics and classification of living organisms
B2 Organisation of the organism
B3 Movement into and out of cells
B4 Biological molecules
B5 Enzymes

First year, 2nd term

B7 Human nutrition
B9 Transport in animals
B11 Gas exchange in humans
B12 Respiration
B13 Excretion in humans

First year, 3rd term

B6 Plant nutrition
B8 Transport in plants
B14 Coordination and response
B15 Drugs
B19 Organisms and their environment

Second year, 1st term

B16 Reproduction
B17 Inheritance
B18 Variation and selection

Second year, 2nd term

B10 Diseases and immunity
B20 Human influences on ecosystems
B21 Biotechnology and genetic modification

Content overview

The information in this section is taken from the Cambridge IGCSE and IGCSE (9–1) Biology syllabuses (0610/0970) for examination from 2023. You should always refer to the appropriate syllabus document for the year of your examination to confirm the details and for more information. The syllabus document is available on the Cambridge International website at www. cambridgeinternational.org.

Learning episode	Learning objectives and learning aims Learning objectives are given which link to the syllabus. Learning aims reflect the intended outcomes of the consolidation and summary sessions.	Syllabus reference	Student Book pages	Practical activity

B1.1
Characteristics of living organisms
	Learning objective
	• Describe the characteristics of living organisms by describing:
	  (a) movement as an action by an organism or part of an organism causing a change of position or place
	  (b) respiration as the chemical reactions in cells that break down nutrient molecules and release energy for metabolism
	  (c) sensitivity as the ability to detect and respond to changes in the internal or external environment
	  (d) growth as a permanent increase in size and dry mass
	  (e) reproduction as the processes that make more of the same kind of organism
	  (f) excretion as the removal of the waste products of metabolism and substances in excess of requirements
	  (g) nutrition as the taking in of materials for energy, growth and development
		1.1	11–13	

B1.2 Concept and uses of classification systems	**Learning objectives** • State that organisms can be classified into groups by the features that they share • Describe a species as a group of organisms that can reproduce to produce fertile offspring • Describe the binomial system of naming species as an internationally agreed system in which the scientific name of an organism is made up of two parts showing the genus and species • Construct and use dichotomous keys based on identifiable features • **Supplement** Explain that classification systems aim to reflect evolutionary relationships • **Supplement** Explain that the sequences of bases in DNA are used as a means of classification • **Supplement** Explain that groups of organisms which share a more recent ancestor (are more closely related) have base sequences in DNA that are more similar than those that share only a distant ancestor	1.2	13–18	Activity: Dichotomous keys

B1.3
Features of organisms
	Learning objectives
	• State the main features used to place animals and plants into the appropriate kingdoms
	• State the main features used to place organisms into groups within the animal kingdom, limited to:
	  (a) the main groups of vertebrates: mammals, birds, reptiles, amphibians, fish
	  (b) the main groups of arthropods: myriapods, insects, arachnids, crustaceans
	• Classify organisms using the features identified above
	• **Supplement** State the main features used to place all organisms into one of the five kingdoms: animal, plant, fungus, prokaryote, protoctist
	• **Supplement** State the main features used to place organisms into groups within the plant kingdom, limited to ferns and flowering plants (dicotyledons and monocotyledons)
	• **Supplement** Classify organisms using the features identified above
	• **Supplement** State the features of viruses, limited to a protein coat and genetic material
		1.3	19–32	Activity: Classifying arthropods
				Activity: Classifying plants

Content overview

			Learning content	Practicals / Demonstrations
B1.4 Consolidation and summary	1	33–37	**Learning aims** • Review the learning points of the topic summarised in the end of topic checklist • Test understanding of the topic content by answering the end of topic questions	
B2.1 Cell structure and size of specimens	2.1	41–53	**Learning objectives** • Describe and compare the structure of a plant cell with an animal cell, limited to: cell wall, cell membrane, nucleus, cytoplasm, chloroplasts, ribosomes, mitochondria, vacuoles • Describe the structure of a bacterial cell, limited to: cell wall, cell membrane, cytoplasm, ribosomes, circular DNA, plasmids • Identify the cell structures listed above in diagrams and images of plant, animal and bacterial cells • Describe the functions of the structures listed above in plant, animal and bacterial cells • State that new cells are produced by division of existing cells • State that specialised cells have specific functions, limited to: (a) ciliated cells – movement of mucus in the trachea and bronchi (b) root hair cells – absorption (c) palisade mesophyll cells – photosynthesis (d) neurones – conduction of electrical impulses (e) red blood cells – transport of oxygen (f) sperm and egg cells (gametes) – reproduction • Describe the meaning of the terms: cell, tissue, organ, organ system and organism as illustrated by examples given in the syllabus • Calculate magnification and size of biological specimens using millimetres as units • State and use the formula: magnification = image size ÷ actual size Supplement Convert measurements between millimetres (mm) and micrometres (μm)	Practical: Using a light microscope Practical: Making a microscope slide
B2.2 Consolidation and summary	2	54–57	**Learning aims** • Review the learning points of the topic summarised in the end of topic checklist • Test understanding of the topic content by answering the end of topic questions	
B3.1 Diffusion	3.1	61–65	**Learning objectives** • Describe diffusion as the net movement of particles from a region of their higher concentration to a region of their lower concentration (i.e. down a concentration gradient), as a result of their random movement • State that the energy for diffusion comes from the kinetic energy of random movement of molecules and ions • State that some substances move into and out of cells by diffusion through the cell membrane • Describe the importance of diffusion of gases and solutes in living organisms • Investigate the factors that influence diffusion, limited to: surface area, temperature, concentration gradient and distance	Demonstration: Diffusion Practical: Diffusion across visking tubing (dialysis tubing) Practical: Temperature and the rate of diffusion Practical: Surface area and the rate of diffusion

Content overview

Topic	Learning objectives		Pages	Practical
B3.2 Osmosis	Learning objectives • Describe the role of water as a solvent in organisms with reference to digestion, excretion and transport • State that water diffuses through partially permeable membranes by osmosis • State that water moves into and out of cells by osmosis through the cell membrane • Investigate osmosis using materials such as dialysis tubing • Investigate and describe the effects on plant tissues of immersing them in solutions of different concentrations • State that plants are supported by the pressure of water inside the cells pressing outwards on the cell wall • Supplement Describe osmosis as the net movement of water molecules from a region of higher water potential (dilute solution) to a region of lower water potential (concentrated solution), through a partially permeable membrane • Supplement Explain the effects on plant cells of immersing them in solutions of different concentrations by using the terms: turgid, turgor pressure, plasmolysis, flaccid • Supplement Explain the importance of water potential and osmosis in the uptake and loss of water by organisms	3.2	66–73	Practical: Osmosis and visking tubing (dialysis tubing) Practical: Measuring osmosis by change in mass Practical: Osmosis and change in shape Practical: Water potential
B3.3 Active transport	Learning objectives • Describe active transport as the movement of particles through a cell membrane from a region of lower concentration to a region of higher concentration (i.e. against a concentration gradient), using energy from respiration • Supplement Explain the importance of active transport as a process for movement of molecules or ions across membranes, including ion uptake by root hairs • Supplement State that protein carriers move molecules or ions across a membrane during active transport	3.3	73–74	
B3.4 Consolidation and summary	Learning aims • Review the learning points of the topic summarised in the end of topic checklist • Test understanding of the topic content by answering the end of topic questions	3	75–77	

B4.1 Carbohydrates, fats and proteins	Learning objectives
	• List the chemical elements that make up: carbohydrates, fats and proteins
	• State that large molecules are made from smaller molecules, limited to:
	  (a) starch, glycogen and cellulose from glucose
	  (b) proteins from amino acids
	  (c) fats and oils from fatty acids and glycerol
		4	80–81	

B4.2 Supplement DNA	Learning objective
	• Describe the structure of a DNA molecule:
	  (a) two strands coiled together to form a double helix
	  (b) each strand contains chemicals called bases
	  (c) bonds between pairs of bases hold the strands together
	  (d) the bases always pair up in the same way: A with T, and C with G (full names are **not** required)
		4	84–85	

B4.3 Food tests	Learning objective
	• Describe the use of:
	  (a) iodine solution test for starch
	  (b) Benedict's solution test for reducing sugars
	  (c) biuret test for proteins
	  (d) ethanol emulsion test for fats and oils
	  (e) DCPIP test for vitamin C
		4	82–84	Practical: Food tests

Content overview

B4.4 Consolidation and summary	Learning aims • Review the learning points of the topic summarised in the end of topic checklist • Test understanding of the topic content by answering the end of topic questions	4	86–87
B5.1 Enzyme structure	Learning objectives • Describe a catalyst as a substance that increases the rate of a chemical reaction and is not changed by the reaction • Describe enzymes as proteins that are involved in all metabolic reactions, where they function as biological catalysts • Describe why enzymes are important in all living organisms in terms of a reaction rate necessary to sustain life • Describe enzyme action with reference to the shape of the active site of an enzyme being complementary to its substrate and the formation of products • Supplement Explain enzyme action with reference to: active site, enzyme-substrate complex, substrate and product • Supplement Explain the specificity of enzymes in terms of the complementary shape and fit of the active site with the substrate	5	90–92
B5.2 Enzyme activity	Learning objectives • Investigate and describe the effect of changes in temperature and pH on enzyme activity with reference to optimum temperature and denaturation • Supplement Explain the effect of changes in temperature on enzyme activity in terms of kinetic energy, shape and fit, frequency of effective collisions and denaturation • Supplement Explain the effect of changes in pH on enzyme activity in terms of shape and fit and denaturation	5	92–97	Practical: Gelatin, enzymes and temperature Practical: Trypsin, milk and temperature Practical: The effect of pH on amylase
B5.3 Consolidation and summary	Learning aims • Review the learning points of the topic summarised in the end of topic checklist • Test understanding of the topic content by answering the end of topic questions	5	98–99

Topic	Learning objectives	Syllabus ref	Pages	Practical / Demonstration

B6.1 Photosynthesis	**Learning objectives**
	• Describe photosynthesis as the process by which plants synthesise carbohydrates from raw materials using energy from light
	• State the word equation for photosynthesis as:
	  carbon dioxide + water → glucose + oxygen
	  in the presence of light and chlorophyll
	• State that chlorophyll is a green pigment that is found in chloroplasts
	• State that chlorophyll transfers energy from light into energy in chemicals, for the synthesis of carbohydrates
	• Outline the subsequent use and storage of the carbohydrates made in photosynthesis, limited to:
	  (a) starch as an energy store
	  (b) cellulose to build cell walls
	  (c) glucose used in respiration to provide energy
	  (d) sucrose for transport in the phloem
	  (e) nectar to attract insects for pollination
	• Investigate the need for chlorophyll, light and carbon dioxide for photosynthesis, using appropriate controls
	• Investigate and describe the effects of varying light intensity, carbon dioxide concentration and temperature on the rate of photosynthesis
	• Investigate and describe the effect of light and dark conditions on gas exchange in an aquatic plant using hydrogencarbonate indicator solution
	• Supplement Identify and explain the limiting factors of photosynthesis in different environmental conditions
	• Supplement State the balanced chemical equation for photosynthesis as:
	  $6CO_2 + 6H_2O \rightarrow C_6H_{12}O_6 + 6O_2$
		6.1	103–112	Practical: Testing leaves for starch
				
				Practical: Chlorophyll and photosynthesis
				
				Practical: Carbon dioxide and photosynthesis
				
				Practical: Light and the rate of photosynthesis

B6.2 Mineral requirements of plants	**Learning objective**
	• Explain the importance of:
	  (a) nitrate ions for making amino acids
	  (b) magnesium ions for making chlorophyll
		6.1	104–105	Practical: Plants and mineral nutrients

B6.3 Leaf structure	**Learning objectives** • State that most leaves have a large surface area and are thin, and explain how these features are adaptations for photosynthesis • Identify in diagrams and images the following structures in the leaf of a dicotyledonous plant: chloroplasts, cuticle, guard cells and stomata, upper and lower epidermis, palisade mesophyll, spongy mesophyll, air spaces, vascular bundles, xylem and phloem • Explain how the structures listed above adapt leaves for photosynthesis	6.2	112–115	Demonstration: Leaf structure
B6.4 Consolidation and summary	**Learning aims** • Review the learning points of the topic summarised in the end of topic checklist • Test understanding of the topic content by answering the end of topic questions	6	116–119	

Content overview

B7.1 Diet	7.1	123–125		

	Learning objectives
	 • Describe what is meant by a balanced diet
	 • State the principal dietary source and describe the importance of:
	   (a) carbohydrates
	   (b) fats and oils
	   (c) proteins
	   (d) vitamins, limited to C and D
	   (e) mineral ions, limited to calcium and iron
	   (f) fibre (roughage)
	   (g) water
	 • State the cause of scurvy and rickets			

B7.2 Digestive system	7.2	125–128		

	Learning objectives
	 • Identify in diagrams and images the main organs of the digestive system, limited to:
	   (a) alimentary canal: mouth, oesophagus, stomach, small intestine (duodenum and ileum) and large intestine (colon, rectum, anus)
	   (b) associated organs: salivary glands, pancreas, liver and gall bladder
	 • Describe the functions of the organs of the digestive system listed above, in relation to:
	   (a) ingestion – the taking of substances, e.g. food and drink, into the body
	   (b) digestion – the breakdown of food
	   (c) absorption – the movement of nutrients from the intestines into the blood
	   (d) assimilation – uptake and use of nutrients by cells
	   (e) egestion – the removal of undigested food from the body as faeces			

B7.3 Physical digestion	7.3	128–130		
	Learning objectives • Describe physical digestion as the breakdown of food into smaller pieces without chemical change to the food molecules • State that physical digestion increases the surface area of food for the action of enzymes in chemical digestion • Identify in diagrams and images the types of human teeth: incisors, canines, premolars and molars • Describe the structure of human teeth, limited to: enamel, dentine, pulp, nerves, blood vessels and cement, and understand that teeth are embedded in bone and the gums • Describe the functions of the types of human teeth in physical digestion of food • Describe the function of the stomach in physical digestion • **Supplement** Outline the role of bile in emulsifying fats and oils to increase the surface area for chemical digestion			

Content overview

Topic	Content		Pages	Practical

B7.4 Chemical digestion	**Learning objectives**
	 • Describe chemical digestion as the breakdown of large insoluble molecules into small soluble molecules
	 • State the role of chemical digestion in producing small soluble molecules that can be absorbed
	 • Describe the functions of enzymes as follows:
	   (a) amylase breaks down starch to simple reducing sugars
	   (b) proteases break down protein to amino acids
	   (c) lipase breaks down fats and oils to fatty acids and glycerol
	 • State where, in the digestive system, amylase, protease and lipase are secreted and where they act
	 • Describe the functions of hydrochloric acid in gastric juice, limited to killing harmful microorganisms in food and providing an acidic pH for optimum enzyme activity
	 • Supplement Describe the digestion of starch in the digestive system:
	   (a) amylase breaks down starch to maltose
	   (b) maltase breaks down maltose to glucose on the membranes of the epithelium lining the small intestine
	 • Supplement Describe the digestion of protein by proteases in the digestive system:
	   (a) pepsin breaks down protein in the acidic conditions of the stomach
	   (b) trypsin breaks down protein in the alkaline conditions of the small intestine
	 • Supplement Explain that bile is an alkaline mixture that neutralises the acidic mixture of food and gastric juices entering the duodenum from the stomach, to provide a suitable pH for enzyme action
		7.4	130–132	Practical: The digestion and absorption of carbohydrates

B7.5 Absorption	**Learning objectives** • State that the small intestine is the region where nutrients are absorbed • State that most water is absorbed from the small intestine but that some is also absorbed from the colon • Supplement Explain the significance of villi and microvilli in increasing the internal surface area of the small intestine • Supplement Describe the structure of a villus • Supplement Describe the roles of capillaries and lacteals in villi	7.5	133–134	
B7.6 Consolidation and summary	**Learning aims** • Review the learning points of the topic summarised in the end of topic checklist • Test understanding of the topic content by answering the end of topic questions	7	135–139	

B8.1 Transport in plants	**Learning objectives**
	 • State the functions of xylem and phloem:
	   (a) xylem – transport of water and mineral ions, and support
	   (b) phloem – transport of sucrose and amino acids
	 • Identify in diagrams and images the position of xylem and phloem as seen in sections of roots, stems and leaves of non-woody dicotyledonous plants
	 • Supplement Relate the structure of xylem vessels to their function, limited to:
	   (a) thick walls with lignin (details of lignification are **not** required)
	   (b) no cell contents
	   (c) cells joined end to end with no cross walls to form a long continuous tube
		8.1	143–145	Practical: Looking at microscope slides of transverse sections of plant tissues

© HarperCollins*Publishers* Ltd 2021

Content overview

B8.2 Water uptake	Learning objectives • Identify in diagrams and images root hair cells and state their functions • State that the large surface area of root hairs increases the uptake of water and mineral ions • Outline the pathway taken by water through the root, stem and leaf as: root hair cells, root cortex cells, xylem, mesophyll cells • Investigate, using a suitable stain, the pathway of water through the above-ground parts of a plant	8.2	145–146	Practical: Water movement through plants
B8.3 Transpiration	Learning objectives • Describe transpiration as the loss of water vapour from leaves • State that water evaporates from the surfaces of the mesophyll cells into the air spaces and then diffuses out of the leaves through the stomata as water vapour • Investigate and describe the effects of variation of temperature and wind speed on transpiration rate • Supplement Explain how water vapour loss is related to: the large internal surface area provided by the interconnecting air spaces between mesophyll cells and the size and number of stomata • Supplement Explain the mechanism by which water moves upwards in the xylem in terms of a transpiration pull that draws up a column of water molecules, held together by forces of attraction between water molecules • Supplement Explain the effects on the rate of transpiration of varying the following factors: temperature, wind speed and humidity • Supplement Explain how and why wilting occurs	8.3	146–150	Demonstration: Transpiration in a potted plant Practical: The rate of transpiration
Supplement B8.4 Translocation	Learning objectives • Supplement Describe translocation as the movement of sucrose and amino acids in phloem from sources to sinks • Supplement Explain why some parts of a plant may act as a source and a sink at different times • Supplement Describe: (a) sources as the parts of plants that release sucrose or amino acids (b) sinks as the parts of plants that use or store sucrose or amino acids	8.4	150–151	
B8.5 Consolidation and summary	Learning aims • Review the learning points of the topic summarised in the end of topic checklist • Test understanding of the topic content by answering the end of topic questions	8	152–154	

	Learning objectives	9.1 9.2	175–183	Practical: Exercise and pulse rate Demonstration: Heart dissection
B9.1 Circulatory systems and the heart	• Describe the circulatory system as a system of blood vessels with a pump and valves to ensure one-way flow of blood • Identify in diagrams and images the structures of the mammalian heart, limited to: muscular wall, septum, left and right ventricles, left and right atria, one-way valves and coronary arteries • State that blood is pumped away from the heart in arteries and returns to the heart in veins • State that the activity of the heart may be monitored by: ECG, pulse rate and listening to sounds of valves closing • Investigate and describe the effect of physical activity on the heart rate • Describe coronary heart disease in terms of the blockage of coronary arteries and state the possible risk factors including: diet, lack of exercise, stress, smoking, genetic predisposition, age and sex • Discuss the roles of diet and exercise in reducing the risk of coronary heart disease • Supplement Describe the single circulation of a fish • Supplement Describe the double circulation of a mammal • Supplement Explain the advantages of a double circulation • Supplement Identify in diagrams and images the atrioventricular and semilunar valves in the mammalian heart • Supplement Explain the relative thickness of: (a) the muscle walls of the left and right ventricles (b) the muscle walls of the atria compared to those of the ventricles • Supplement Explain the importance of the septum in separating oxygenated and deoxygenated blood • Supplement Describe the functioning of the heart in terms of the contraction of muscles of the atria and ventricles and the action of the valves • Supplement Explain the effect of physical activity on the heart rate			

17

Content overview

B9.2 Blood vessels and blood	**Learning objectives**
	• Describe the structure of arteries, veins and capillaries, limited to: relative thickness of wall, diameter of the lumen and the presence of valves in veins
	• State the functions of capillaries
	• Identify in diagrams and images the main blood vessels to and from the:
	  (a) heart, limited to: vena cava, aorta, pulmonary artery and pulmonary vein
	  (b) lungs, limited to: pulmonary artery and pulmonary vein
	  (c) kidney, limited to: renal artery and renal vein
	• List the components of blood as: red blood cells, white blood cells, platelets and plasma
	• Identify red and white blood cells in photomicrographs and diagrams
	• State the functions of the following components of blood:
	  (a) red blood cells in transporting oxygen, including the role of haemoglobin
	  (b) white blood cells in phagocytosis and antibody production
	  (c) platelets in clotting (details are not required)
	  (d) plasma in the transport of blood cells, ions, nutrients, urea, hormones and carbon dioxide
	• State the roles of blood clotting as preventing blood loss and the entry of pathogens
	• Supplement Explain how the structure of arteries and veins is related to the pressure of the blood that they transport
	• Supplement Explain how the structure of capillaries is related to their functions
	• Supplement Identify, in diagrams and images, the main blood vessels to and from the liver as: hepatic artery, hepatic veins and hepatic portal vein
	• Supplement Identify lymphocytes and phagocytes in photomicrographs and diagrams
	• Supplement State the functions of:
	  (a) lymphocytes – antibody production
	  (b) phagocytes – engulfing pathogens by phagocytosis
	• Supplement Describe the process of clotting as the conversion of fibrinogen to fibrin to form a mesh
		9.3
		9.4
			184–191	Demonstration: looking at prepared slides of blood vessels and blood

B9.3 Consolidation and summary	**Learning aims** • Review the learning points of the topic summarised in the end of topic checklist • Test understanding of the topic content by answering the end of topic questions	9	192–195	

B10.1 Disease, defence and hygiene	**Learning objectives**
	• Describe a pathogen as a disease-causing organism
	• Describe a transmissible disease as a disease in which the pathogen can be passed from one host to another
	• State that a pathogen is transmitted:
	  (a) by direct contact, including through blood and other body fluids
	  (b) indirectly, including from contaminated surfaces, food, animals and air
	• Describe the body defences, limited to: skin, hairs in the nose, mucus, stomach acid and white blood cells
	• Explain the importance of the following in controlling the spread of disease:
	  (a) a clean water supply
	  (b) hygienic food preparation
	  (c) good personal hygiene
	  (d) waste disposal
	  (e) sewage treatment (details of the stages of sewage treatment are **not** required)
		10	199–203	

Cambridge IGCSE Biology Teacher's Guide

	Learning objectives		

Supplement **B10.2** The immune system	Learning objectives
	• **Supplement** Describe active immunity as defence against a pathogen by antibody production in the body
	• **Supplement** State that each pathogen has its own antigens, which have specific shapes
	• **Supplement** Describe antibodies as proteins that bind to antigens leading to direct destruction of pathogens or marking of pathogens for destruction by phagocytes
	• **Supplement** State that specific antibodies have complementary shapes which fit specific antigens
	• **Supplement** Explain that active immunity is gained after an infection by a pathogen or by vaccination
	• **Supplement** Outline the process of vaccination:
	  (a) weakened pathogens or their antigens are put into the body
	  (b) the antigens stimulate an immune response by lymphocytes which produce antibodies
	  (c) memory cells are produced that give long-term immunity
	• **Supplement** Explain the role of vaccination in controlling the spread of diseases
	• **Supplement** Explain that passive immunity is a short-term defence against a pathogen by antibodies acquired from another individual, including across the placenta and in breast milk
	• **Supplement** Explain the importance of breast-feeding for the development of passive immunity in infants
	• **Supplement** State that memory cells are not produced in passive immunity
	• **Supplement** Describe cholera as a disease caused by a bacterium which is transmitted in contaminated water
	• **Supplement** Explain that the cholera bacterium produces a toxin that causes secretion of chloride ions into the small intestine, causing osmotic movement of water into the gut, causing diarrhoea, dehydration and loss of ions from the blood
		204–209	10

B10.3 Consolidation and summary	Learning aims • Review the learning points of the topic summarised in the end of topic checklist • Test understanding of the topic content by answering the end of topic questions	210–213	10
B11.1 Gas exchange system	Learning objectives • Describe the features of gas exchange surfaces in humans, limited to: large surface area, thin surface, good blood supply and good ventilation with air • Identify in diagrams and images the following parts of the breathing system: lungs, diaphragm, ribs, intercostal muscles, larynx, trachea, bronchi, bronchioles, alveoli and associated capillaries • **Supplement** Identify in diagrams and images the internal and external intercostal muscles • **Supplement** State the function of cartilage in the trachea • **Supplement** Explain the role of the ribs, the internal and external intercostal muscles and the diaphragm in producing volume and pressure changes in the thorax leading to the ventilation of the lungs • **Supplement** Explain the role of goblet cells, mucus and ciliated cells in protecting the breathing system from pathogens and particles	217–222	11	Demonstration: Dissection of sheep lungs Demonstration: Modelling breathing

Content overview

	Learning objectives			Practical
B11.2 The effects of gas exchange	Learning objectives • Investigate the differences in composition between inspired and expired air using limewater as a test for carbon dioxide • Describe the differences in composition between inspired and expired air, limited to: oxygen, carbon dioxide and water vapour • Investigate and describe the effects of physical activity on the rate and depth of breathing • Supplement Explain the differences in composition between inspired and expired air • Supplement Explain the link between physical activity and the rate and depth of breathing in terms of: an increased carbon dioxide concentration in the blood, which is detected by the brain, leading to an increased rate and greater depth of breathing	11	222–225	Practical: The effect of exercise on breathing Demonstration: Limewater
B11.3 Consolidation and summary	Learning aims • Review the learning points of the topic summarised in the end of topic checklist • Test understanding of the topic content by answering the end of topic questions	11	226–227	
B12.1 Aerobic respiration	Learning objectives • State the uses of energy in living organisms, including: muscle contraction, protein synthesis, cell division, active transport, growth, the passage of nerve impulses and the maintenance of a constant body temperature • Describe aerobic respiration as the chemical reactions in cells that use oxygen to break down nutrient molecules to release energy • State the word equation for aerobic respiration as: glucose + oxygen → carbon dioxide + water • Supplement State the balanced chemical equation for aerobic respiration as: $C_6H_{12}O_6 + 6O_2 \rightarrow 6CO_2 + 6H_2O$	12.1 12.2	231–233	
B12.2 Anaerobic respiration	Learning objectives • Describe anaerobic respiration as the chemical reactions in cells that break down nutrient molecules to release energy without using oxygen • State that anaerobic respiration releases much less energy per glucose molecule than aerobic respiration • State the word equation for anaerobic respiration in yeast as: glucose → alcohol + carbon dioxide • Investigate and describe the effect of temperature on respiration in yeast • State the word equation for anaerobic respiration in muscles during vigorous exercise as: glucose → lactic acid • Supplement State the balanced chemical equation for anaerobic respiration in yeast as: $C_6H_{12}O_6 \rightarrow 2C_2H_5OH + 2CO_2$ • Supplement State that lactic acid builds up in muscles and blood during vigorous exercise causing an oxygen debt • Supplement Outline how the oxygen debt is removed after exercise, limited to: (a) continuation of fast heart rate to transport lactic acid in the blood from the muscles to the liver (b) continuation of deeper and faster breathing to supply oxygen for aerobic respiration of lactic acid (c) aerobic respiration of lactic acid in the liver	12.1 12.3	233–237	Practical: Effect of temperature on respiration in yeast
B12.3 Consolidation and summary	Learning aims • Review the learning points of the topic summarised in the end of topic checklist • Test understanding of the topic content by answering the end of topic questions	12	238–239	

© HarperCollins Publishers Ltd 2021

Topic	Learning objectives		Practical / Demonstration
B13.1 Excretion	Learning objectives • State that carbon dioxide is excreted through the lungs • State that the kidneys excrete urea and excess water and ions • Identify in diagrams and images the kidneys, ureters, bladder and urethra • Supplement Describe the role of the liver in the assimilation of amino acids by converting them to proteins • Supplement State that urea is formed in the liver from excess amino acids • Supplement Describe deamination as the removal of the nitrogen-containing part of amino acids to form urea • Supplement Explain the importance of excretion, limited to toxicity of urea	13	243–244	
Supplement B13.2 Kidney function	Learning objectives • Supplement Explain why volume and concentration of urine vary in different conditions. • Supplement Describe the structure of a kidney including the cortex, medulla and ureter. • Supplement Describe the structure of a kidney tubule. • Supplement Describe the role of the glomerulus in filtration. • Supplement Describe the reabsorption of glucose, water and salts in the kidney tubule, which concentrates urea in urine.	13	245–247	Practical: Kidney dissection Demonstration: Excretion
B13.3 Consolidation and summary	Learning aims • Review the learning points of the topic summarised in the end of topic checklist • Test understanding of the topic content by answering the end of topic questions	13	248–249	
B14.1 The human nervous system	Learning objectives • Describe sense organs as groups of receptor cells responding to specific stimuli: light, sound, touch, temperature and chemicals • State that electrical impulses travel along neurones • Describe the mammalian nervous system in terms of: (a) the central nervous system (CNS) consisting of the brain and the spinal cord (b) the peripheral nervous system (PNS) consisting of the nerves outside of the brain and spinal cord • Describe the role of the nervous system as coordination and regulation of body functions • Identify in diagrams and images sensory, relay and motor neurones • Describe a simple reflex arc in terms of: receptor, sensory neurone, relay neurone, motor neurone and effector • Describe a reflex action as a means of automatically and rapidly integrating and coordinating stimuli with the responses of effectors (muscles and glands) • Describe a synapse as a junction between two neurones • Supplement Describe the structure of a synapse, including the presence of vesicles containing neurotransmitter molecules, the synaptic gap and receptor proteins • Supplement Describe the events at a synapse as: (a) an impulse stimulates the release of neurotransmitter molecules from vesicles into the synaptic gap (b) the neurotransmitter molecules diffuse across the gap (c) neurotransmitter molecules bind with receptor proteins on the next neurone (d) an impulse is then stimulated in the next neurone • Supplement State that synapses ensure that impulses travel in one direction only	14.1	254–259	Demonstration: Response to light Practical B14.1a Speed of reaction

Content overview

	Learning objectives			Demonstration:
B14.2 The human eye	• Identify in diagrams and images the structures of the eye, limited to: cornea, iris, pupil, lens, retina, optic nerve and blind spot • Describe the function of each part of the eye, limited to: (a) cornea – refracts light (b) iris – controls how much light enters the pupil (c) lens – focuses light on to the retina (d) retina – contains light receptors, some sensitive to light of different colours (e) optic nerve – carries impulses to the brain • Explain the pupil reflex, limited to changes in light intensity and pupil diameter • Supplement Explain the pupil reflex in terms of the antagonistic action of circular and radial muscles in the iris • Supplement Explain accommodation to view near and distant objects in terms of the contraction and relaxation of the ciliary muscles, tension in the suspensory ligaments, shape of the lens and refraction of light • Supplement Describe the distribution of rods and cones in the retina of a human • Supplement Outline the function of rods and cones, limited to: (a) greater sensitivity of rods for night vision (b) three different kinds of cones, absorbing light of different colours, for colour vision • Supplement Identify in diagrams and images the position of the fovea and state its function	14.2	259–263	Dissection of the eye
B14.3 Human hormones	• Describe a hormone as a chemical substance, produced by a gland and carried by the blood, which alters the activity of one or more specific target organs • Identify in diagrams and images specific endocrine glands and state the hormones they secrete, limited to: (a) adrenal glands and adrenaline (b) pancreas and insulin (c) testes and testosterone (d) ovaries and oestrogen • Describe adrenaline as the hormone secreted in 'fight or flight' situations and its effects, limited to: (a) increased breathing rate (b) increased heart rate (c) increased pupil diameter • Compare nervous and hormonal control, limited to speed of action and duration of effect • Supplement State that glucagon is secreted by the pancreas • Supplement Describe the role of adrenaline in the control of metabolic activity, limited to: (a) increasing the blood glucose concentration (b) increasing heart rate	14.3	263–265	

	Learning objectives / aims		Pages	Practical / Demonstration
B14.4 Homeostasis	**Learning objectives** • Describe homeostasis as the maintenance of a constant internal environment • State that insulin decreases blood glucose concentration • Supplement Explain the concept of homeostatic control by negative feedback with reference to a set point • Supplement Describe the control of blood glucose concentration by the liver and the roles of insulin and glucagon • Supplement Outline the treatment of Type 1 diabetes • Supplement Identify in diagrams and images of the skin: hairs, hair erector muscles, sweat glands, receptors, sensory neurones, blood vessels and fatty tissue • Supplement Describe the maintenance of a constant internal body temperature in mammals in terms of: insulation, sweating, shivering and the role of the brain • Supplement Describe the maintenance of a constant internal body temperature in mammals in terms of vasodilation and vasoconstriction of arterioles supplying skin surface capillaries	14.4	265–269	Demonstration: Temperature control Practical: Investigating a model for sweating
B14.5 Tropic responses	**Learning objectives** • Describe gravitropism as a response in which parts of a plant grow towards or away from gravity • Describe phototropism as a response in which parts of a plant grow towards or away from the direction of the light source • Investigate and describe gravitropism and phototropism in shoots and roots • Supplement Explain phototropism and gravitropism of a shoot as examples of the chemical control of plant growth • Supplement Explain the role of auxin in controlling shoot growth, limited to: (a) auxin is made in the shoot tip (b) auxin diffuses through the plant from the shoot tip (c) auxin is unequally distributed in response to light and gravity (d) auxin stimulates cell elongation	14.5	270–274	Demonstration: Gravitropism Practical: Investigating gravitropism Practical: Investigating phototropism
B14.6 Consolidation and summary	**Learning aims** • Review the learning points of the topic summarised in the end of topic checklist • Test understanding of the topic content by answering the end of topic questions	14	275–279	
B15.1 Drugs and medicinal drugs	**Learning objectives** • Describe a drug as any substance taken into the body that modifies or affects chemical reactions in the body • Describe the use of antibiotics for the treatment of bacterial infections • State that some bacteria are resistant to antibiotics which reduces the effectiveness of antibiotics • State that antibiotics kill bacteria but do not affect viruses • Supplement Explain how using antibiotics only when essential can limit the development of resistant bacteria such as MRSA	15.1 15.2	282–285	
B15.2 Consolidation and summary	**Learning aims** • Review the learning points of the topic summarised in the end of topic checklist • Test understanding of the topic content by answering the end of topic questions	15	286–287	

23

Content overview

Section		Page	Practical

B16.1 Sexual and asexual reproduction	**Learning objectives**
	• Describe asexual reproduction as a process resulting in the production of genetically identical offspring from one parent
	• Identify examples of asexual reproduction in diagrams, images and information provided
	• Describe sexual reproduction as a process involving the fusion of the nuclei of two gametes to form a zygote and the production of offspring that are genetically different from each other
	• Describe fertilisation as the fusion of the nuclei of gametes
	• Supplement Discuss the advantages and disadvantages of asexual reproduction:
	  (a) to a population of a species in the wild
	  (b) to crop production
	• Supplement State that nuclei of gametes are haploid and that the nucleus of a zygote is diploid
	• Supplement Discuss the advantages and disadvantages of sexual reproduction:
	  (a) to a population of a species in the wild
	  (b) to crop production
	
	16.1
	16.2
		292–296	Practical: Taking plant cuttings

16.2 Sexual reproduction in plants	**Learning objectives** • Identify in diagrams and images and draw the following parts of an insect-pollinated flower: sepals, petals, stamens, filaments, anthers, carpels, style, stigma, ovary and ovules • State the functions of the structures listed above • Identify in diagrams and images and describe the anthers and stigmas of a wind-pollinated flower • Distinguish between the pollen grains of insect-pollinated and wind-pollinated flowers • Describe pollination as the transfer of pollen grains from an anther to a stigma • State that fertilisation occurs when a pollen nucleus fuses with a nucleus in an ovule • Describe the structural adaptations of insect-pollinated and wind-pollinated flowers • Investigate and describe the environmental conditions that affect germination of seeds, limited to the requirement for: water, oxygen and a suitable temperature • Supplement Describe self-pollination as the transfer of pollen grains from the anther of a flower to the stigma of the same flower or a different flower on the same plant • Supplement Describe cross-pollination as the transfer of pollen grains from the anther of a flower to the stigma of a flower on a different plant of the same species • Supplement Discuss the potential effects of self-pollination and cross-pollination on a population, in terms of variation, capacity to respond to changes in the environment and reliance on pollinators • Supplement Describe the growth of the pollen tube and its entry into the ovule followed by fertilisation (details of production of endosperm and development are **not** required) 16.3	297–307	Practical: B16.2a Flower dissection Practical: B16.2b Conditions for germination

B16.3 Sexual reproduction in humans	Learning objectives • Identify on diagrams and state the functions of the following parts of the male reproductive system: testes, scrotum, sperm ducts, prostate gland, urethra and penis • Identify on diagrams and state the functions of the following parts of the female reproductive system: ovaries, oviducts, uterus, cervix and vagina • Describe fertilisation as the fusion of the nuclei from a male gamete (sperm) and a female gamete (egg cell) • Explain the adaptive features of sperm, limited to: flagellum, mitochondria and enzymes in the acrosome • Explain the adaptive features of egg cells, limited to: energy stores and the jelly coat that changes at fertilisation • Compare male and female gametes in terms of: size, structure, motility and numbers • State that in early development, the zygote forms an embryo which is a ball of cells that implants into the lining of the uterus • Identify on diagrams and state the functions of the following in the development of the fetus: umbilical cord, placenta, amniotic sac and amniotic fluid • Supplement Describe the function of the placenta and umbilical cord in relation to the exchange of dissolved nutrients, gases and excretory products between the blood of the mother and the blood of the fetus • Supplement State that some pathogens and toxins can pass across the placenta and affect the fetus	16.4	307–312
B16.4 Sexual hormones in humans	Learning objectives • Describe the roles of testosterone and oestrogen in the development and regulation of secondary sexual characteristics during puberty • Describe the menstrual cycle in terms of changes in the ovaries and in the lining of the uterus • Supplement Describe the sites of production of oestrogen and progesterone in the menstrual cycle and in pregnancy • Supplement Explain the role of hormones in controlling the menstrual cycle and pregnancy, limited to FSH, LH, progesterone and oestrogen	16.5	312–314
B16.5 Sexually transmitted infections	Learning objectives • Describe a sexually transmitted infection (STI) as an infection that is transmitted through sexual contact • State that human immunodeficiency virus (HIV) is a pathogen that causes an STI • State that HIV infection may lead to AIDS • Describe the methods of transmission of HIV • Explain how the spread of STIs is controlled	16.6	314–315
B16.6 Consolidation and summary	Learning aims • Review the learning points of the topic summarised in the end of topic checklist • Test understanding of the topic content by answering the end of topic questions	16	316–321

Content overview

B17.1 Chromosomes, genes and proteins	Learning objectives	17.1	326–330	B17.1b The structure of proteins
	Learning objectives • State that chromosomes are made of DNA, which contains genetic information in the form of genes • Define a gene as a length of DNA that codes for a protein • Define an allele as an alternative form of a gene • **Supplement** State that the sequence of bases in a gene determines the sequence of amino acids used to make a specific protein (knowledge of the details of nucleotide structure is **not** required) • **Supplement** Explain that different sequences of amino acids give different shapes to protein molecules • **Supplement** Explain that DNA controls cell function by controlling the production of proteins, including enzymes, membrane carriers and receptors for neurotransmitters • **Supplement** Explain how a protein is made, limited to: (a) the gene coding for the protein remains in the nucleus (b) messenger RNA (mRNA) is a copy of a gene (c) mRNA molecules are made in the nucleus and move to the cytoplasm (d) the mRNA passes through ribosomes (e) the ribosome assembles amino acids into protein molecules (f) the specific sequence of amino acids is determined by the sequence of bases in the mRNA (g) (knowledge of the details of transcription or translation is **not** required) • **Supplement** Explain that most body cells in an organism contain the same genes, but many genes in a particular cell are not expressed because the cell only makes the specific proteins it needs • **Supplement** Describe a haploid nucleus as a nucleus containing a single set of chromosomes • **Supplement** Describe a diploid nucleus as a nucleus containing two sets of chromosomes • **Supplement** State that in a diploid cell, there is a pair of each type of chromosome and in a human diploid cell there are 23 pairs			

	Learning objectives			Practical:
B17.2 Inheriting characteristics	Learning objectives • Describe the inheritance of sex in humans with reference to X and Y chromosomes • Describe inheritance as the transmission of genetic information from generation to generation • Describe genotype as the genetic make-up of an organism and in terms of the alleles present • Describe phenotype as the observable features of an organism • Describe homozygous as having two identical alleles of a particular gene • State that two identical homozygous individuals that breed together will be pure-breeding • Describe heterozygous as having two different alleles of a particular gene • State that a heterozygous individual will not be pure-breeding • Describe a dominant allele as an allele that is expressed if it is present in the genotype • Describe a recessive allele as an allele that is only expressed when there is no dominant allele of the gene present in the genotype • Interpret pedigree diagrams for the inheritance of a given characteristic • Use genetic diagrams to predict the results of monohybrid crosses and calculate phenotypic ratios, limited to 1 : 1 and 3 : 1 ratios • Use Punnett squares in crosses which result in more than one genotype to work out and show the possible different genotypes • Supplement Explain how to use a test cross to identify an unknown genotype • Supplement Describe codominance as a situation in which both alleles in heterozygous organisms contribute to the phenotype • Supplement Explain the inheritance of ABO blood groups: phenotypes are A, B, AB and O blood groups and alleles are I^A, I^B and I^o • Supplement Describe a sex-linked characteristic as a feature in which the gene responsible is located on a sex chromosome and that this makes the characteristic more common in one sex than in the other • Supplement Describe red-green colour blindness as an example of sex linkage • Supplement Use genetic diagrams to predict the results of monohybrid crosses involving codominance or sex linkage and calculate phenotypic ratios	17.4	332–345	Practical: B17.2 Investigating monohybrid inheritance
Supplement B17.3 Mitosis and meiosis	Learning objectives • Supplement Describe mitosis as nuclear division giving rise to genetically identical cells (details of the stages of mitosis are **not** required) • Supplement State the role of mitosis in growth, repair of damaged tissues, replacement of cells and asexual reproduction • Supplement State that the exact replication of chromosomes occurs before mitosis • Supplement State that during mitosis, the copies of chromosomes separate, maintaining the chromosome number in each daughter cell • Supplement Describe stem cells as unspecialised cells that divide by mitosis to produce daughter cells that can become specialised for specific functions • Supplement State that meiosis is involved in the production of gametes • Supplement Describe meiosis as a reduction division in which the chromosome number is halved from diploid to haploid resulting in genetically different cells (details of the stages of meiosis are **not** required)	17.2 17.3	330–331	

Content overview

B17.4 Consolidation and summary	17	346–350	Learning aims • Review the learning points of the topic summarised in the end of topic checklist • Test understanding of the topic content by answering the end of topic questions
B18.1 Variation	18.1	365–369	Learning objectives • Describe variation as differences between individuals of the same species • State that continuous variation results in a range of phenotypes between two extremes; examples include body length and body mass • State that discontinuous variation results in a limited number of phenotypes with no intermediates; examples include ABO blood groups, seed shape in peas and seed colour in peas • State that discontinuous variation is usually caused by genes only and continuous variation is caused by both genes and the environment • Investigate and describe examples of continuous and discontinuous variation • Describe mutation as genetic change • State that mutation is the way in which new alleles are formed • State that ionising radiation and some chemicals increase the rate of mutation • Supplement Describe gene mutation as a random change in the base sequence of DNA • Supplement State that mutation, meiosis, random mating and random fertilisation are sources of genetic variation in populations
B18.2 Adaptive features	18.2	369–371	Learning objectives • Describe an adaptive feature as an inherited feature that helps an organism to survive and reproduce in its environment • Interpret images or other information about a species to describe its adaptive features • Supplement Explain the adaptive features of hydrophytes and xerophytes to their environments

Content overview

| B18.3 Selection | Learning objectives • Describe natural selection with reference to: (a) genetic variation within populations (b) production of many offspring (c) struggle for survival, including competition for resources (d) a greater chance of reproduction by individuals that are better adapted to the environment than others (e) these individuals pass on their alleles to the next generation • Describe selective breeding with reference to: (a) selection by humans of individuals with desirable features (b) crossing these individuals to produce the next generation (c) selection of offspring showing the desirable features • Outline how selective breeding by artificial selection is carried out over many generations to improve crop plants and domesticated animals and apply this to given contexts • Supplement Describe adaptation as the process, resulting from natural selection, by which populations become more suited to their environment over many generations • Supplement Describe the development of strains of antibiotic resistant bacteria as an example of natural selection • Supplement Outline the differences between natural and artificial selection | 18.3 | 372–377 | Practical: A model of natural selection |
| B18.4 Consolidation and summary | Learning aims • Review the learning points of the topic summarised in the end of topic checklist • Test understanding of the topic content by answering the end of topic questions | 18 | 378–381 | |

Content overview

B19.1 Energy flow, food chains and food webs	Learning objectives	19.1 19.2	386–399	
	• State that the Sun is the principal source of energy input to biological systems			
	• Describe the flow of energy through living organisms, including light energy from the Sun and chemical energy in organisms, and its eventual transfer to the environment			
	• Describe a food chain as showing the transfer of energy from one organism to the next, beginning with a producer			
	• Construct and interpret simple food chains			
	• Describe a food web as a network of interconnected food chains and interpret food webs			
	• Describe a producer as an organism that makes its own organic nutrients, usually using energy from sunlight, through photosynthesis			
	• Describe a consumer as an organism that gets its energy by feeding on other organisms			
	• State that consumers may be classed as primary, secondary, tertiary and quaternary according to their position in a food chain			
	• Describe a herbivore as an animal that gets its energy by eating plants			
	• Describe a carnivore as an animal that gets its energy by eating other animals			
	• Describe a decomposer as an organism that gets its energy from dead or waste organic material			
	• Use food chains and food webs to describe the impact humans have through overharvesting of food species and through introducing foreign species to a habitat			
	• Draw, describe and interpret pyramids of numbers and pyramids of biomass			
	• Discuss the advantages of using a pyramid of biomass rather than a pyramid of numbers to represent a food chain			
	• Describe a trophic level as the position of an organism in a food chain, food web or ecological pyramid			
	• Identify the following as the trophic levels in food webs, food chains and ecological pyramids: producers, primary consumers, secondary consumers, tertiary consumers and quaternary consumers			
	• Supplement Draw, describe and interpret pyramids of energy			
	• Supplement Discuss the advantages of using a pyramid of energy rather than pyramids of numbers or biomass to represent a food chain			
	• Supplement Explain why the transfer of energy from one trophic level to another is often not efficient			
	• Supplement Explain, in terms of energy loss, why food chains usually have fewer than five trophic levels			
	• Supplement Explain why it is more energy efficient for humans to eat crop plants than to eat livestock that have been fed on crop plants			

Content overview

B19.2 Nutrient cycles	Learning objectives • Describe the carbon cycle, limited to: photosynthesis, respiration, feeding, decomposition, formation of fossil fuels and combustion • Supplement Describe the nitrogen cycle with reference to: (a) decomposition of plant and animal protein to ammonium ions (b) nitrification (c) nitrogen fixation by lightning and bacteria (d) absorption of nitrate ions by plants (e) production of amino acids and proteins (f) feeding and digestion of proteins (g) deamination (h) denitrification • Supplement State the roles of microorganisms in the nitrogen cycle, limited to: decomposition, nitrification, nitrogen fixation and denitrification (generic names of individual bacteria, e.g. *Rhizobium*, are **not** required)	19.3	399–405	
B19.3 Populations	Learning objectives • Describe a population as a group of organisms of one species, living in the same area, at the same time • Describe a community as all of the populations of different species in an ecosystem • Describe an ecosystem as a unit containing the community of organisms and their environment, interacting together • Identify and state the factors affecting the rate of population growth for a population of an organism, limited to food supply, competition, predation and disease • Identify the lag, exponential (log), stationary and death phases in the sigmoid curve of population growth for a population growing in an environment with limited resources • Interpret graphs and diagrams of population growth • Supplement Explain the factors that lead to each phase in the sigmoid curve of population growth, making reference, where appropriate, to the role of limiting factors	19.4	405–410	Practical: Investigating change in population size
B19.4 Consolidation and summary	Learning aims • Review the learning points of the topic summarised in the end of topic checklist • Test understanding of the topic content by answering the end of topic questions	19	411–417	
B20.1 Food supply	Learning objectives • Describe how humans have increased food production, limited to: (a) agricultural machinery to use larger areas of land and improve efficiency (b) chemical fertilisers to improve yields (c) insecticides to improve quality and yield (d) herbicides to reduce competition with weeds (e) selective breeding to improve production by crop plants and livestock • Describe the advantages and disadvantages of large-scale monocultures of crop plant • Describe the advantages and disadvantages of intensive livestock production	20.1	421–424	

Content overview

B20.2 Habitat destruction	Learning objectives
	• Describe biodiversity as the number of different species that live in an area
	• Describe the reasons for habitat destruction, including:
	  (a) increased area for housing, crop plant production and livestock production
	  (b) extraction of natural resources
	  (c) freshwater and marine pollution
	• State that through altering food webs and food chains, humans can have a negative impact on habitats
	• Explain the undesirable effects of deforestation as an example of habitat destruction, to include: reducing biodiversity, extinction, loss of soil, flooding and increase of carbon dioxide in the atmosphere
		20.2	424–427	

B20.3 Pollution	Learning objectives
	• Describe the effects of untreated sewage and excess fertiliser on aquatic ecosystems
	• Describe the effects of non-biodegradable plastics, in both aquatic and terrestrial ecosystems
	• Describe the sources and effects of pollution of the air by methane and carbon dioxide, limited to: the enhanced greenhouse effect and climate change
	• Supplement Explain the process of eutrophication of water, limited to:
	  (a) increased availability of nitrate and other ions
	  (b) increased growth of producers
	  (c) increased decomposition after death of producers
	  (d) increased aerobic respiration by decomposers
	  (e) reduction in dissolved oxygen
	  (f) death of organisms requiring dissolved oxygen in water
		20.3	428–435	Demonstration: Measuring dissolved oxygen in water

Content overview

	Learning objectives		Pages	Practical

B20.4 Conservation	Learning objectives
	• Describe a sustainable resource as one which is produced as rapidly as it is removed from the environment so that it does not run out
	• State that some resources can be conserved and managed sustainably, limited to forests and fish stocks
	• Explain why organisms become endangered or extinct, including: climate change, habitat destruction, hunting, overharvesting, pollution and introduced species
	• Describe how endangered species can be conserved, limited to:
	  (a) monitoring and protecting species and habitats
	  (b) education
	  (c) captive breeding programmes
	  (d) seed banks
	• Supplement Explain how forests can be conserved using: education, protected areas, quotas and replanting
	• Supplement Explain how fish stocks can be conserved using: education, closed seasons, protected areas, controlled net types and mesh size, quotas and monitoring
	• Supplement Describe the reasons for conservation programmes, limited to:
	  (a) maintaining or increasing biodiversity
	  (b) reducing extinction
	  (c) protecting vulnerable ecosystems
	  (d) maintaining ecosystem functions, limited to nutrient cycling and resource provision, including food, drugs, fuel and genes
	• Supplement Describe the use of artificial insemination (AI) and in vitro fertilisation (IVF) in captive breeding programmes
	• Supplement Explain the risks to a species if its population size decreases, reducing genetic variation (knowledge of genetic drift is **not** required)
		20.4	435–442	

B20.5 Consolidation and summary	Learning aims • Review the learning points of the topic summarised in the end of topic checklist • Test understanding of the topic content by answering the end of topic questions	20	443–447	

B21.1 Biotechnology	Learning objectives
	• State that bacteria are useful in biotechnology and genetic modification due to their rapid reproduction rate and their ability to make complex molecules
	• Describe the role of anaerobic respiration in yeast during the production of ethanol for biofuels
	• Describe the role of anaerobic respiration in yeast during bread-making
	• Describe the use of pectinase in fruit juice production
	• Investigate and describe the use of biological washing powders that contain enzymes
	• Supplement Discuss why bacteria are useful in biotechnology and genetic modification, limited to:
	  (a) few ethical concerns over their manipulation and growth
	  (b) the presence of plasmids
	• Supplement Explain the use of lactase to produce lactose-free milk
	• Supplement Describe how fermenters can be used for the large-scale production of useful products by bacteria and fungi, including insulin, penicillin and mycoprotein
	• Supplement Describe and explain the conditions that need to be controlled in a fermenter, including: temperature, pH, oxygen, nutrient supply and waste products
		21.1
		21.2
			451–459	Practical: Extracting juice from fruit
				Practical: Biological detergents
				Practical: Immobilising enzymes
				Practical: Making lactose-free milk

Content overview

B21.2 Genetic engineering	**Learning objectives**
	• Describe genetic modification as changing the genetic material of an organism by removing, changing or inserting individual genes
	• Outline examples of genetic modification:
	  (a) the insertion of human genes into bacteria to produce human proteins
	  (b) the insertion of genes into crop plants to confer resistance to herbicides
	  (c) the insertion of genes into crop plants to confer resistance to insect pests
	  (d) the insertion of genes into crop plants to improve nutritional qualities
	• **Supplement** Outline the process of genetic modification using bacterial production of a human protein as an example, limited to:
	  (a) isolation of the DNA making up a human gene using restriction enzymes, forming sticky ends
	  (b) cutting of bacterial plasmid DNA with the same restriction enzymes, forming complementary sticky ends
	  (c) insertion of human DNA into bacterial plasmid DNA using DNA ligase to form a recombinant plasmid
	  (d) insertion of recombinant plasmids into bacteria (specific details are **not** required)
	  (e) multiplication of bacteria containing recombinant plasmids
	  (f) expression in bacteria of the human gene to make the human protein
	• **Supplement** Discuss the advantages and disadvantages of genetically modifying crops, including soya, maize and rice
		21.3	459–464

B21.3 Consolidation and summary	**Learning aims** • Review the learning points of the topic summarised in the end of topic checklist • Test understanding of the topic content by answering the end of topic questions	21	465–469

B1 Characteristics and classification of living organisms

Introduction

This section revises and extends knowledge from previous work on the seven characteristics that are displayed by living organisms. It forms the basis for many other topics covering the separate characteristics in more detail later in the course. It also covers the classification of organisms, which will be referred to in many other parts of the course.

Links to other topics

Topics	Essential background knowledge	Useful links
2 Organisation of the organism		2.1 Cell structure and size of specimens
6 Plant nutrition		6.1 Photosynthesis
7 Human nutrition		7.1 Diet
12 Respiration		12.1 Respiration 12.2 Aerobic respiration 12.3 Anaerobic respiration
13 Excretion in humans		13.1 Excretion in humans
14 Coordination and response		14.1 Coordination and response 14.2 Sense organs 14.5 Tropic responses
16 Reproduction		16.3 Sexual reproduction in plants 16.4 Sexual reproduction in humans

Topic overview

B1.1	**Characteristics of living organisms** A sorting activity will help students to revise their knowledge of the characteristics of living organisms from previous work.
B1.2	**Concept and uses of classification systems** This learning episode will help students to understand the term *species*, and to understand the concept of the binomial system for naming organisms and why it is used. Supplement This learning episode will also help students to extend their understanding of classification, based on the use of DNA. In this learning episode, students will use a dichotomous key to identify organisms. They will also have the opportunity to construct a dichotomous key for organisms.
B1.3	**Features of organisms** This learning episode will help students to identify the key features used to classify animals and plants, and to group animals within the animal kingdom. Supplement This learning episode also provides opportunities to learn the key features of the five kingdoms of organisms, and also the key features for classifying some plants. Students will also be introduced to the basic structure of viruses.
B1.4	**Consolidation and summary** This learning episode quickly recaps the ideas encountered in this section and allows time for students to answer the end of topic questions in the Student Book.

Career Links

These are some scientific careers that focus on this area of biology but careers in many other fields use the knowledge and skills gained studying science. A knowledge of the characteristics and classification of living organisms underpins many careers in biology. **Curators** in natural history museums and botanical gardens need to know the taxonomy of the organisms that they study. **Conservation biologists** use their understanding of the characteristics of different organisms to help protect the organisms and their habitats.

Starting points

The Student Book section opener puts the ideas in the section into context and sets the scene. It also allows students to acknowledge and value their prior learning, and provides a benchmark against which future learning can be compared.

The questions provide a structure for introducing the section and can be used in a number of different ways:

- You could ask students to consider the questions as an introductory homework task.
- You could put students into groups to share their own ideas and understanding and then to report back to the whole class.
- Students could be given access to the Internet, preferably with a tight timescale, to find out the information required.

You could then use a spider chart or other form of wall chart to summarise everybody's ideas.

Recording these initial ideas allows you to retain them for reference as the individual topics are developed. In this way, your students' progress in learning can be readily acknowledged.

Learning episode B1.1 Characteristics of living organisms

Learning objectives

- Describe the characteristics of living organisms by describing:
  (a) movement as an action by an organism or part of an organism causing a change of position or place
  (b) respiration as the chemical reactions in cells that break down nutrient molecules and release energy for metabolism
  (c) sensitivity as the ability to detect and respond to changes in the internal or external environment
  (d) growth as a permanent increase in size and dry mass
  (e) reproduction as the processes that make more of the same kind of organism
  (f) excretion as the removal of the waste products of metabolism and substances in excess of requirements
  (g) nutrition as the taking in of materials for energy, growth and development

Common misconceptions

Students are likely to have several misconceptions that will be dealt with in more detail in later topics. For example, they may think that: plants do not respire and only carry out photosynthesis; excretion and egestion mean the same thing; breathing and respiration share the same meaning. Briefly correct any misconceptions where necessary, but return to them later to make sure they are sorted properly when the section is covered in detail.

Resources

Student Book pages 11–13

Worksheet B1.1 Characteristics of living organisms

Approach

1. Recap knowledge from previous work

Introduce the topic by giving students two minutes to write down the characteristics shown by all living organisms. Then ask them to compare what they have written with another student to identify the similarities and differences in their answers.

Take examples from around the class to collate a class list of characteristics. Encourage discussion within the class for each new suggestion. Add any suggestions to the list that the students agree on.

2. Characteristics of living organisms

Give students Worksheet B1.1 to work on either individually, or in pairs, to help them clarify the seven key characteristics of living organisms. They could use the cards to create their own list of characteristics, or you could give them the characteristics and ask them to identify which is shown in each example. Students could add one more plant and one more animal example of their own for each characteristic.

3. Create a mnemonic (optional)

Ask students to work on their own or in pairs to create a mnemonic (word made up of a series of letters to help you to remember something) for the initials of the seven characteristics, in whichever order they like.

4. Consolidation: Create a crossword

Ask students to create the clues for a crossword that includes all seven characteristics of living organisms.

Answers

Page 13

1. a) Any suitable answers for human, such as:
    movement, walking; respiration, combination of oxygen with glucose to release energy, carbon dioxide and water; sensitivity, vision; growth, increase in height; reproduction, having a baby; excretion, producing urine; nutrition, eating food.
    b) Any suitable answers for a specific animal, such as:
    movement, crawling; respiration, combination of oxygen with glucose to release energy, carbon dioxide and water; sensitivity, smell; growth, increase in length; reproduction, producing young; excretion, losing carbon dioxide through respiratory surface; nutrition, eating food.
    c) Any suitable answers for a plant, such as:
    movement, growing towards light; respiration, combination of oxygen with glucose to release energy, carbon dioxide and water; sensitivity, detecting direction of light; growth, increase in height; reproduction, producing seeds; excretion, diffusion of waste products out of leaf for photosynthesis (oxygen) and respiration (carbon dioxide); nutrition, taking in nutrients from soil and making glucose by photosynthesis.
2. movement – to reach best place to get food or other conditions favourable for growth
    respiration – to release energy from food that can be used for all life processes
    sensitivity – to detect changes in the environment
    growth – to increase in size until large/mature enough for reproduction
    reproduction – to pass genes on to next generation
    excretion – to remove harmful substances from body
    nutrition – to take in substances needed by the body for growth and reproduction

Worksheet B1.1

1. movement/sensitivity	2. respiration/excretion	3. excretion
4. respiration/excretion	5. reproduction	6. growth
7. sensitivity	8. movement	9. nutrition
10. nutrition	11. reproduction	12. sensitivity
13. growth	14. sensitivity	15. reproduction

Worksheet B1.1 Characteristics of living organisms

Cut out the cards and then sort them into piles to identify which key characteristic of living organisms they are examples of.

Note that some examples may match more than one characteristic, so try to identify all characteristics matched by each example.

1. The leaves of some plants track the Sun's movement across the sky during the day.	2. At night, a plant gives out carbon dioxide.	3. Humans produce urine every day, which contains waste products from the body as well as any water that the body doesn't need.
4. The air that you breathe out of your lungs contains more carbon dioxide than the air that you breathe in.	5. Humans give birth to live babies, but birds and reptiles lay eggs from which the young hatch.	6. During the first year of life a human baby may triple its birth weight.
7. A *Mimosa* plant responds to touch by closing the leaflets.	8. A student runs a 400 m race.	9. The green colouring of plant leaves is chlorophyll, which plants use to make glucose and starch.
10. A healthy diet includes a good balance of protein, fruit and vegetables.	11. Flowering plants produce seeds that can be grown to produce new plants.	12. If you receive a shock, your heart usually starts to beat more quickly.
13. Trees add another ring of wood around their trunks each year.	14. A houseplant on a windowsill grows towards the light.	15. A bacterium in ideal conditions can divide into two every 20 minutes.

Learning episode B1.2 Concept and uses of classification systems

Learning objectives

- State that organisms can be classified into groups by the features that they share
- Describe a species as a group of organisms that can reproduce to produce fertile offspring
- Describe the binomial system of naming species as an internationally agreed system in which the scientific name of an organism is made up of two parts showing the genus and species
- Construct and use dichotomous keys based on identifiable features
- **Supplement** Explain that classification systems aim to reflect evolutionary relationships
- **Supplement** Explain that the sequences of bases in DNA are used as a means of classification
- **Supplement** Explain that groups of organisms which share a more recent ancestor (are more closely related) have base sequences in DNA that are more similar than those that share only a distant ancestor

Common misconceptions

Classification can be a confusing topic because, although we classify on obvious features, many features of an organism are adapted, and therefore modified, to help it survive in a particular environment. This can result in some species within a group having features more commonly associated with other groups. For example, some species of fish, amphibians and reptiles give birth to live young, a feature more commonly associated with mammals. It is important in this learning episode, and the next one, that students realise this and look for more than one main morphological or anatomical feature of a species to help identify it.

Resources

Student Book pages 13–18

Supplement Worksheet B1.2a Evolutionary relationships

Supplement Worksheet B1.2b Classification using DNA

Worksheet B1.2c Dichotomous keys

B1.2 Technician's notes

Resources for a class practical (see Technician's notes)

Approach

1. Introducing species

Ask students for examples of names of species. Write a selection on the board, encouraging as wide a range as possible of animals, plants, vertebrates, non-vertebrates, etc. Then give students a minute to think of a definition of the term *species* and to write it down on a piece of paper. Ask students to compare their definition with that of a neighbour, and to work together to improve the definition. Ask pairs then to work in fours to produce a better definition, and then in eights if the class is large. Then ask a student from each group to give their agreed definition.

Compare the definitions to identify their similarities and differences. With higher-ability students, discuss the limitations of each definition in terms of how it could be tested. Then compare the definitions with the one given on page 13 of the Student Book. Point out that, although it is difficult to produce a definition for the term that can be proved, it is a useful working concept for much of biological work.

2. Binomial naming

Give students the common name of a flower, such as *African daisy*, and either provide a range of examples of different plants that bear that name or give students the opportunity to research examples of this. (Wikipedia offers seven genera that may bear the common name of *African daisy*, but there will be others.) Ask students to suggest why they all have that same common name (e.g. they all bear the features of a *daisy* [a flower with a large central yellow section surrounded by many thin petals] and all originate from Africa). Students could then research some binomial names for *African daisy* species.

If preferred, use an animal name that covers many species, such as *worm*. However, note that this will cover different classificatory groups, including roundworms, flatworms and nematodes. Adjust the questioning to suit the example but to cover the same learning.

Ask students to suggest what problems might be caused in identification as a result of using common names, such as in a horticulture reference that explains how best to grow and care for plants.

They should use the binomial names to explain the importance of identifying each species uniquely. For example, as different plants need different growing conditions, advice for one species of *African daisy* may not be suitable for another. This could be extended to consider other situations in which unique identification of a species may be important (such as in breeding, or in the treatment of infection by a particular parasite, e.g. malaria that may be caused by different species of *Plasmodium*).

Supplement **3. Evolutionary relationships**
Worksheet B1.2a provides some images of the front limb of four species of mammal. The first question on the sheet asks students to compare the diagrams in terms of similarities and differences.

The second question asks why there are both similarities and differences. If students are unsure of the answer, discuss the effect of the environment on the shape of the body, which they should be able to answer from earlier work on adaptations to the environment. For example, the cat's front limb is adapted for running, the whale flipper for swimming, etc. Explain that all mammals evolved from an ancestor that had a similar structure to their limb. Students should then consider how the limbs evolved from the ancestor's limb and how this accounts for their similarities.

Supplement **4. Classification using DNA**
Details of DNA structure, function and mutation are covered in Sections 4, 17 and 18. For this activity students need only appreciate that DNA carries genetic information coded for by the sequence of millions of copies of the bases A, T, G and C. The base sequences in individuals of the same species are almost, but not quite, identical. The DNA base sequences are inherited by offspring from their parents. Changes occur due to faults in copying the sequences during reproduction. The more times a sequence is copied, the more changes can occur. This is why there are greater differences in the sequences between individuals of distantly related species than individuals of closely related species.

To help students gain some appreciation of this, give them Worksheet B1.2b, which contains six very small sequences of DNA. Students should compare the sequences and put them in order of how different they are compared with the sequence from species 1. When scientists carry out real comparisons, they are working with many thousands of letters in the code, which is why the comparison is generally done by computer!

The comparison should put the species into this order, starting with 1: 1, 4, 6, 2, 5, 3. Use this to ask questions such as:

- Which species is most distantly related to species 1? [answer: 3]
- Which species is most closely related to species 3? [answer: 5]

Students should then combine what they have learned in this, and the previous task, to try to explain why using the base sequence in DNA can give a more accurate result than body structure in classification.

5. Dichotomous keys

Explain to students that a dichotomous key uses yes/no style questions to help identify organisms.

Worksheet B1.2c has a dichotomous key for identifying some herbivores found on the African savannah, such as on the Serengeti Plain. You will need to provide students with pictures of each of the named herbivores (zebra, eland, oryx, Thompson's gazelle, wildebeest, buffalo, impala), but make sure the pictures do not show the names of the animals.

Students could work on their own, or in small groups, to encourage discussion of the features and questions. When students have used the key, ask them to compare their results with those of other students and to discuss any differences. Students could consider which questions were most useful, and which caused the most difficulty. For the latter, they could try to think of other questions that might have been more useful.

Ask students to use the content of Worksheet B1.2c and convert it to the same format as the identification key of fruits on page 17 of the Student Book. They could use the pictures provided for the previous task to complete the key.

If students are confident in the way the key is constructed, ask them to use the herbivore pictures (or another set of pictures of organisms) to create their own key. This is not a simple task – it needs careful thought to identify appropriate characteristics to separate out each species. Students could then compare their keys by trying them out on each other. Encourage discussion on which method works best and why, so that students can draw general conclusions on the best approaches.

6. Consolidation: describing a new species

Ask students to imagine they have been shown a new organism that has never been seen before. Give them a few minutes to think about how they would work out whether it was a new species, and how that species might be named. Take examples around the class to elicit the importance of comparing features to known species to identify its relationships and therefore what genus name it should be given.

Technician's notes

The following resources are needed for the class practical on dichotomous keys:

Pictures of the following animals: zebra, eland, oryx, Thompson's gazelle, wildebeest, buffalo, impala
The pictures should be of males and should not show the names of the animals.

Answers

Page 15

1. A group of organisms that share many features and that can interbreed and produce fertile offspring.
2. Organisms are grouped according to how similar they are. The more similar their features, the more closely they are grouped, e.g. into species or genus rather than order or class.
3. Any suitable example, such as *Homo sapiens*, showing the two parts of the name, described as the genus name (the first part) and species name (the second part). Each species has a different binomial name.
4. Supplement It can be used to identify evolutionary relationships and it can help identify which species need conservation.

Supplement Page 16

1. The sequence of bases in DNA is more similar in organisms that are closely related than in organisms that are more distantly related. So, organisms with a similar DNA sequence have evolved from a more recent ancestor than those with DNA that is more different.

Page 17 Science in Context: Identification in the field

Different sex; different age.

Page 17

1. A key that identifies organisms by using questions, where each question has only two possible answers. So with each question a group of different organisms is divided into two groups.
2. Any two suitable answers, such as: the key may not include sufficient differences between groups to place an organism at species level accurately; individuals vary within a species, so it may be difficult to decide whether an individual does or does not have a particular feature; it might be the wrong time of year to identify particular features, for example, plants don't have flowers at certain times of the year.

Page 18

1. So that the user of the key can follow through the questions easily.
2. Any suitable solution that splits the three animals into a group of two and a single animal using one yes/no question, followed by another yes/no question that splits the group of two into two separate species. For example:

Does it have straight horns?   yes – gazelle no – go to next question

Does it have horns twisted in a spiral shape?   yes – kudu   no – wildebeest

Supplement Worksheet B1.2a

1. Similarities: they are all quite long; those of the cat and human clearly bend at the 'elbow'; they all have a similar arrangement of bones, one in the upper 'arm', two in the lower 'arm', several in the 'wrist' and several that make up separate 'fingers'. Differences: they are very different shapes and are adapted for different purposes – human for grasping and manipulating things, cat for running, whale for swimming, bat for flying; in the human, the fingers are separate and long; in the cat the 'fingers/toes' are short and angled for running; in the whale the upper bones are very short and there are many more 'finger bones'; in the bat, the 'fingers' are very extended to support the membrane of the wing.
2. Characteristics are inherited from parents/ancestors. The ancestor of all these species would have had the same basic organisation of bones in their front limb. Adaptations to different kinds of environment have resulted in the evolution into very different forms (morphs) in the different species shown.

Supplement Worksheet B1.2a Evolutionary relationships

The diagram shows the outline and bone structure of the front limb of four species of mammal. The same bones are shaded the same tone in each diagram.

Look at the diagrams and answer the questions.

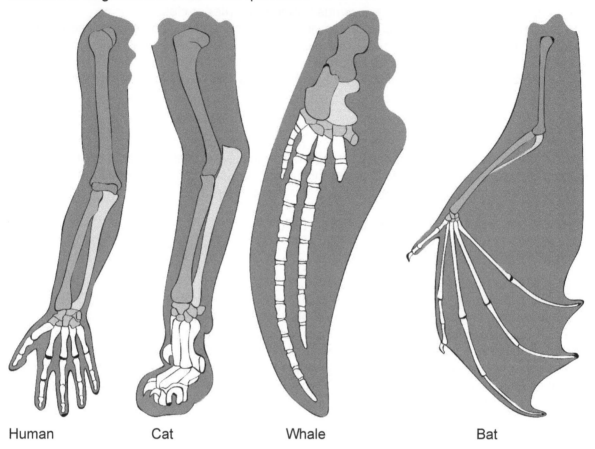

Human   Cat   Whale   Bat

1.   Compare these front limbs. How are they similar and how are they different?
2.   Suggest reasons why there are similarities and differences between the limbs. (You will need to consider adaptation to the environment and inheritance to answer this question fully.)

Supplement Worksheet B1.2b Classification using DNA

The diagrams below show some very short pieces of DNA. Each piece has been taken from the same part of the genetic information in six different, but related, species.

Compare the pieces of DNA for differences. For species 2 to 6, count how many differences each has compared to species 1.

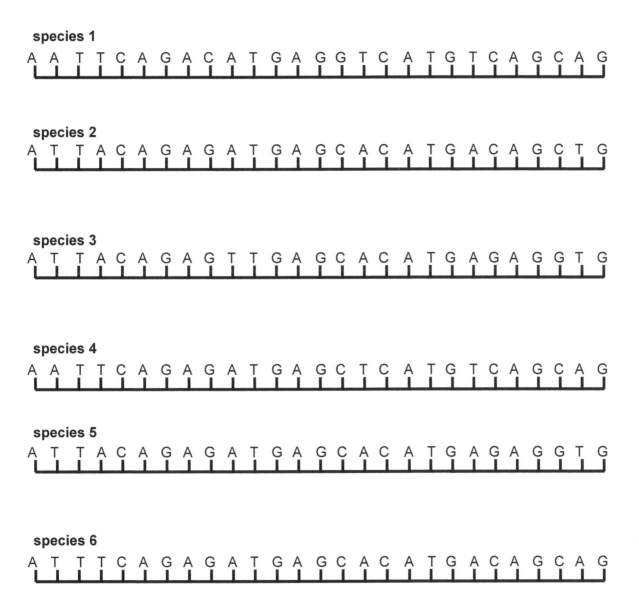

species 1

A A T T C A G A C A T G A G G T C A T G T C A G C A G

species 2

A T T A C A G A G A T G A G C A C A T G A C A G C T G

species 3

A T T A C A G A G T T G A G C A C A T G A G A G G T G

species 4

A A T T C A G A G A T G A G C T C A T G T C A G C A G

species 5

A T T A C A G A G A T G A G C A C A T G A G A G G T G

species 6

A T T T C A G A G A T G A G C A C A T G A C A G C A G

Worksheet B1.2c Dichotomous keys

There are many herbivores on the savannah (grass plains) of Africa.

You will be given pictures of some of these herbivores. Use the key below to identify them.

1. Has horns?	yes	go to 2
	no	**zebra**
2. Has tight spiral horns?	yes	**eland**
	no	go to 3
3. Has white markings on the face?	yes	go to 4
	no	go to 5
4. Has long horns?	yes	**oryx**
	no	**Thompson's gazelle**
5. Has black stripes across its back?	yes	**wildebeest**
	no	go to 6
6. Is a large, heavy animal?	yes	**buffalo**
	no	**impala**

Learning episode B1.3 Features of organisms

Learning objectives

- State the main features used to place animals and plants into the appropriate kingdoms
- State the main features used to place organisms into groups within the animal kingdom, limited to:
  (a) the main groups of vertebrates: mammals, birds, reptiles, amphibians, fish
  (b) the main groups of arthropods: myriapods, insects, arachnids, crustaceans
- Classify organisms using the features identified above
- Supplement State the main features used to place all organisms into one of the five kingdoms: animal, plant, fungus, prokaryote, protoctist
- Supplement State the main features used to place organisms into groups within the plant kingdom, limited to ferns and flowering plants (dicotyledons and monocotyledons)
- Supplement Classify organisms using the features identified above
- Supplement State the features of viruses, limited to a protein coat and genetic material

Common misconceptions

Students are likely to have the general understanding that there are plants (green things with leaves), animals (things that have legs and move) and 'other things' (for which they have no special form of classification except they are 'not' plants or animals). They may need more questioning to help them to distinguish between the 'other things' and to focus on the key features that are used to classify the main groups.

Resources

Student Book pages 19–32

Supplement Worksheet B1.3a The five kingdoms

Worksheet B1.3b Classifying vertebrates

B1.3 Technician's notes

Resources for a class practical (see Technician's notes)

Approach

1. Introducing classification
Give students two minutes to write down the key features of plants and animals. Take examples from around the class and ask why each is a key feature. Encourage discussion to highlight differences in understanding. Students could then test their understanding by collecting, or drawing, images of a range of organisms in each group, summarising the key features that the organisms in each group share. Make sure that students include as wide a range of animals and plants as possible, although they are not expected to classify them beyond 'plant' or 'animal' at this point.

Some students may suggest organisms that do not fit into either of these groups (e.g. bacteria, fungi), in which case ask them to suggest why they do not fit. These other groups are covered elsewhere in this learning episode.

Supplement 2. The five kingdoms
Give students Worksheet B1.3a, which contains a table of characteristics for the five kingdoms of organisms. Ask them to complete the table to show which groups of organisms have each characteristic. Encourage them to add detail to their responses, rather than just a tick or cross.

The completed table is given in the answers section.

3. Classifying vertebrates
Give students Worksheet B1.3b, which shows a selection of vertebrates. Ask students to cut out the images and to group them into the five main groups of vertebrates. They may have difficulty with this,

because the images do not all clearly show the main features of the group. Further research will probably be needed to complete the task. Students could use the feature information on pages 24–27 of the Student Book to help them.

The task could be extended by asking students to find additional examples of each group. Encourage them to find as wide a range of examples as possible. Students should justify their grouping by identifying at least one key feature of the group for each example. It might be appropriate to point out that the variety of forms is related to adaptations of each kind of organism to their environment, and that this can lead to organisms of different groups that live in similar environments showing similar forms, for example dolphins and fish, or bats and birds.

4. Classifying arthropods

Many students may have no understanding of the term arthropod. However, they are likely to have encountered examples from several different groups within the classification. Provide a selection of unlabelled images or real-life examples of different arthropods, for example, arachnids such as spiders or scorpions, myriapods such as centipedes or millipedes, insects such as butterflies or wasps, and crustaceans such as crabs or crayfish. First, ask students what the organisms all have in common. They should be able to identify that all the examples have a hard outer skeleton (exoskeleton) and that the skeleton is jointed to allow movement. Contrast the exoskeleton with the internal skeleton of vertebrates. Challenge students to consider the advantages and disadvantages of the two types of skeleton. (For example, an exoskeleton provides greater protection against attack, but limits growth and must be shed periodically for growth to continue.)

Then ask students to identify features that could be used to separate the organisms into different groups. This is challenging to do without support if the examples vary a lot in their features. Students could use pages 27–29 in their Student Book to help them. If there is time, students could research other examples for each group. They could sketch their examples, pointing out the main features for identification.

Supplement 5. Classifying plants

Gather a range of plants that include some ferns, some dicotyledons (for example, from the daisy, pea, deadnettle and rose families) and some monocotyledons (from the grass, lily and orchid families). If possible, choose plants that have flowers or spore cases (sporangia) visible, and also include some tree species. Alternatively, give students a selection of images instead of plant material.

Ask students to sort the species according to any features they can identify. They should aim to find one characteristic that will separate the examples into three main groups, and then find other characteristics that appear unique to each group.

Students are most likely to separate the groups initially into ferns and flowering plants on general structure, particularly the presence of spore cases or flowers. They should then be able to separate the flowering plants into two main groups based on leaf or vein structure, as monocotyledons usually have long strap-like leaves with parallel veins, whereas the dicotyledons have branching leaf veins and a wide range of leaf shapes. Other distinctive features that students might spot is that monocotyledon flowers have parts (e.g. petals, sepals, stamens) in multiples of three, whereas dicotyledon flowers have parts in multiples of four or five.

If there is time, students could investigate related areas of interest, such as:

- other features that are characteristic of each group (this includes pollen structure, number of seed leaves in the seed and arrangement of vascular bundles in the stem)
- what *dicotyledon* means and how it differs from *monocotyledon*
- which group the most important crop plants belong, and why they are so valuable.

Supplement 6. Viruses

Describe the basic structure of a virus and then ask students to suggest why they are not part of any of the five kingdoms. Students work in groups to discuss whether they should be regarded as living organisms. (Refer back to the characteristics of life from learning episode B1.1.)

Technician's notes

Be sure to check the latest safety notes on these resources before proceeding.

The following resources are needed for the activity on classifying arthropods:

Selection of images or real-life examples of a wide range of arthropods: for example, arachnids such as spiders or scorpions, myriapods such as centipedes or millipedes, insects such as butterflies or wasps, and crustaceans such as crabs or crayfish.

If using real-life examples, make sure no venomous animals are included and that all animals are treated with care and returned safely to their original environment as soon as possible after the lesson.

The following resources are needed for the activity on classifying plants:

Samples of flowering plants that include some ferns, some dicotyledons (for example, from the daisy, pea, deadnettle and rose families) and some monocotyledons (from the grass, lily and orchid families). If possible, choose plants that have spore cases or flowers and also include some tree species.

Alternatively, a selection of images of plant material could be used.

Answers

Page 20

1. Plant cells may contain chloroplasts, but animal cells do not.
2. Plants are usually not able to move around freely, but many animals can
3. It is a plant because only plant cells contain chloroplasts.

Page 21 Science in Context: Mushrooms and toadstools

As they are carried by the wind, many spores will land in places where they cannot grow. However, producing very large numbers of spores means that it is likely that at least some will land in suitable conditions to grow.

Page 23 Science in Context: Malaria

Unless the mosquito has fed on someone with the *Plasmodium* parasite in their blood, then the mosquito cannot pass it on to the next person they feed on.

Supplement Page 23

1. animals, plants, fungi, protoctists and prokaryotes; each kingdom with a suitable example.
2. a) Cell walls, cannot move around
   b) no chloroplasts, do not photosynthesise
3. Some protoctists contain chloroplasts and can photosynthesise as some plant cells do; others do not have chloroplasts and feed on other organisms, so are more like single animal cells.
4. Prokaryotes

Page 27 Science in Context: Duck-billed platypus

Advantage: mother does not need to carry around extra weight / may be able to have greater number of young in one brood.

Disadvantage: parent(s) need to stay close to incubate eggs/keep them warm/protect eggs.

Page 29

1. Table like the following:

Group	Key body features	Fertilisation	Production of young
fish	scaly skin, streamlined shape, fins and tail, gills	external	from eggs
birds	feathers and wings, constant body temperature	internal	from eggs
mammals	hair, mammary glands, constant body temperature	internal	live birth
amphibians	moist skin, metamorphosis between very different young and adult forms	external	from eggs
reptiles	tough scaly skin, varying body temperature	internal	from eggs

2. It is a vertebrate (backbone) and a bird because only birds have feathers.
3. They all have a tough exoskeleton and paired jointed legs.
4. Myriapods have many segments and many legs. Insects have a three-part body with six legs and often two pairs of wings. Arachnids have a two-part body with eight legs. Crustaceans usually have a three-part body with two pairs of antennae on their heads, and they may have swimming legs on the abdomen as well as real legs on the thorax.

Supplement Page 31

1. Similarities: both have chloroplasts and are plants, both have roots and leaves. Differences: ferns reproduce using spores, flowering plants reproduce using flowers and seeds.
2. Dicotyledon plants have two cotyledons in the seed; monocotyledon plants only have one. Monocotyledons have long, strap-like leaves with parallel veins; dicotyledons have broad leaves of many shapes with branching veins.
3. Ferns

Page 32 Science in Context: HIV and AIDS

May not be able to destroy the HIV viruses without damaging the human cells they are in.

Supplement Page 32

1. Similar: contain genetic material; different: only have a protein coat, not a cell membrane or other features of cells of living organisms.

Supplement Worksheet B1.3a

Kingdom	Food source?	Cell wall?	Nucleus in cell?	Single- or multi-celled?	Other?
Plants	make own food using light energy from Sun	yes, cellulose	yes	multi-celled	contain green chloroplasts
Animals	eat other organisms	no	yes	multi-celled	move about
Fungi	extracellular digestion of dead or living plant or animal tissue	yes, chitin	yes	some single-celled, some multi-celled	made of mycelium formed by hyphae
Prokaryotes	some use photosynthesis, some feed off dead or living tissue	yes, variable	no – DNA lies free in cytoplasm	single-celled	much smaller than other cells
Protoctists	some use photosynthesis, some feed on living organisms or dead matter	some	yes	Mostly single-celled, some multi-celled	single-celled but larger and more complex than prokaryotes

Worksheet B1.3b

mammals: gorilla, bat, goat

birds: condor, stork

reptiles: crocodile, gecko, python

amphibians: frog, salamander

fish: angel fish, conger eel

Supplement Worksheet B1.3a The five kingdoms

Use the Student Book and your own research to help you complete the table below to show some key characteristics of each group of organisms.

Add details where you can, rather than just a 'yes/no' answer.

Kingdom	Food source?	Cell wall?	Nucleus in cell?	Single- or multi-celled?	Other?
Plants					
Animals					
Fungi					
Prokaryotes					
Protoctists					

Worksheet B1.3b Classifying vertebrates

Examples of vertebrates (animals with backbones) are shown below. Cut out the examples and group them using their main features to show whether they are mammals, birds, reptiles, amphibians or fish. You may need to do some research to help you group them correctly.

For each group, identify at least two main features shown by all the vertebrates in that group. If you have time, look for other examples of each group. Try to find as great a variety of examples as you can.

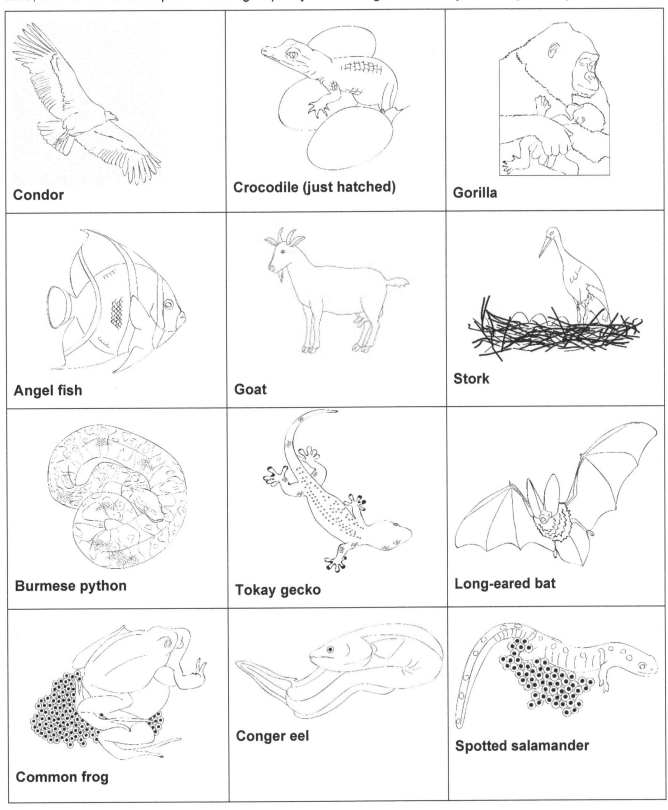

Condor

Crocodile (just hatched)

Gorilla

Angel fish

Goat

Stork

Burmese python

Tokay gecko

Long-eared bat

Common frog

Conger eel

Spotted salamander

Learning episode B1.4 Consolidation and summary

Learning aims

- Review the learning points of the topic summarised in the end of topic checklist
- Test understanding of the topic content by answering the end of topic questions

Resources

Student Book pages 33–37

Approach

Ask students to work with a partner to make a list of key words from this topic. They could then work together to produce a spider diagram showing how the different concepts are linked. They could compare their list with the list of key terms given on page 33 of the Student Book. Discuss the checklist on pages 33–34 and use questioning to see how much of the content they are comfortable with.

Students could make flashcards of the key content and then use the flashcards to quiz each other on the information.

Ask students to work individually through the end of topic questions on pages 35–37 of the Student Book without looking at the text. As they work, walk around the classroom observing their answers and asking questions as necessary to find out which questions are causing difficulties.

After a set period, ask the students to stop working and discuss any areas of difficulty you observed as you walked round the class. Students should complete any unanswered questions for homework, but you should stress that they should try to answer the questions without looking at the text, so that they can see how much they have remembered.

Answers

End of topic questions mark scheme

The marks available for a question can indicate the level of detail you need to provide in your answer.

Question	Correct answer	Marks
1	Movement – changing position or place; respiration – chemical reactions that release energy; sensitivity – detecting and responding to changes in the environment; growth – increasing in size; reproduction – making more individuals; excretion – removal of waste products; nutrition – obtaining food.	7 marks (1 mark for each correct answer)
2	Nutrition and respiration in animals, photosynthesis and respiration in plants.	1 mark 1 mark
3	Dry mass is the mass of all the materials used to make the cells and tissues of the body. Water content in the body varies as water is gained and lost, so wet mass is not reliable.	1 mark 1 mark
4	The crystal increases in size as more of the substance in solution attaches to the crystal, but this is not true growth because the substance can be lost to solution too, so it is not a permanent increase. The crystal also does not show any of the other six characteristics of life.	1 mark 1 mark
5	Animals are not 'more alive' just because they have to move around to get their food, etc. Plants must remain attached to the ground because that is where they get support, water and nutrients, but they move parts of their body. Both plants and animals show all the life processes, so are equally 'alive'.	1 mark 1 mark
6	The tree may not noticeably grow during winter, although there may be some cell growth. Respiration and therefore excretion may still occur, but at a very slow rate (as gas exchange continues slowly through the bark). Nutrition (photosynthesis), movement, reproduction and sensitivity may not take place during winter. As long as the tree can return to a state in which it can carry out all these processes (when leaves grow, during the rest of the year), it is still alive.	1 mark 1 mark 1 mark 1 mark
7 a)	The unique part is *leo*, which is the species name.	1 mark
7 b)	The *Panthera* part is the genus name, which is shared with other closely related species.	1 mark 1 mark
8 a)	They look very alike.	1 mark
8 b)	They would breed and produce fertile offspring.	1 mark
Supplement 8 c)	Compare their DNA sequences. If they are very similar then they could be the same species.	1 mark 1 mark
9	You would have to study it carefully to identify all the key features that could be used for putting it into a group. You would then compare those features with the key features of each of the classificatory groups and decide which group it matched best.	1 mark 1 mark

Supplement 10 a)	They share large parts of their DNA sequence because they share a relatively recent ancestor.	1 mark
Supplement 10 b)	The order of similarity (starting with the most similar) is chimpanzee, mouse, chicken, fly.	1 mark
	This suggests that chimpanzees and humans shared an ancestor more recently in the past than humans and mice, and with mice more recently than with birds or with insects.	1 mark
11	Is the fruit broadly spherical in shape? – other shape	1 mark
	Is the outer surface soft or rigid? – soft	1 mark
	Is the flesh sweet or starchy? – sweet	1 mark
12 a)	Any two suitable advantages such as: - simplicity – looking at only one thing at a time - does not need complicated equipment so can be used in the field.	1 mark each max. 2 marks
12 b)	Any suitable disadvantage, such as the feature under consideration must be present in all organisms from a specific group.	1 mark

13	Any suitable key that uses number of petals, visibility/number/arrangement of stamens, visibility/number/arrangement of stigmas or colour, as features to separate the flowers first into groups and then into single species using yes/no questions. For example:
	a) Are the petals dark red? yes – go to b
	  no – go to c
	b) Are the stigmas and stamens visible? yes – **hibiscus**
	  no – **rose**
	c) Does the flower have more than six petals? yes – **daisy**
	  no – **poppy**
		1 mark for dichotomous questions
		
		1 mark for key that is effective

14 a)	ii) is the most useful question	1 mark
14 b)	There are other organisms that have wings, e.g. bats and insects, and others that lay eggs, e.g. reptiles.	1 mark
15 a)	Having wings is common to many groups and is not a suitable feature to use for classification.	1 mark
	The wings are very different in structure, one being formed by stretched skin, one by feathers.	1 mark
15 b)	They are both vertebrates because they each have a backbone.	1 mark
	Bats are mammals because they have hair, birds have feathers.	1 mark
16	B	1 mark
17	They both have chloroplasts in their cells / can photosynthesise,	1 mark
	and they are both multicellular.	1 mark
18	Gives birth to live young/young in the womb supported by placenta,	1 mark
	young fed on milk made in mammary glands.	1 mark
19 a)	Octopus and bee.	1 mark
	Dog and shark.	1 mark
19 b)	Octopus and bee not vertebrates (no backbone), dog and shark are vertebrates (have backbone).	1 mark
Supplement 20	viruses, prokaryotes, protoctists	2 marks

Supplement 21	Animals	Plants	Fungi	Protoctists	Prokaryotes	5 marks
	eukaryote cell / has nucleus	eukaryote cell / has nucleus	eukaryote cell / has nucleus	eukaryote cell / has nucleus	prokaryote cell / no nucleus	(1 mark for identifying the features of each kingdom)
	multicellular	multicellular	multicellular	Mostly single-celled, some multi-celled	single-celled	
	able to move	chloroplasts	hyphae / mycelium		smaller / less complex than other cells	
Supplement 22 a)	Have cell walls and a central vacuole in their cells, like plants.					1 mark
	Don't move around as animals usually do (at some point in their life).					1 mark
Supplement 22 b)	Cannot photosynthesise as plants do / get food by digesting dead or living plant or animal material.					1 mark
	Different cellular structure from plants e.g. formed of hyphae and mycelium.					1 mark
Supplement 23 a)	reproduction					1 mark
Supplement 23 b)	movement – viruses cannot move on their own					1 mark
	respiration – viruses do not break down food to release energy					1 mark
	sensitivity – viruses do not sense and respond to stimuli					1 mark
	growth – viruses do not grow in the sense of increasing their mass					1 mark
	excretion – viruses do not produce substances, so they cannot excrete					1 mark
	nutrition – viruses do not take in nutrients for conversion into other substances					1 mark
Supplement 23 c)	Viruses should be classified as living organisms because they are able to reproduce, which is one of the life processes.					1 mark
						1 mark
Supplement 23 d)	Viruses should not be classified as living organisms because they only show reproduction when they take over the reproductive processes in another cell,					1 mark
	and they do not show any of the other life processes.					1 mark
	Total:					74 marks

B2 Organisation of the organism

Introduction

In this section, students will learn to recognise the structures found in cells, and relate the structures that they see in plant and animal cells to their functions. In addition, students will learn how to calculate the magnification of an object under a microscope.

This section revises another fundamental building block of knowledge that students should refer to in later work in other sections. It revisits and extends knowledge from earlier work on the organisation of cells into tissues, organs and organ systems.

Links to other topics

Topics	Essential background knowledge	Useful links
1 Characteristics and classification of living organisms	1.2 Concept and uses of classification systems	
3 Movement into and out of cells		3.1 Diffusion 3.2 Osmosis 3.3 Active transport
6 Plant nutrition		6.1 Photosynthesis 6.2 Leaf structure
7 Human nutrition		7.2 Digestive system 7.5 Absorption
8 Transport in plants		8.2 Water uptake 8.3 Transpiration 8.4 Translocation
9 Transport in animals		9.2 Heart 9.3 Blood vessels 9.4 Blood
12 Respiration		12.2 Aerobic respiration 12.3 Anaerobic respiration
13 Excretion in humans		13.1 Excretion in humans
14 Coordination and response		14.1 Coordination and response 14.2 Sense organs 14.3 Hormones 14.4 Homeostasis 14.5 Tropic responses
16 Reproduction		16.3 Sexual reproduction in plants 16.4 Sexual reproduction in humans 16.5 Sexual hormones in humans
17 Inheritance		17.1 Chromosomes, genes and proteins 17.2 Mitosis 17.3 Meiosis

Topic overview

B2.1	**Cell structure and size of specimens**
	This learning episode covers section 2.1: *Cell structure* and section 2.2: *Size of specimens* from the syllabus.
	The learning episode gives students the opportunity to develop their microscope skills while studying the similarities and differences between animal cells and plant cells. They also find out about the structure of bacterial cells.
	Students discover a range of cells, tissues, organs and organ systems and how they work together to carry out their functions in organisms.
	Supplement Students will convert measurements between millimetres and micrometres.
B2.2	**Consolidation and summary**
	This learning episode quickly recaps on the ideas encountered in this section and allows time for students to answer the end of topic questions in the Student Book.

Careers links

These are some scientific careers that focus on this area of biology but careers in many other fields use the knowledge and skills gained studying science. **Histopathologists** diagnose and study diseases in tissues and organs, using microscopes to look at cell samples taken from patients. **Vascular scientists** and technicians work with doctors to assist with the diagnosis of disorders of the circulatory system.

Starting points

The Student Book section opener puts the ideas in the section into context and sets the scene. It also allows students to acknowledge and value their prior learning, and provides a benchmark against which future learning can be compared.

The questions provide a structure for introducing the section and can be used in a number of different ways:

- You could ask students to consider the questions as an introductory homework task.
- You could put students into groups to share their own ideas and understanding and then to report back to the whole class.
- Students could be given access to the Internet, preferably with a tight timescale, to find out the information required.

You could then use a spider chart or other form of wall chart to summarise everybody's ideas.

Recording these initial ideas allows you to retain them for reference as the individual topics are developed. In this way, your students' progress in learning can be readily acknowledged.

Learning episode B2.1 Cell structure and size of specimens

Learning objectives

- Describe and compare the structure of a plant cell with an animal cell, limited to: cell wall, cell membrane, nucleus, cytoplasm, chloroplasts, ribosomes, mitochondria, vacuoles
- Describe the structure of a bacterial cell, limited to: cell wall, cell membrane, cytoplasm, ribosomes, circular DNA, plasmids
- Identify the cell structures listed above in diagrams and images of plant, animal and bacterial cells
- Describe the functions of the structures listed above in plant, animal and bacterial cells
- State that new cells are produced by division of existing cells
- State that specialised cells have specific functions, limited to:
  1. ciliated cells – movement of mucus in the trachea and bronchi
  2. root hair cells – absorption
  3. palisade mesophyll cells – photosynthesis
  4. neurones – conduction of electrical impulses
  5. red blood cells – transport of oxygen
  6. sperm and egg cells (gametes) – reproduction
- Describe the meaning of the terms: cell, tissue, organ, organ system and organism as illustrated by examples given in the syllabus
- State and use the formula: magnification = image size ÷ actual size
- Calculate magnification and size of biological specimens using millimetres as units
- **Supplement** Convert measurements between millimetres (mm) and micrometres (μm)

Common misconceptions

Some students may find it hard to grasp the tiny scale of cells and their components. It may help if they look at a piece of small newsprint or a miniature letter drawn with a sharp pencil under the microscope before viewing cells.

Resources

Student Book pages 41–53

Worksheet B2.1a Using a light microscope

Worksheet B2.1b Preparing a microscope slide

Worksheet B2.1c Organs of the human body

B2.1 Technician's notes

Resources for class practicals (see Technician's notes)

Approach

1. Introducing cells

Ask students to remember what they can about cells, and then ask one student to come to the board and sketch the outline of one cell. Ask other students to suggest other features that could be included, but encourage discussion first on whether the suggested feature is found in *all* cells and is therefore worthy of being included in the sketch. All features drawn should be labelled.

Make sure that the sketch includes the cell membrane, cytoplasm and nucleus containing DNA. Students should draw a fully labelled sketch of a cell with these features in their workbooks.

Ask students if they know how new cells are formed. They will look at cell division in detail later in the course (Section 17).

Take the opportunity to briefly revise SI units of length (based on the metre), and ask which unit might be most appropriate for measuring the length of a cell. Students should be able to respond with *millimetre*. Make sure they understand the relationship between millimetres and metres.

Supplement Ask students which unit would be more useful for measuring structures inside cells, such as the nucleus or even smaller structures. Introduce the *micrometre* and its symbol, μm, and its relationship to millimetres and metres (1000 micrometres = 1 millimetre, 1 micrometre = 1×10^{-6} m).

2. Using a light microscope

If your students are not already familiar with using a light microscope, this activity is a good opportunity to introduce it. Worksheet B2.1a provides basic instructions on viewing a prepared slide through a light microscope. It might help to demonstrate how to set up and focus, first using the low-power objective lens and then the high-power objective lens, to avoid the risk of crashing high-power objectives into slides (see Technician's notes). Students may need reminding of this several times in order to develop good technique.

SAFETY INFORMATION: Remind students always to focus first using the low-power objective, starting with it near to the slide, then moving it upwards to avoid crashing the lens into the slide. Also remind them to use the fine-focusing knob only while working with the high-power objective lens. If using natural light to see the specimen, students should not focus the sun's rays directly through the objective and eyepiece.

Any prepared slide of plant or animal tissue where it is possible to see a simple internal structure in cells will be suitable. Ideally you should show a labelled diagram or photograph on the board, to help students interpret what they see. Do not show tissues with complex structures, such as muscle cells, as this may be confusing.

Students should be expected to draw clearly one or two labelled cells from each slide. Remind them to use a sharp pencil for their drawings, and to draw only what they see even if it is not as clear and obvious as any drawing they have seen elsewhere. Students should not use shading in their drawings.

Step 11 on the worksheet involves students calculating the magnification of their drawings. Some students may need help with this. Use page 52 of the Student Book to introduce the formula for magnification: magnification = image size ÷ actual size. Students could also view a transparent plastic ruler or a simple graticule under the microscope to calculate the field of view, and use that to estimate the size of the cell they have drawn.

Supplement Expect students to use units of micrometres in their answers.

3. Preparing a microscope slide

Worksheet B2.1b provides instructions for making stained slides of plant cells (onion) and animal cells (human cheek cells). You may need to demonstrate how to place a coverslip on a slide to avoid trapping air bubbles (see Technician's notes).

Students often over- or under-stain material, so they may need more than one attempt at making a slide of each kind. If time is limited, have slides that you prepared earlier ready for display to the class in case they are unsuccessful.

Supplement This practical could be extended to look at cells from other sources, such as banana parenchyma or moss leaf. In each case, students should apply their skills and understanding of cell structure to interpret what they see under the microscope.

SAFETY INFORMATION: Before working with their own cheek cells, remind students of safety protocols and make sure that pots of disinfectant are available for discarded cotton buds, slides and coverslips. Demonstrate to students how to use a scalpel/knife safely (see Technician's notes). Check your school guidelines as using cheek cells may be prohibited. Animal liver cells could be used as an alternative or follow local advice.

While working with the high-power objective lens, use the fine-focusing knob only. If using natural light to see the specimen, remind students to make sure they do not focus the sun's rays directly through the objective and eyepiece. Handle coverslips with care – they are fragile and very sharp if broken. Students should only handle slides of their own cheek cells, and should place slides into disinfectant when they have finished. Remind students to take care when using a scalpel or knife.

4. Functions of animal and plant cell structures

Use pages 41–43 of the Student Book to introduce animal and plant cell structures and functions. Students should make notes and use these to test each other on the functions of the cell wall, cell membrane, nucleus, cytoplasm, chloroplasts, ribosomes, mitochondria and vacuole.

5. Bacterial cell structures

Use page 44 of the Student Book to introduce bacterial cell structures and functions. Students should make notes and use these to test each other on the functions of the cell wall, cell membrane, cytoplasm, ribosomes, circular DNA and plasmids.

6. Cell organisation

Organ system	Organ	Tissues	Cells

On the board, sketch a table like the one shown below. Make sure it has at least four rows.

Ask students to name organ systems found in the human body, and add each name to a different row in the first column. They should then suggest the names of any organs in each system, tissues in each organ and cells in each tissue.

Alternatively, students could copy the table outline and complete it as far as possible individually or in pairs. Take examples from around the class and check that there are no errors. Keep the table for the end of this learning episode, so that students can add further examples from what they have learned.

7. Organs of the human body

Worksheet B2.1c provides a human body outline on which students can draw the positions of some organs. They should include annotated labels to explain what each organ does.

8. Cell fact cards

Ask students to carry out research on the cell (or cells) of their choice to produce a 'cell fact file'. They should choose any cell found in animals or plants, such ciliated cells, root hair cells, palisade mesophyll cells, neurones, red blood cells, sperm or egg cells. They should draw a large, labelled diagram describing the cell structure and explaining how this is related to its function.

They can present their research in a poster or using presentation software.

9. Consolidation: chain reaction quiz

Give students a few minutes to write down as many quick quiz questions on cell structures and their functions as they can. They should also jot down the answers. Ask one student to read out a question, and choose another to answer it. If they answer correctly, they then read out one of their own questions. If they do not answer correctly, choose another student to answer. Continue round the class until most cell structures and functions have been covered.

Technician's notes

Be sure to check the latest safety notes on these resources before proceeding.

The following resources are needed for the class practical, B2.1a, per student or pair:

light microscope with low-power (e.g. ×4) and high-power (e.g. ×10) objective lenses
prepared slides of plant and/or animal tissue in which cell structures are easily visible (suitable examples are palisade cells in a leaf and squamous epithelial cells (cheek cells))
Check your school guidelines as using cheek cells may be prohibited. Animal liver cells could be used as an alternative or follow local advice.

The following resources are needed for the class practical, B2.1b, per student or pair:

eye protection
light microscope with low-power (e.g. ×4) and high-power (e.g. ×10) objectives, microscope slides and coverslips
mounted needle or blunt seeker
forceps
pipette
methylene blue stain
iodine solution – iodine in potassium iodide solution
onion – fewer onions are needed if a layer is peeled off for each student or pair
sharp knife or scalpel and cutting board
clean cotton buds, at least 1 per student; paper towel
pot of disinfectant, e.g. 1% Virkon
transparent plastic ruler or simple graticule

SAFETY INFORMATION: Cotton buds should be kept in disinfectant for long enough for them to be sterilised after students have used them. Please refer to the manufacturer's instructions. Scalpel blades break if they are used to cut whole onions, causing a risk of injury. Knives are more appropriate for cutting up onions, or students could be supplied with sections of onions from which to prepare slides.

Check your school guidelines as using cheek cells may be prohibited. Animal liver cells could be used as an alternative or follow local advice.

Answers

Page 44 Science in Context: Artificial cells

They can be made so their structure is perfectly suited to the job they are needed for.

Page 45 Science in Context: Bacterial plasmids

Organisms in the other kingdoms do not have plasmids. They can only transfer genes through reproduction, and this can only happen within the same species.

Pages 45–46 Developing Practical Skills

1. a) Place the slide on the microscope stage under a low-power objective lens. Use the stage clips to hold the slide in place. Make sure the light from the source passes through the slide to the objective lens (but not focused sunlight). Use the coarse-focusing knob to bring the lens close to the slide and then focus by moving it upwards to bring the slide into as sharp a focus as possible. Then use the fine-focusing knob to get the best focus.

b) The specimen should always be focused using low power before moving the high-power objective into position. Only the fine-focusing knob should be used to focus on high power. This should avoid crashing the objective into the slide because the high-power objective is longer than the low-power objective.
c) Make sure that the sunlight is not shining directly on the mirror so that it is focused through the microscope into the eye, as this can damage the eye.
2. a) Careful drawing with clean pencil lines of a white blood cell, with no shading, labelled to show the nucleus, cell
membrane (not really visible at this magnification but understood to be at the edge of the cell) and cytoplasm.

b) magnification = eyepiece ($\times$ 4) $\times$ objective ($\times$ 20) = $\times$ 80

3. The cells are animal cells. They do not have features of plant cells, e.g. cell wall, large vacuole.

Page 46

1. a) Drawing should be drawn with thin, clear pencil lines, no crossing out and no shading, to show the outline of the cell in the photograph and the central shape.
   b) Diagram should be labelled to show nucleus, cytoplasm and cell membrane.
2. a) cell wall, large vacuole, chloroplast
   b) cell wall, circular DNA, plasmids
3. a) chloroplast
   b) large vacuole
   c) cell wall
4. Mitochondria release energy during aerobic respiration. Ribosomes are where new proteins are formed.
5. Because they are too small to be seen properly with a light microscope.

Page 47 Science in Context: Stem cells

The stem cells could divide and specialise forming new neurones (nerve cells) to replace those that have been damaged.

Page 50

1. a) Lining some tubes in animal organs, such as the respiratory tract of humans; the cilia on the outside of the cells help move substances along inside the tubes.
   b) Throughout the body, conduct electrical impulses around the body.
   c) In blood; carry oxygen around attached to haemoglobin inside the cell.
   d) Near the tips of plant roots; have long cell extensions to increase surface area for absorption of substances into the root.
   e) In plant leaves; carry out most of the photosynthesis in plants.
2. Sperm cells are small, and have a tail(flagellum) for movement.
   Mitochondria provide energy for movement of the tail, and the acrosome contains enzymes which digest the egg cell membrane so the sperm nucleus can enter the egg cell for fertilisation.
   Egg cells are large and contain a lot of cytoplasm to provide nutrients for the fertilised cell during the early stages of division.

Page 51

1. a) Any suitable two, such as: muscle tissue, nervous tissue, bone tissue.
   b) Any suitable two, such as: heart, liver, brain.
   c) Any suitable two, such as: nervous system, digestive system, circulatory system.
2. a) Any suitable two, such as: palisade cells, root hair cells.
   b) Any suitable two, such as: leaf, root.

Page 53

1. 2 mm
   The magnification is image size ÷ actual size = 10 ÷ 0.5 = ×20
   So second image size = magnification × actual size = 20 × 0.1 = 2 mm
2. If you are not using a suitable magnification for the specimen you are looking at you may not be able to see what you want to. (It is most useful to start by focusing on a lower magnification and then moving up to the magnification you want to use.)
3. actual size = image size ÷ magnification = 2.5 mm ÷ 100
4. 0.025 mm = 25 μm

Worksheet B2.1a Using a light microscope

The light microscope is a very useful tool for biologists. It can be used to look at living specimens and sections of dead material that have been stained with coloured dyes to show up the cells more clearly. In this practical you will practise using a light microscope with a prepared slide.

Apparatus

light microscope with low-power and high-power objectives   prepared slide

SAFETY INFORMATION

Always focus first using the low-power objective lens, starting with it near to the slide, then moving it upwards to avoid crashing the lens into the slide.

While working with the high-power objective, use the fine-focusing knob only.

If using natural light to see the specimen, make sure you do not focus the sun's rays directly through the objective and eyepiece.

Method

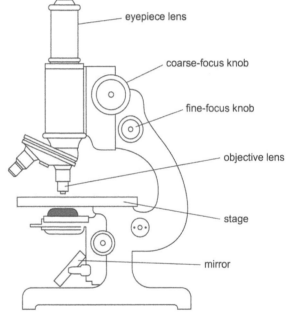

1. Check the magnification of the eyepiece and write it down. Also check the magnification of the low-power and high-power objectives and write them down.
2. Make sure the low-power objective is in the lowest position.
3. Place the prepared slide on the microscope stage and hold it firmly with the stage clips. Move it so the specimen on the slide is directly below the eyepiece.
4. Look through the eyepiece, and use the coarse-focus knob to bring the specimen into focus as best you can. (Note: practise working with both eyes open, as this will be more comfortable over long periods of working.)
5. Use the fine-focus knob to focus clearly on the specimen. You may want to move the slide around a little at this point, to bring the most interesting part of the slide into the middle of the field of view (the area you can see). If so, make sure you refocus with the fine-focus knob.
6. Draw the area of interest in your book, using a sharp pencil to make a neat drawing without any shading. Try to label what you can see.
7. Choose a part of the slide that you want to look at in more detail and make sure it is in the centre of the field of view and properly focused. Then move the high-power objective in to replace the low-power objective.
8. **Use the fine-focusing knob only** to focus the specimen clearly.
9. Draw the area of interest, e.g. one cell, and label it clearly.
10. Before removing the slide from the microscope stage, replace the high-power objective with the low-power objective in the lowest position.
11. Calculate the magnification of your drawings by multiplying the magnifying power of the objective with the magnifying power of the eyepiece lens, and add the magnifications to your drawings.

Worksheet B2.1b Preparing a microscope slide

Sometimes in your work you will need to prepare your own slides. In this practical you will take samples of plant and animal cells to prepare slides, and then stain them to show key structures more clearly. You will then use a microscope to look at your slides. If you are not confident when using the microscope, use Worksheet B2.1a: Using a light microscope, to guide you.

Apparatus

microscope with low- and high-power objective lenses

microscope slides and coverslips

mounted needle

forceps

pipette

water

methylene blue stain

iodine solution

onion

sharp knife or scalpel and cutting board

clean cotton buds

filter paper or paper towel

pot of disinfectant

SAFETY INFORMATION

While working with the high-power objective, use the fine-focusing knob only.

If using natural light to see the specimen, make sure you do not focus the sun's rays directly through the objective and eyepiece.

Handle coverslips with care – they are fragile.

Only handle slides of your own cheek cells, and place slides into disinfectant when you have finished.

Take care when using a knife. Follow your teacher's instructions.

Wear eye protection.

Method

Plant cells

1. Peel a thick layer from an onion, and find the very fine sheet of cells inside that layer. Use the forceps to carefully remove the sheet of cells and then cut a small piece about 1 cm square. Place this in the middle of a clean microscope slide and add one drop of iodine solution to the specimen using a pipette.
2. Place one edge of a coverslip to one side of the specimen and support the opposite edge with a mounted needle. Carefully lower the coverslip over the specimen so that no air bubbles are trapped.

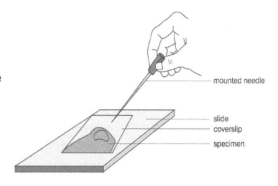

Page 1 of 2

3. Use the filter paper to gently blot any excess liquid from the slide.
4. Place the slide on the microscope stage and view it first with a low-power objective lens and then a high-power objective lens. Draw one or two cells from what you can see and label the structures that are visible.

Animal cells

1. Use a clean cotton bud to wipe the inside of your cheek in your mouth.
2. Wipe the cotton bud across the middle of a clean microscope slide to cover an area of about 1 cm square. **Then put the cotton bud straight into the pot of disinfectant.**
3. Place one drop of methylene blue stain on the smear on the microscope.
4. Place one edge of a coverslip to one side of the smear and support the opposite edge with a mounted needle. Carefully lower the coverslip over the smear so that no air bubbles are trapped.
5. Use the filter paper to gently blot any excess liquid from the slide.
6. Place the slide on the microscope stage and view it first with a low-power objective lens and then a high-power objective lens. Draw one or two cells from what you can see and label the structures that are visible.

Analyse and interpret data

1. Calculate the magnification at low power and high power of your microscope. An example is shown in the table below. Add these magnifications to your drawings.

Eyepiece magnification	Objective magnification	Overall magnification
×10	×4	×40

2. What was the effect of the stain? What did it help you to see more clearly?
3. Compare the structures of the animal cells and plant cells that you looked at. Describe any similarities and any differences.

Worksheet B2.1c Organs of the human body

Use the outline of the human body below to mark the positions of the major organs including: stomach, small intestine, large intestine, liver, pancreas, heart, lungs, kidneys.

Draw the organs in the right place and try to make them the right size and shape.

Label the organs and add short notes to say what you think each organ does.

Compare your diagram with your neighbour's. What has your neighbour drawn well? What will they need to improve?

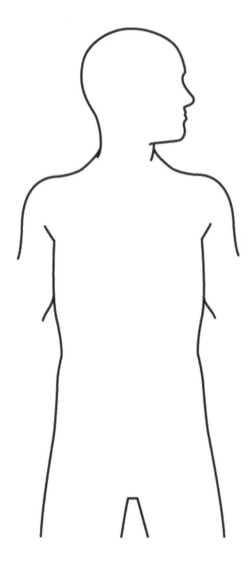

68

Learning episode B2.2 Consolidation and summary

Learning aims

- Review the learning points of the topic summarised in the end of topic checklist
- Test understanding of the topic content by answering the end of topic questions

Resources

Student Book pages 54–57

Approach

Introduce the learning episode

Ask students to work with a partner to make a list of key words from this topic. They could then work together to produce a spider diagram showing how the different concepts are linked. They could compare their list with the list of key terms given on page 54 of the Student Book. Discuss the checklist on pages 54–55 and use questioning to see how much of the content they are comfortable with.

Students could make flashcards of the key content and then use the flashcards to quiz each other on the information.

Develop the learning episode

Ask students to work individually through the end of topic questions on pages 56–57 of the Student Book without looking at the text. As they work, walk around the classroom observing their answers and asking questions as necessary to find out which questions are causing difficulties.

Finish the learning episode

After a set period, ask the students to stop working and discuss any areas of difficulty you observed as you walked round the class. Students should complete any unanswered questions for homework, but you should stress that they should try to answer the questions without looking at the text, so that they can see how much they have remembered.

Answers

End of topic questions mark scheme

The marks available for a question can indicate the level of detail you need to provide in your answer.

Question	Correct answer	Marks
1	Table similar to the following, showing the same information: | **Plant cell**	**Animal cell**
---	---
has a nucleus	has a nucleus
surrounded by cell membrane	surrounded by cell membrane
surrounded by cellulose cell wall	no cell wall
cytoplasm fills cell	cytoplasm fills cell
usually large central vacuole	no large central vacuole
may contain chloroplasts	no chloroplasts		12 marks
2	C	1 mark
3 a)	Structure that contains the genetic material and controls cell division.	1 mark
3 b)	Structure surrounding a cell that controls what enters and leaves the cell.	1 mark
3 c)	Material inside the cell that contains the cell structures and in which many chemical reactions take place.	1 mark
4 a)	Animal cells are surrounded by a <u>cell membrane</u> that controls what enters	1 mark

	and leaves the cell.	
4 b)	<u>Some</u> plant cells contain chloroplasts.	1 mark
4 c)	<u>Only</u> plant cells contain a large central vacuole in the middle of the cell.	1 mark
5	As red blood cells have no nucleus, there is nothing to control cell division. So they cannot divide like many other cells to replace damaged cells and must be replaced with new cells from the bone marrow.	1 mark 1 mark

6	Any three suitable systems, such as:					1 mark for each column for 3 suitable systems, to a max. of 15
	Organ system	**Function**	**Organs in this system**	**Tissues in these organs**	**Cells in these tissues**	
	respiratory	gas exchange between body and atmosphere	lungs	epithelial tissue lining tubes and air spaces	epithelial cells	
	nervous	receives stimuli from surroundings and coordinates responses	brain spinal cord nerves	nervous tissue	neurones / nerve cells	
	circulatory	carries substances around the body	heart blood vessels	muscle tissue epithelial tissue blood	muscle cells epithelial cells blood cells	

7	By having specialised cells, the body of a multicellular organism can develop in many different ways, such as developing muscles for rapid movement or producing flowers for reproduction. This makes a huge range of different body designs possible and means that the cells, tissues, organs and systems can carry out the life processes more efficiently.	1 mark 1 mark 1 mark
8	organism, organ system, organ, tissue, cell	3 marks
9 a)	Group of similar cells that carry out a similar function.	1 mark
9 b)	Several tissues that are grouped together to carry out a particular function.	1 mark
9 c)	Several organs that work together to carry out a particular function.	1 mark
Supplement 10	image size = magnification × actual size = 100 × 0.1 mm = 10 mm = 10 000 μm	1 mark 1 mark 1 mark
	Total:	48 marks

B3 Movement into and out of cells

Introduction

This section covers the topics of diffusion, osmosis and active transport to help students understand how substances can move into and out of cells. Students will need to apply this knowledge in later topics in relation to processes such as gas exchange, nutrition and transport.

Note that a number of practical activities are suggested for the activities on diffusion and osmosis. You may wish to select what can be fitted into your lessons – this need only be one or two practical activities.

Links to other topics

Topics	Essential background knowledge	Useful links
1 Characteristics and classification of living organisms	1.1 Characteristics of living organisms	
2 Organisation of the organism	2.1 Cell structure and size of specimens	
6 Plant nutrition		6.1 Photosynthesis 6.2 Leaf structure
7 Human nutrition		7.5 Absorption
8 Transport in plants		8.2 Water uptake 8.3 Transpiration
9 Transport in animals		9.4 Blood
11 Gas exchange in humans		11.1 Gas exchange in humans
12 Respiration		12.2 Aerobic respiration
13 Excretion in humans		13.1 Excretion in humans
16 Reproduction		16.4 Sexual reproduction in humans

Topic overview

B3.1	**Diffusion** This learning episode introduces students to the concept of diffusion in general, and later to diffusion across membranes. Students will link the energy for diffusion to the kinetic energy of particles, and investigate factors that affect the rate of diffusion.
B3.2	**Osmosis** This learning episode provides students with a range of opportunities to investigate osmosis across partially permeable membranes. Supplement Students are given the opportunity to link their learning on osmosis to water potential, and are expected to interpret the effect of solutions on living cells in terms of water potential and turgor pressure.

B3.3	**Active transport**
	This short learning episode helps students to distinguish active transport from the passive processes of diffusion and osmosis.
	Supplement Students are given the opportunity to discuss the process of active transport in particular situations.
B3.4	**Consolidation and summary**
	This learning episode quickly recaps on the ideas encountered in this section. Students answer the end of topic questions in the Student Book.

Careers links

These are some scientific careers that focus on this area of biology but careers in many other fields use the knowledge and skills gained studying science. Dialysis therapy requires collaboration between many different people. For example, **biomedical engineers** design the dialysis machines and software to run them, **clinical engineers** manage the day-to-day operation of the machines, and **renal nurses** look after patients undergoing dialysis.

Starting points

The Student Book section opener puts the ideas in the section into context and sets the scene. It also allows students to acknowledge and value their prior learning, and provides a benchmark against which future learning can be compared.

- The questions provide a structure for introducing the section and can be used in a number of different ways:
- You could ask students to consider the questions as an introductory homework task.
- You could put students into groups to share their own ideas and understanding and then to report back to the whole class.
- Students could be given access to the Internet, preferably with a tight timescale, to find out the information required.

You could then use a spider chart or other form of wall chart to summarise everybody's ideas.

Recording these initial ideas allows you to retain them for reference as the individual topics are developed. In this way, your students' progress in learning can be readily acknowledged.

Learning episode B3.1 Diffusion

Learning objectives

- Describe diffusion as the net movement of particles from a region of their higher concentration to a region of their lower concentration (i.e. down a concentration gradient), as a result of their random movement
- State that the energy for diffusion comes from the kinetic energy of random movement of molecules and ions
- State that some substances move into and out of cells by diffusion through the cell membrane
- Describe the importance of diffusion of gases and solutes in living organisms
- Investigate the factors that influence diffusion, limited to: surface area, temperature, concentration gradient and distance

Common misconceptions

Many students confuse the term *diffusion* with simple movement and think that once diffusion has stopped, the particles have stopped moving too. This can lead to difficulties in understanding the process. It is worth reinforcing the point that diffusion is *net movement*, by repeating this definition every time you use the word 'diffusion' early on in this work.

A video clip or some kinaesthetic work (with the students acting as particles) can also help, as suggested below.

Resources

Student Book pages 61–65

Worksheet B3.1a Diffusion across visking tubing (dialysis tubing)

B3.1a Technician's notes

Worksheet B3.1b Temperature and the rate of diffusion

B3.1b Technician's notes

Worksheet B3.1c Surface area and the rate of diffusion

B3.1c Technician's notes

Resources for class practicals (see Technician's notes)

Approach

The practical work on diffusion in Worksheets B3.1a and B3.1b uses identification tests for glucose and starch. These are not covered by the syllabus until Section 4 in relation to food tests. If you wish to use Worksheets B3.1a and B3.1b to focus on the general factors that affect diffusion, you will need to briefly explain the tests for glucose and starch beforehand. Details of these are given on Worksheet B4.3.

Alternatively, you could retain these worksheets to use in Section 7 to focus on the role of diffusion in the human alimentary canal.

1. Introduction to diffusion: demonstration

Place a crystal of potassium manganate(VII) (potassium permanganate) into a beaker of water and ask students to watch and describe what happens. Or put a slice of slice of coloured root vegetable that will leach colour easily, such as beetroot, into a beaker of water and ask students to watch and describe what happens (see Technician's notes).

SAFETY INFORMATION: Potassium manganate(VII) is harmful and should be handled with tongs or disposable gloves. Dispose of the solution carefully to avoid splashes, as the colour stains permanently.

Alternatively, spray a small amount of strong perfume at one side of the room and ask students to put their hand up when they smell the perfume.

Ask them to think at the level of particles and to try and imagine what is happening in the beaker or room. If possible, show them a video of Brownian motion. You will need to explain that the dots they see are not molecules, but larger particles that are being constantly jiggled by the movement of molecules that are too small to see.

Explain that particles are always in a state of movement as a result of the energy they have – the greater the temperature, the more energy they have and so the more they move (this should reinforce learning in physics).

Ask students to consider the effect of distance on diffusion by asking how quickly particles reach a specified distance. Extrapolate this to the importance of distance for diffusion of substances within an organism. Students should appreciate that the shorter the diffusion distance, the more rapidly particles will reach their destination.

(Note that the examples given may also involve mass flow of gas or solution, e.g. by convection, more than diffusion, because diffusion is only effective quickly over a very short distance. However, the examples do help students to visualise that particles are always in constant movement, and so helps with understanding of diffusion.)

2. Moving as particles

You can model the effect of the colour, or perfume, spreading using students as particles. Choose a small number of sensible students who are wearing similar coloured tops (or provide hats or something to distinguish them from other students). Within a fairly limited area, ask students to walk in straight lines, and when they come up against another student or a boundary edge, to change direction like billiard balls. After a few minutes, ask your key group of students to raise their hands, so that everyone can see how they have become spread out.

Give students a few minutes to work in pairs or small groups to find an explanation for the movement of colour through the solution, or perfume, across the room. Then ask for a selection of answers from different groups and discuss any differences. The key point is that the movement out from the area of high concentration is a result of the random movement of the particles.

It is also worth asking whether any particles of colour/perfume are moving back towards the area of high concentration. As a result of random motion, some will be. However, there will be more particles moving from the area of higher concentration to the area of lower concentration, and this is what we notice.

Then ask, when the colour is evenly distributed in the beaker, or the perfume in the air within the room, what happens to the particles. It is essential students understand that the movement continues but that, as there is no difference in concentration between areas, we cannot perceive any change.

Introduce the idea of *net movement*, that is the sum of movements in different directions, and go on to use this to define the term *diffusion*.

3. Diffusion and cell membranes

In biology we are most interested in the effect of diffusion as a mechanism for getting substances into and out of cells. The practical work for this is covered in Worksheet B3.1a in relation to diffusion in general. However, this practical work could be retained until Section 7 for use in the context of absorption of digested food materials from the human alimentary canal, where it depends on knowledge of the tests for starch and glucose.

The practical uses visking tubing (dialysis tubing) as a model for a cell membrane, and acts as a good model for understanding that some molecules (e.g. glucose) can cross cell membranes easily, and therefore can be redistributed in the system as a result of diffusion, whereas others (e.g. starch) cannot. Students often find visking tubing difficult to use. If you prefer, you could carry out the practical as a demonstration, asking students to predict what they think will happen and to explain their predictions using what they know about diffusion and partially permeable membranes (see Technician's notes).

SAFETY INFORMATION: Demonstrate how to connect the visking tubing (dialysis tubing) to the cut-off syringe. Students should wear eye protection when using iodine solution and Benedict's solution. Remind students to take care when heating solutions in the water bath for the glucose test (see Technician's notes).

4. Factors affecting the rate of diffusion

Students need to understand the main factors that affect the rate of diffusion of substances. They will need to apply this information later to their understanding of processes such as gas exchange, nutrition and excretion.

Distance: The effect of distance on diffusion was covered in the introductory task. This effect will be covered again in relation to diffusion of food molecules into the blood from the small intestine in Section 7: *Human nutrition,* and *The diffusion of gases between cells, the blood and the lungs*. It will also be covered in Section 9: *Transport in animals*, Section 11: *Gas exchange in humans*, and *The diffusion of substances in the kidneys in excretion* and Section 13: *Excretion in humans*.

Temperature*:* Worksheet B3.1b gives students the opportunity to plan an experiment into the effect of temperature on the rate of diffusion using a visking tubing (dialysis tubing) system. Discuss the prediction with them, and if they seem unsure, remind them about the effect of increasing temperature on the energy of particles. If there is time, and their plan is acceptable, students could carry out their plan.

Encourage students to carry out two or more repeats themselves. Please note that if students do carry out two repeats at three different temperatures, several hours will be needed for the practical. If there is a limited amount of time available for the practical, students could pool their results, instead of doing the repeats. If you decide to do this, draw students' attention to the possibility of bringing in uncertainties by sharing results with neighbours.

Surface area*:* Worksheet B3.1c gives students the opportunity to investigate the rate of diffusion into visking tubing (dialysis tubing) of different lengths, to explore the effect of surface area on the rate of diffusion. Technically, what is happening in this experiment is osmosis, as the substance that crosses into the tubing through the membrane is water. As osmosis is not covered until the next Learning episode, it may be best to ignore this point for now. However, the model is useful as it can be applied to any substance that diffuses through a partially permeable membrane.

Students should find that the tube of longest length absorbs most water because it has the largest surface area over which diffusion can take place. More able students could calculate the surface area of each piece of tubing and use this in their analysis, but using length as a proxy for surface area is sufficient as long as students are aware of its relationship to surface area.

For both these worksheets, check that that students have drawn up a suitable table for their results before they carry out the tests.

Concentration gradient: For the effect of concentration gradient on the rate of diffusion, ask students to draw a cartoon, or develop a model that helps to explain why the rate of diffusion is greater when the concentration gradient is greater. This will provide an opportunity to reinforce what they have learned about diffusion generally. Take time to check students' understanding of the basic principles at this point, as this will underpin a lot of later work.

SAFETY INFORMATION: Students should wear eye protection when using iodine solution and Benedict's solution. Remind students to take care when heating solutions in a hot water bath.

Technician's notes

Be sure to check the latest safety notes on these resources before proceeding.

The following resources are needed for the demonstration in the introduction:

250 cm³ beaker of water
crystal of potassium manganate(VII) *or* slice of a root vegetable that leaches colour easily, such as beetroot

The following resources are needed for the class practical, B3.1a, per group:

5 cm^3 10% glucose solution (100 g dm^{-3}) and 5 cm^3 10% starch solution
15 cm length of visking tubing (dialysis tubing) and elastic band, to attach visking tubing to syringe end
sawn-off end of 10 cm^3 plastic syringe (the plunger end)
100 cm^3 beaker
tap water
iodine solution in dropper bottle
spotting tile
Benedict's solution + water bath at 80 °C and heatproof tongs
boiling tube
teat pipette
eye protection

In this practical, students investigate the diffusion of substances across a partially permeable membrane using visking tubing (dialysis tubing). The visking tubing could be prepared ahead of the lesson, as students may have difficulty with it.

Students should find that glucose diffuses through the tubing into the water, but not starch. This is because the glucose molecules are small enough to pass through the holes in the membrane, and are therefore free to diffuse from the region of higher concentration (inside the tubing) to the area of lower concentration (water outside the tubing). The starch molecules are too large to move through the holes in the membrane, so are not subject to diffusion.

Note that dry visking tubing (dialysis tubing) needs to be soaked for a few minutes in water before use. This can be done ahead of the lesson.

The following resources are needed for the class practical B3.1b, per group:

5 cm^3 10% glucose solution (100 g dm^{-3})
15 cm length of visking tubing (dialysis tubing)
sawn-off end of 10 cm^3 plastic syringe (the plunger end)
elastic band, to attach visking tubing to syringe end
100 cm^3 small beaker
tap water
Benedict's solution + water bath at 80 °C and heatproof tongs
boiling tube
teat pipette
eye protection

This practical can be used as a planning exercise or, if there is time, students may be allowed to carry out their plans after approval by you. The method is based on the one used on Worksheet B3.1a. If students have not had the chance to carry that out, they should be given the worksheet to help them develop their planned method.

If students carry out their investigation, they should find that the higher the temperature, the faster the rate of diffusion (the sooner that glucose is detected in the water). This is because at a higher temperature a molecule has more kinetic energy and so moves faster, increasing the chances of it reaching the membrane and passing through one of the holes to the other side.

The equipment list details what students will need. They may suggest other equipment in their plans. Plans should be checked for the equipment suggested, and alternatives offered if some suggestions are not available.

The following resources are needed for the class practical B3.1c, per group:

3 × 12cm lengths of visking tubing (dialysis tubing)
15 ml sugar solution, 5% (use sucrose, not glucose and use fresh solution)
ruler
3 paper clips
beaker containing about 150 ml of distilled water
dropping pipette
tongs
access to an electronic balance
clock or watch
3 sheets of paper towel, for removing excess water

This practical investigates the rate of diffusion of water (osmosis, but see point in notes above)using visking tubing (dialysis tubing) to represent cells of different surface area.

Visking tubing should be wet to soften it. Students may find it difficult to tie a knot in the visking tubing. The tubing could be provided with a knot already tied at one end.

Answers

Page 63 Science in Context: Kidney failure and haemodialysis

Substances that need to remain in the blood will be at a similar concentration in the dialysis fluid so they do not diffuse out of the patient's blood. Substances that do need to be removed from the patient's blood will be at a much lower concentration in the dialysis fluid so they will diffuse from the patient's blood to the dialysis fluid.

Page 65

1. Any answer that means the same as the following:
   net movement – the sum of movement in all the different directions possible
   diffusion – the sum of the movement of particles from an area of high concentration to an area of lower concentration in a solution or across a partially permeable membrane.
2. Passive, because no energy is provided by the cell for it to happen.
3. Only particles that are small enough to pass through the membrane can diffuse. Larger molecules cannot diffuse through the membrane.

Worksheet B3.1a Diffusion across visking tubing (dialysis tubing)

Visking tubing (dialysis tubing) is an artificial membrane. Like cell membranes, it is partially permeable so that small molecules will move through it but large molecules will not. We can use visking tubing to model which molecules can enter cells by diffusion and which cannot.

In this practical, you will use visking tubing to investigate whether glucose and starch molecules can diffuse through the membrane. You will describe the tests for starch and glucose that you will use in this practical. An alternative to using Benedict's solution to test for glucose is to use Clinistix®, which change colour in the presence of glucose.

Apparatus

glucose solution	starch solution	small beaker
sawn-off end of plastic syringe	elastic band	iodine solution in dropper bottle
tap water	pipettes	
spotting tile	boiling tube	
eye protection	visking tubing (dialysis tubing)	

Benedict's solution + water bath at 80 °C and heatproof tongs, or Clinistix®

SAFETY INFORMATION

Wear eye protection when using iodine solution and Benedict's solution.

Take care when heating solutions in the water bath for the glucose test.

Method

boiling tube

visking tubing

water

1. Put on eye protection and gloves.
2. Tie a tight knot in one end of the visking tubing (dialysis tubing) and attach the other end to the sawn-off syringe end with an elastic band.
3. Half fill the boiling tube with water.
4. In a small beaker, mix about 5 cm³ glucose solution with a similar amount of starch solution.
5. Carefully fill the visking tubing with the starch/glucose mixture, and wash the outside of the tubing with fresh water to remove any spills.
6. Suspend the tubing in the boiling tube of water as shown in the diagram.
7. After 15 minutes, use a pipette to remove some of the water surrounding the visking tubing in the boiling tube. Test the water for glucose. Record your results.
8. Test some more of the water surrounding the visking tubing for starch. Record your results.

Handling experimental observations and data

9. Describe your results.
10. Explain your results using your scientific knowledge about membranes.
11. Explain what your results suggest about the diffusion of glucose and/or starch through partially permeable membranes.

Worksheet B3.1b Temperature and the rate of diffusion

Many factors can affect rate of diffusion, including temperature, concentration gradient and surface area.

Using the method in Worksheet B3.1a, with only glucose solution in the tubing, plan an investigation into the effect of temperature on the rate of diffusion. The apparatus list contains some of the equipment you need, but you may add other items.

You will need to consider:

* the range of temperatures you will test

* the number of repeats of each temperature you will use to get reliable results

* how frequently you will test each tube to give reliable results of the rate of diffusion

* how you will decide the end point of each test

* apart from the safety notes below, what other risks should be considered with this method and how they should be handled.

When you have written your plan, show it to your teacher. When your plan has been checked, your teacher will let you know if you can carry out your investigation.

Apparatus

glucose solution	visking tubing (dialysis tubing)	sawn-off end of plastic syringe
elastic band	small beaker	tap water
boiling tube	pipettes	eye protection

Benedict's solution + water bath at 80 °C and heatproof tongs, or Clinistix®

SAFETY INFORMATION
Wear eye protection when using iodine solution and Benedict's solution.
Take care when heating solutions in a hot water bath.

Prediction

1. Write a prediction for your experiment. Explain your prediction using your scientific knowledge.

Handling experimental observations and data

2. Record your results in a suitable table.
3. Identify any anomalous values in repeat experiments and, ignoring those, calculate an average time for each temperature.
4. Use your averaged results to draw a suitable graph.
5. Describe any patterns in your results.
6. Explain any patterns in your results using your scientific knowledge.

Planning and evaluating investigations

7. Describe any problems that you had in carrying out your experiment and how they may have affected your results.

Worksheet B3.1c Surface area and the rate of diffusion

The organs of large, multi-celled organisms where diffusion takes place, such as the lungs and small intestines in humans, have adaptations that increase their surface area. In this practical, you will investigate the effect of surface area on the rate of diffusion of a substance through cell membranes of different surface areas.

Apparatus

visking tubing (dialysis tubing), three lengths

sugar solution

dropping pipette

3 paper clips, ruler

beaker of distilled water

stopwatch or clock

access to an electronic balance

eye protection

disposable gloves or tongs

paper towel

SAFETY INFORMATION
Wear eye protection

Method

1. Put on eye protection.
2. Wet the visking tubing (dialysis tubing) to make it pliable then tie a knot close to one end of it. Rub the other end between your fingers to open the tube. Do this to all three pieces of tubing.
3. Using a dropping pipette, measure 3 ml of sugar solution into one length of visking tubing. Excluding as much air as possible, roll up the open end of the tubing until there is 4 cm between the knot and the rolled-up part. Hold the rolled-up part in place with a paper clip, taking care not to puncture the tube.
4. Repeat this procedure for the other two pieces of visking tubing but roll one up to a length of 6 cm and the other to 8 cm.
5. Dip the three lengths of visking tubing in the beaker of distilled water, carefully remove excess water and weigh them. Note the weight (mass) of each tube.
6. Place the three lengths of visking tubing in the beaker of distilled water and start the timer. After 3 minutes, remove all three tubes, carefully remove excess water and reweigh them.
7. Calculate the change in weight (mass) of each tube and record this in a suitable table.
8. Repeat steps 6 and 7 two more times.

Handling experimental observations and data

9. Use your final results to draw a suitable chart or graph.
10. Describe any patterns you see in your data.
11. Explain the implications of your results for living organisms.
12. Explain why it was important to repeat the measurements as in step 7.

Learning episode B3.2 Osmosis

Learning objectives

- Describe the role of water as a solvent in organisms with reference to digestion, excretion and transport
- State that water diffuses through partially permeable membranes by osmosis
- State that water moves into and out of cells by osmosis through the cell membrane
- Investigate osmosis using materials such as dialysis tubing
- Investigate and describe the effects on plant tissues of immersing them in solutions of different concentrations
- State that plants are supported by the pressure of water inside the cells pressing outwards on the cell wall
- Supplement Describe osmosis as the net movement of water molecules from a region of higher water potential (dilute solution) to a region of lower water potential (concentrated solution), through a partially permeable membrane
- Supplement Explain the effects on plant cells of immersing them in solutions of different concentrations by using the terms: turgid, turgor pressure, plasmolysis, flaccid
- Supplement Explain the importance of water potential and osmosis in the uptake and loss of water by organisms

Common misconceptions

As with diffusion, the important concept is *net movement*, not simply the movement of the water particles.

This section can be very confusing because we usually talk in terms of the concentration of *solute* molecules in a solution, not the concentration of *water* molecules. If you talk in terms of the concentration of water molecules, then it should be clear to students that osmosis is just diffusion in terms of water molecules. This is particularly important in living systems, as water is essential for life – hence it is given a special name.

Expect students to be precise in their definition of which concentration they are talking about, water or solute, in their discussions and writing.

It is essential that students appreciate the difference between the relatively elastic, but weak, cell membrane and the inelastic, strong plant cell wall, otherwise they will have difficulty in understanding the importance of water in the support of plants (or in the bursting of animal cells).

Resources

Student Book page 66–73

Worksheet B3.2a Osmosis and visking tubing (dialysis tubing)

3.2a Technician's notes

Worksheet B3.2b Measuring osmosis by change in mass

3.2b Technician's notes

Worksheet B3.2c Osmosis and change in shape

3.2c Technician's notes

Supplement Worksheet B3.2d Water potential

Resources for class practicals (see Technician's notes)

Approach

1. Introduction

Show students two potted plants, one that has been well-watered and one that is wilting. (Alternatively show pictures of a well-watered plant and a wilted plant.) Give students a short while to suggest why the plants are different, and to suggest how the wilted plant could recover. Take examples from around the class to get answers related to watering. Demonstrate recovery of the wilted plant by adding water and comparing the difference at the end of the lesson (see Technician's notes).

Ask students to think about what other ways water is important to living organisms. Students write down own ideas and then share with a partner and then with the whole class. Lead students to the idea that its roles in digestion, excretion, transport and so on, depend upon water being a solvent.

2. Osmosis

Prepare two eggs before the lesson by placing the unboiled eggs in vinegar (or other dilute acid) for 3–4 days to dissolve the shell. Then place each egg in a conical flask and cover one egg with a concentrated sugar solution and the other with water. After 24 hours, this should result in one shrunken egg that can be removed easily from the flask, and one large 'full' egg that cannot be extracted easily from the flask.

Show these to students and explain that the egg is like a large cell, with the membrane surrounding the egg acting as a cell membrane. Ask them to suggest how you produced the difference between the eggs.

Alternatively, show them some photos of plasmolysed (shrunken) and turgid (full) plant cells and ask them to describe and explain any differences in the movement of substances in the cells.

If necessary, remind students that water molecules are small enough to pass through cell membranes. Introduce the term *osmosis* as the special case of diffusion that considers the movement of water molecules.

The investigation in Worksheet B3.2a could be used as a demonstration, or class practical, to help students understand the effect of partially permeable membranes on osmosis and help them to visualise what effect osmosis can have on living cells (see Technician's notes).

SAFETY INFORMATION: Demonstrate how to connect the visking tubing (dialysis tubing) to the glass tubing (see Technician's notes).

Supplement 3. Defining osmosis

Using the definition of diffusion that they learned earlier, ask students to produce a definition for the term *osmosis*. This should raise the opportunity to discuss the need to distinguish between the concentration of solute molecules and concentration of water molecules in a solution when talking or writing about osmosis.

4. Factors affecting osmosis

Worksheets B3.2b and B3.2c provide two opportunities for students to investigate the effect of concentration of solution on osmosis in living systems. Worksheet B3.2b uses a straightforward method, and uses change in mass to measure the gain or loss of water by cells. Worksheet B3.2c is more challenging and uses the change in curvature of pieces of plant stem as a result of differential absorption of water by inner and outer cell layers. It also gives students the opportunity to produce a dilution sequence of solutes for the investigation.

For Worksheet B3.2c, the stems need to be left for at least 2 hours, so it may be more appropriate to leave them until the next lesson.

If necessary, discuss with students how they are going to record their results in a table and what is the most suitable graph for them to draw. Worksheet B3.2b also asks students to consider what errors and uncertainties there may be in their results.

Supplement Give students opportunities to apply their knowledge of the effect of temperature and surface area on diffusion to their understanding of osmosis. This could be done by simple questioning, or by asking students to draw, or describe, what happens when cells are placed in solutions of different temperature, or the problems for larger organisms that need to take in water.

SAFETY INFORMATION: Students should take care with sharp knives. Students should wash their hands after handling plant material.

Supplement **5. Water potential**

Worksheet B3.2d offers students questions that will give them practice in the understanding of water potential. This is best used after they have read the section on water potential on pages 66–67 of the Student Book to test how well they have understood the concept.

Comparing negative numbers can be confusing, so consider providing more examples like the one on the worksheet for question 4 if students seem to be having difficulty with this. If students have difficulty understanding that a water potential of e.g. -2.1 (MPa) is higher than -2.5 (MPa), then ask them to think about temperatures below zero: e.g. -2.1 °C is higher (warmer) than -2.5 °C.

6. Consolidation

Return to the potted plants shown in the introduction and ask students to describe the change in the wilted plant. (Over a period of about 1 hour, it should show significant recovery.) Ask students to sketch a cell from the plant at the start of the lesson and the same cell at the end of the lesson. They should annotate their sketches to show what has changed.

Supplement Expect students to include the terms *turgor pressure*, *water potential*, *plasmolysis* or *flaccid* in their annotations.

Technician's notes

Be sure to check the latest safety notes on these resources before proceeding.

The following resources are needed for the demonstrations in tasks 1 and 2:

two similar potted plants, one well-watered and the other unwatered for long enough to be wilted but not dead (soft-leaved and soft-stemmed plants are best, e.g. *Impatiens* (busy Lizzie))
two unboiled eggs placed in vinegar (or other dilute acid) at least 4 days before the lesson and left for 3–4 days to dissolve the shell, then 24 hours before the lesson, place each egg in a conical flask and cover one egg with a concentrated sugar solution and the other with water alternatively, photomicrographs of plasmolysed and turgid plant cells

The following resources are needed for the class practical B3.2a, per group:

plastic/glass tube about 10 cm long, fairly wide bore so that the solution can be pipetted into it
15 cm length of visking tubing (dialysis tubing)
10% starch solution or other solution of substance that will not diffuse through the visking tubing
water
100 cm³ beaker or boiling tube
clamp + stand
wire or elastic band to attach tubing to plastic/glass tube
teat pipette
marker pen

This investigation tests understanding of osmosis. It can be carried out as a class demonstration, or as a practical by students working in groups (see Technician's notes). The set-up for the apparatus is shown on the worksheet. The key point is to get students to predict what will happen to the level of liquid in the plastic/glass tube after a few hours, and to explain their prediction.

Plastic tubing should be used if possible as there is less danger of it shattering when students attach the visking tubing (dialysis tubing) to it.

The equipment will need to be set up and left for a few hours, or preferably overnight, to show a significant change in liquid level.

After this time, the level of liquid should have risen in the glass tube. This is because the concentration of water molecules in the starch solution inside the visking tubing is lower than in the water in the beaker, and the starch molecules do not diffuse through the membrane. The net movement of water molecules (osmosis) will be into the visking tubing, increasing the volume of the solution, so liquid level rises.

The following resources are needed for the class practical B3.2b, per group:

5 boiling tubes + rack, 100 cm^3 beaker
10 cm^3 coloured fruit juice or fruit cordial
distilled water
glass rod
paper towels
fresh root vegetable, for example a potato (see note below)
cork borer
sharp knife
weighing balance (must be capable of weighing to 0.001 g)
10 cm^3 measuring cylinder
marker pen

This investigation uses change in mass of tissue samples to measure the effect on osmosis of solutions of different concentration. Any suitable plant tissue can be used, but root vegetables are particularly useful due to their volume. Provide whole vegetables, from which students can cut their own samples. Alternatively take cores from the vegetables as close to the lesson as possible and store them in water.

Any suitable solution can be used, but choose one that is cheap and safe to handle. The worksheet suggests the use of a coloured fruit cordial, such as blackcurrant, as students will be able to see differences in concentration and are less likely to muddle tubes.

The worksheet expects students to carry out the dilutions themselves. Alternatively, these could be prepared ahead of the lesson to save time.

The discs will need to be left in solution for at least two hours to get significant changes in mass. Some students may need help with the calculations in the first step of the analysis on the worksheet.

Students should find that discs gain mass or lose mass (water) depending on the extent to which the concentration of cell cytoplasm differs from the solution in which they are placed. Mass loss is a result of the movement of water out of the cell when placed in more concentrated solutions, and mass gain is the result of the movement of water into the cell when placed in more dilute solutions.

The following resources are needed for the class practical B3.2c, per group:

3 boiling tubes and rack
10 cm^3 0.5 mol dm^{-3} sodium chloride solution
distilled water
glass rod
3 Petri dishes
fresh hollow plant stalks, e.g. dandelion, rhubarb
sharp knife or scalpel
marker pen

This investigation links to the Developing Practical Skills box on page 69 in the Student Book. It is similar to the previous investigation, but uses a different method of changing stem curvature to show the effects of osmosis. It also gives students an opportunity to create and calculate a dilution sequence.

If dandelions are not available, any hollow-stalked plant with soft stalks may be suitable. It is advisable to test other plants with a trial run to make sure that they produce measurable differences in the solution concentrations offered.

The plant stalks must be fresh, or stored in water or a dilute salt solution until use to prevent them drying out.

Students should find that the stalks change in curvature as a result of osmosis, and because the outer cells of the stalk cannot absorb water whereas the inner cells can. Fig. 3.12 on page 69 of the Student Book shows results from an experiment like this.

Stalks in solutions that are more dilute (have a higher concentration of water molecules) than the cytoplasm in the cells will curve with the outer layer on the inside of the curve and inner layer outside. Stalks in solutions that are more concentrated (have a lower concentration of water molecules) than the cytoplasm in the cells will curve with the inner layer on the inside of the curve and outer layer outside.

The greater the difference from the concentration of cell cytoplasm, the greater the curvature (although the pressure caused by the cell wall will prevent the curvature in distilled water going beyond a certain point).

Answers

Page 65 Developing Practical Skills

1. So there is no red pigment on the outside, so all pigment that colours the water has diffused out of the cells.
2. a) Repeat the investigation using cubes of different sizes.
   b) Repeat the investigation using water of different temperatures.
   c) Repeat the investigation using cubes of different colour intensities.
   d) Repeat the investigation using beakers of different sizes and measuring the time taken for all the water to turn a certain shade of pink.
3. So they all contain the same concentration of pigment.
4. Have a 'standard' – e.g. a pink solution or a coloured piece of card to use as a comparison.

Page 67 Developing Practical Skills

1. To make sure none of the liquid on the outside of the tubing was included in the weight.
2. So the tubing did not lose water by evaporation.
3. The weight of the visking tubing (dialysis tubing) and contents would increase during the investigation.
4. Water will move by osmosis from the water in the beaker to the more concentrated solution inside the visking tubing.

Page 69 Developing Practical Skills

1. Plan should include:

- cutting stems with sharp knife to give accurate size and reduce damage to cells near the cut
- using a ruler or similar to measure the lengths and widths of the pieces when cutting
- placing cut pieces into water for storage until investigation begins, to prevent cells drying out
- accurate preparation of solutions of different concentrations given in diagram
- quickly placing pieces into each solution so they all have the same time in solution
- ideally, use of several pieces of stem in each solution (two are shown in the diagrams, but more might be better) so that results from different pieces can be averaged to give a more accurate result (as living things can vary).

2. The diagrams show that in solutions of concentration less than 0.25 M, the inner cells enlarge and can cause the strips to curve with the outer surface on the inside of the curve. In 0.25 M there is little change in the strips, and in the most concentrated solution (0.5 M), the strips curve more with the outer layer on the outside of the curve.
3. Osmosis results when there is a concentration gradient for water across a partially permeable membrane. In the solutions of concentration less than 0.25 M, the inner cells gain water. As the outer cells do not absorb as much water, because of the waxy layer, the increase in size/turgidity of the inner cells causes the strips to curve with the inner layer on the outer side of the curve. In the most concentrated solution, the inner cells have lost water as a result of osmosis and become smaller, and so the strips have curved even more than they normally do, with the outer layer on the outside of the curve.
4. The normal concentration of cell cytoplasm of dandelion cells appears to be about 0.25 M because these strips changed least, suggesting there was little osmosis between the cells and solution and so there is only a small concentration gradient between them.

Page 70

1. It is a solvent and many substances can dissolve in it.
2. In humans, substances dissolve in blood so they can be transported around the body.
3. The net movement of water molecules from a region of higher water potential to a region of lower water potential, through a partially permeable membrane.

4. a) Both involve the passive movement of molecules as the result of a concentration gradient.
   b) Osmosis only considers the movement of water molecules; diffusion considers the solute molecules.
5. The strong cell wall prevents more water entering a plant cell than there is space for in the cell (i.e. when the cell is full of water). The cell wall gives cells that are full of water a specific shape and a rigidity, and this helps to support the plant, keeping it upright.

Pages 71–72 Science in Context: Stomata

Making sugar lowers the water potential inside the guard cells so they gain water by osmosis from the surrounding cells.

Page 72 Science in Context: Overhydration

The water containing dissolved sugar would have a lower water potential than pure water. Therefore, there would not be as great a water potential gradient between the water and the person's blood. Therefore, water would not enter the blood – and in turn the brain – so quickly.

Supplement Page 73

1. a) A plant cell that is not full of water.
   b) A plant cell that is full of water.
   c) The removal of water from a plant cell so that the cell membrane surrounding the cytoplasm pulls away from the cell wall.
   d) The pressure on the cell wall caused by the water in the cytoplasm that prevents more water entering the plant cell.
2. The water potential of the cells inside the plant root is lower than the water potential of the soil water surrounding the root, so water moves down the water potential gradient into the root.
3. Diagram should show water molecules leaving the red blood cell as a result of osmosis and entering the solution.
   Labels should indicate a water potential gradient from the cell to the solution and indicate that the loss of water results in the cell shrinking.

Page 74 Science in Context: Metabolic poisons

Osmosis is a passive process and does not need energy from respiration.

Supplement Worksheet B3.2d

1. The tendency for a solution to absorb water.
2. 0 (MPa)
3. Negative, because a solution always has a greater potential to absorb water than pure water.
4. a) B, because -2.1 is nearer to zero than -2.5
   b) from B to A
   c) Because osmosis always occurs down the water potential gradient (or any similar answer).

Worksheet B3.2a Osmosis and visking tubing (dialysis tubing)

Visking tubing (dialysis tubing) is an artificial partially permeable membrane. You can use visking tubing to investigate the effect of the concentration of a solution on osmosis. This is because water molecules are small enough to pass through the tubing membrane. The solute used must be made of molecules that are too large to diffuse through the membrane, so in this practical you will use starch.

Apparatus

plastic/glass tube

visking tubing (dialysis tubing)

starch solution

water

beaker or boiling tube

clamp + stand

wire or elastic band to attach visking tubing to plastic/glass tube

pipette

marker pen

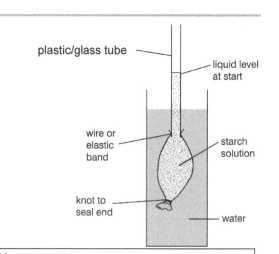

SAFETY INFORMATION

Connect the visking tubing (dialysis tubing) *to the plastic/glass tube in the way that your teacher shows you.*

Method

1. Knot one end of the visking tubing (dialysis tubing), then attach it to the end of the plastic/glass tube using a piece of wire or elastic band. Support the plastic/glass tube in a clamp.
2. Use the pipette to carefully add starch solution to the visking tubing through the plastic/glass tube. Keep adding starch solution until about 1 cm is visible in the plastic/glass tube. Mark the level of the solution with a marker pen.
3. Wash any spills of starch solution from the outside of the tubing. Then suspend the tubing in a beaker or boiling tube of water supported by a clamp, so that the visking tubing is completely covered in water, as shown in the diagram.
4. Leave the apparatus for several hours or overnight. Then record any difference in the level of solution in the glass tube.

Prediction

5. Write a prediction for this experiment. Explain your prediction using your science knowledge.

Handling experimental observations and data

6. Compare the level of solution before and after the experiment. How did it change?
7. Explain any change using your science knowledge.
8. Explain the importance of your findings to living organisms.
9. Explain why it was important to use a solution of a substance that cannot diffuse through visking tubing for this investigation.

Worksheet B3.2b Measuring osmosis by change in mass

When cells take in water by osmosis, they increase in mass. You can use this to investigate the effect of concentration of solution on osmosis.

Apparatus

5 boiling tubes + rack

100 cm^3 beaker

coloured fruit cordial

distilled water

glass rod

paper towels

fresh root vegetable, for example, a potato

cork borer

sharp knife

weighing balance

10 cm^3 measuring cylinder

marker pen

SAFETY INFORMATION
Take care with sharp knives.

Method

Preparing solutions of different concentration

1. Pour 5 cm^3 fruit cordial into the first tube. Label this tube *Full concentration*.
2. Pour 5 cm^3 fruit cordial into the second tube. Add 5 cm^3 distilled water and stir the solution with a dry glass rod. Label this tube *0.5 concentration*.
3. Pour 5 cm^3 of the 0.5 concentration solution into an empty tube. Add 5 cm^3 distilled water and stir the solution with a dry glass rod. Label this tube *0.25 concentration*.
4. Pour 5 cm^3 of the 0.25 concentration solution into an empty tube. Add 5 cm^3 distilled water and stir the solution with a dry glass rod. Pour 5 cm^3 into a beaker, leaving 5 cm^3 in the tube. Label this tube *0.125 concentration*.
5. Pour 5 cm^3 distilled water into the final empty tube. Label this tube *Pure water*.

Preparing the discs

6. Use the cork borer to extract cores from the root. Cut the cores into discs that are about 3 mm deep.
7. When you have five discs, blot them dry on a paper towel and weigh them on the balance. Record their total mass and then place the discs into one of the tubes.
8. Repeat step 7 until all the tubes contain five discs.
9. Leave the discs for at least 2 hours, or overnight. After this time, remove all the discs from one tube, blot them dry on a paper towel, then measure their mass again. Record the results and repeat for the discs from each of the remaining tubes.

Page 1 of 2

Handling experimental observations and data

10. Record your results in a suitable table.
11. For each tube, calculate the change in mass of the discs, and then the percentage change in mass.
    To calculate the percentage change in mass, use the following formula:
    percentage change in mass = × 100%
    Record this in a column in your table.
12. Draw a suitable graph or chart to display your results.
13. Describe any patterns in your results.
14. Explain any pattern in your results using your knowledge of osmosis.
15. Explain why you calculated the percentage changes in mass, and not just used the change in mass.

Planning and evaluating investigations

16. Describe any problems that you had with your investigation and explain how these may have affected your results.
17. Explain what could have been done to reduce their effect on the results.
18. What errors or uncertainties do you think there may have been in your results from this activity?

Page 2 of 2

Worksheet B3.2c Osmosis and change in shape

In the stalks of many plants, the outer cells are unable to take in water because they are covered with a waterproof layer that stops the plant losing water in normal conditions. However, the cells inside the stalk are not protected, so they will take in water due to osmosis until the cell walls prevent further expansion. You can use this difference to investigate osmosis in the cells of pieces of hollow stalk, such as from a dandelion plant.

Apparatus

3 boiling tubes	0.5 mol/dm³ sodium chloride solution
distilled water	glass rod
5 Petri dishes	fresh plant stalks
sharp knife or scalpel	marker pen

SAFETY INFORMATION

Take care with sharp knives.

Wash your hands after handling the plant stalks.

Method

Preparing the dilutions

1. Use the pen to label the tubes from 1 to 3, and label the Petri dishes from 1 to 5.
2. Measure 5 cm³ sodium chloride solution into dish 1.
3. Measure 5 cm³ sodium chloride solution into tube 1. Add 5 cm³ distilled water and mix with a clean dry glass rod. Measure 5 cm³ of the solution in tube 1 into tube 2 and pour the rest into dish 2.
4. Add 5 cm³ distilled water to tube 2 and mix the solution with a clean, dry glass rod. Measure 5 cm³ of the solution in tube 2 into tube 3 and pour the rest into dish 3.
5. Add 5 cm³ distilled water to tube 3 and mix the solution with a clean, dry glass rod. Measure 5 cm³ of the solution into dish 4.
6. Measure 5 cm³ distilled water into dish 5.

Preparing the plant stalks

7. Cut across a plant stalk to produce a tube about 5 cm long.
8. Cut along the cylinder and open it out to produce a flat sheet.
9. Cut the flat sheet into thin strips, about 3 mm wide along their length.
10. Place two strips into each dish and leave them for about 5 minutes.
11. Note any changes in curvature between the strips. Record your results in words or pictures.

Handling experimental observations and data

12. Describe any patterns in your results.
13. Use your knowledge of osmosis to explain any pattern in your results.

Planning and evaluating investigations

14. Describe any problems that you had with your investigation, and explain what could have been done to reduce their effect on the results.
15. Suggest how the method could be adapted to estimate the concentration of the cell cytoplasm in tissue from different plants.

Supplement Worksheet B3.2d Water potential

1. Write a definition for water potential.
2. What is the water potential of pure water?
3. Does a solution have a positive or a negative water potential? Explain your answer.
4. The diagram shows the water potential of two cells in a plant.
   (Water potential is measured in MPa, megapascals.)

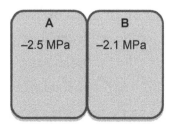

a) Which plant has the higher water potential? Explain your answer.
b) In which direction will osmosis take place?
c) Explain your answer to part b).

Learning episode B3.3 Active transport

Learning objectives

- Describe active transport as the movement of particles through a cell membrane from a region of lower concentration to a region of higher concentration (i.e. against a concentration gradient), using energy from respiration
- Supplement Explain the importance of active transport as a process for movement of molecules or ions across membranes, including ion uptake by root hairs
- Supplement State that protein carriers move molecules or ions across a membrane during active transport

Common misconceptions

This is an opportunity to reinforce the fact that diffusion and osmosis are passive processes that result from the random movement of particles, whereas active transport is the deliberate selection of certain molecules by a cell for transport across the cell membrane.

Resources

Student Book page 73–74

Approach

1. Introduction to active transport

Show students a plant in a pot and explain that you water the plant every one or two days, but occasionally you add some plant 'food' to the water. Ask them to suggest what would happen if you did not water the plant. Use the discussion to briefly revise what students have learned about osmosis.

Then ask what would happen if the plant did not receive any plant 'food' occasionally. (Make sure students do not confuse 'plant food', which provides the mineral nutrients that plants need to convert carbohydrates from photosynthesis into other chemicals needed for cell growth and function, with 'animal food', which supplies both the energy and nutrient needs of an animal.)

Tell students that the concentration of solutes inside the root cells is greater than the concentration of solutes in the soil water. Ask them to suggest what problems this causes for plants. They should remember from their work on diffusion that the concentration gradient is in the wrong direction for root cells to absorb what they need from soil water by diffusion.

Introduce the idea that moving molecules from an area of lower concentration to an area of higher concentration needs energy, like pushing something uphill. The energy for this work comes from respiration (which is covered in more detail in Section 12: *Respiration*).

Supplement 2. Why do organisms need active transport?

Ask students to read through the text on pages 73–74 in the Student Book and make notes to explain why certain substances move across cell membranes as a result of active transport, and why cells need special protein carrier molecules in their membranes for this purpose.

3. Consolidation of work on movement of substances

Give each student three small pieces of paper and ask them to write *diffusion* on one and *osmosis* on another and *active transport* on the third. Then ask a series of questions to which one or more of these words is the answer. Ask students to hold up the appropriate answer(s) for each question. For example:

- Which of the processes involves the random movement of molecules? [diffusion and osmosis]
- Which process only involves water molecules? [osmosis]
- Which of the processes involves molecules other than water? [diffusion and active transport]
- Which process requires energy to happen? [active transport]

This will help you to identify any weaknesses in learning by individual students.

Answers

Page 74

1. The movement of particles through a cell membrane from a region of lower concentration to a region of higher concentration (i.e. against a concentration gradient), using energy from respiration.
1. Supplement Uptake of nitrate ions by root cells in plants which plants need but are in higher concentration inside plant cells than in soil water; uptake of glucose from digested food in the small intestine by cells of the villi in humans, to ensure all the glucose is absorbed.
2. Supplement Energy from respiration makes it possible for protein carrier molecules to transport particles from one side of a membrane to the other against the concentration gradient.

Learning episode B3.4 Consolidation and summary

Learning aims

- Review the learning points of the topic summarised in the end of topic checklist
- Test understanding of the topic content by answering the end of topic questions

Resources

Student Book pages 75–77

Approach

Introduce the learning episode

Ask students to work with a partner to make a list of key words from this topic. They could then work together to produce a spider diagram showing how the different concepts are linked. They could compare their list with the list of key terms given on page 75 of the Student Book. Discuss the checklist on pages 75–76 and use questioning to see how much of the content they are comfortable with.

Students could make flashcards of the key content and then use the flashcards to quiz each other on the information.

Develop the learning episode

Ask students to work individually through the end of topic questions on page 77 of the Student Book without looking at the text. As they work, walk around the classroom observing their answers and asking questions as necessary to find out which questions are causing difficulties.

Finish the learning episode

After a set period, ask the students to stop working and discuss any areas of difficulty you observed as you walked round the class. Students should complete any unanswered questions for homework, but you should stress that they should try to answer the questions without looking at the text, so that they can see how much they have remembered.

Answers

End of topic questions mark scheme

The marks available for a question can indicate the level of detail you need to provide in your answer.

Question	Correct answer	Marks
1	C	1 mark
2 a)	diffusion	1 mark
2 b)	neither	1 mark
2 c)	osmosis	1 mark
3	The salt forms a very concentrated solution on the slug's surface,	1 mark
	so water leaves its body by osmosis.	1 mark
4	Plants are supported by the pressure of water inside cells.	1 mark
	Water is a solvent,	1 mark
	so is involved in transport / excretion/ moving substances around / out of the body,	1 mark
		1 mark

Question	Correct answer	Marks
	chemical reactions / processes such as nutrient absorption happen in solution (Allow alternative marks for other valid answers, e.g. sweating)	
Supplement 5	Turgid cells contain as much water as they can,	1 mark
	so the cell contents press against the cell wall making the cells rigid/firm supporting the plant.	1 mark
	Flaccid cells have lost water,	1 mark
	so are no longer rigid/firm and the plant will droop/wilt.	1 mark
Supplement 6. a)	ii) more than	1 mark
Supplement 6 b)	to provide energy for the active transport of mineral ions, such as nitrates, into the roots	1 mark
	Total:	16 marks

Introduction

This short section covers some of the important molecules in living organisms: carbohydrates, fats, proteins and DNA.

This learning provides the information needed for later sections on biological systems, particularly nutrition and inheritance. If preferred, the content could be saved for these later sections where it can be introduced in context, for example protein structure in Section 5: *Enzymes,* Section 7: *Human nutrition* or Section 17: *Inheritance.*

Links to other topics

Topics	Essential background knowledge	Useful links
2 Organisation of the organism	2.1 Cell structure and size of specimens	
3 Movement into and out of cells		3.1 Diffusion 3.2 Osmosis 3.3 Active transport
5 Enzymes		5.1 Enzymes
6 Plant nutrition		6.1 Photosynthesis 6.2 Leaf structure
7 Human nutrition		7.1 Diet 7.3 Physical digestion 7.4 Chemical digestion 7.5 Absorption
8 Transport in plants		8.1 Xylem and phloem 8.4 Translocation
9 Transport in animals		9.4 Blood
10 Disease and immunity		10.1 Diseases and immunity
13 Excretion in humans		13.1 Excretion in humans
17 Inheritance		17.1 Chromosomes, genes and proteins

Topic overview

B4.1	**Carbohydrates, fats and proteins**
	In this learning episode, students learn about the basic structure of biological molecules and the elements from which they are made. This knowledge forms an essential basis for Section 7: *Human nutrition*.
Supplement **B4.2**	**DNA**
	This short learning episode introduces the shape of the DNA molecule. The relationship between this and genes is covered in Section 17, so you may prefer to wait until then and include this as part of Learning episode 17.1 *Chromosomes, genes and proteins*.
B4.3	**Food tests**
	This learning episode provides practical work on identifying starch, proteins, reducing sugars, lipids and vitamin C in foods. This knowledge links closely to Learning episode 7.1 *Diet*.
B4.4	**Consolidation and summary**
	This learning episode quickly recaps on the ideas encountered in the section. Students can answer the end of topic questions in the Student Book.

Careers links

These are some scientific careers that focus on this area of biology but careers in many other fields use the knowledge and skills gained studying science. **Food scientists** and **food technologists** use a knowledge of biological molecules in foods to design and plan the manufacture of food and drinks, making sure that they meet guidelines e.g. for sugar content.

Starting points

The Student Book section opener puts the ideas in the section into context and sets the scene. It also allows students to acknowledge and value their prior learning, and provides a benchmark against which future learning can be compared.

The questions provide a structure for introducing the section and can be used in a number of different ways:
- You could ask students to consider the questions as an introductory homework task.
- You could put students into groups to share their own ideas and understanding and then to report back to the whole class.
- Students could be given access to the Internet, preferably with a tight timescale, to find out the information required.

You could then use a spider chart or other form of wall chart to summarise everybody's ideas.

Recording these initial ideas allows you to retain them for reference as the individual topics are developed. In this way, your students' progress in learning can be readily acknowledged.

Learning episode B4.1 Carbohydrates, fats and proteins

Learning objectives

- List the chemical elements that make up: carbohydrates, fats and proteins
- State that large molecules are made from smaller molecules, limited to:
  (a) starch, glycogen and cellulose from glucose
  (b) proteins from amino acids
  (c) fats and oils from fatty acids and glycerol

Common misconceptions

This section anticipates some of the content in Section 7: *Human nutrition,* and some students may be confused about the source of elements that form the molecules of our body. If they are uncertain, prompt for the correct answer by asking how nutrients get into our bodies.

Resources

Student Book pages 80–81

B4.1 Technician's notes

Worksheet B4.1 Elements in the human body

Resources for a class practical (see Technician's notes)

Approach

1. Introduction

Write the words *carbohydrate*, *protein* and *fat* on the board, and give students a few minutes to think of three really interesting facts that they remember about these substances. Take examples of facts from around the class and ask students to use the key points to construct a concept map about these biological molecules.

If any student mentions the word *lipid* to cover fats (lipids that are solid at room temperature, for example, butter or lard) and oils (lipids that are liquid at room temperature), explain what it means. Students may come across the term in any research they do. However, tell students that this term is beyond the core syllabus.

Students could add any other words they think are suitable to the concept map. Keep the concept map for the end of this learning episode.

2. Elements in the human body

Worksheet B4.1 gives some information about the elements in the human body. Students should use the information to draw a pie chart and answer questions that reinforce their learning of elements present in the major groups of biological molecules.

Some students may need help with creating the pie chart to answer Question 1. Question 4 may confuse some students because it considers elements, not the molecules they form. Make sure any confusion is clarified before moving on.

3. The structure of carbohydrates, fats and proteins

Use pages 80–81 of the Student Book to introduce the different basic units of the three groups of food molecules. Students could use 'popper beads' or similar (for example, chemical models) to reinforce their understanding of the structure of carbohydrates and proteins. Use identical beads for making a 'carbohydrate', explaining that starch, glycogen and cellulose simply have different arrangements of identical glucose molecules. Use a selection of different beads for making a 'protein'. Students could suggest a different way of modelling to show how different fats and oils all contain glycerol, but each contain different fatty acids.

Students could make models using modelling putty if 'popper beads' or chemical models are not available.

4. Consolidation

Ask students to return to their concept maps from the introduction and add to, or amend, them using what they have learned in this learning episode. Take examples of changes from around the class to make sure all the key points are included.

Technician's notes

Be sure to check the latest safety notes on these resources before proceeding.

The following resources are needed for task 3, per group:

popper beads or similar

Answers

Page 81

1. a) fatty acids and glycerol
   b) simple sugars
   c) amino acids
2. Protein is formed from amino acids and carbohydrate is formed from simple sugars; carbohydrates are often made from one kind of simple sugar and proteins are made from many different kinds of amino acid.

Worksheet B4.1

1. Pie chart correctly drawn – students may or may not include a section for other elements (3.8%).
2. oxygen – water, carbohydrates, fats, proteins; carbon – carbohydrates, fats, proteins; hydrogen – water, carbohydrates, fats, proteins; nitrogen – proteins; sulfur – proteins.
3. 96.2%, because there are small amounts of other elements too
4. In food and drink.

Worksheet B4.1 Elements in the human body

The table shows the proportion of key elements in the human body.

Element	Proportion (%)
oxygen	65
carbon	18
hydrogen	10
nitrogen	3
sulfur	0.2

1. Use the information in the table to draw a pie chart in the space below. Remember to label the sections of the pie chart with the names of the elements.

2. A key biological molecule is water (H_2O). The main groups of biological molecules are carbohydrates, fats and proteins. Annotate your pie chart to show which of the main groups of biological molecules contain each element. (Page 82 in the Student Book may help you.)

3. Use the table to calculate the total proportion of these elements in your body. Explain why it is not 100%.

4. Suggest how these elements got into your body to become part of one of the biological molecules.

Supplement Learning episode B4.2 DNA

Learning objectives

- Supplement Describe the structure of a DNA molecule:
    - (a) two strands coiled together to form a double helix
    - (b) each strand contains chemicals called bases
    - (c) bonds between pairs of bases hold the strands together
    - (d) the bases always pair up in the same way: A with T, and C with G (full names are **not** required)

Common misconceptions

Students may be confused by the fact that the bases on a single strand of DNA may be in any order, but that the base sequence on one strand defines the sequence on the complementary strand.

Resources

Student Book pages 84–85

Supplement Worksheet B4.2 DNA in the cell

Approach

1. Introduction

Organise students into pairs or small groups and give them a few minutes to exchange what they know about DNA and decide on a few key points. Then ask for one student from each group to feed one of those key points back to the class. Allow brief discussion of anything students seem unclear about, but avoid going into any depth – explain that students will be doing this later in the course. Instead bring together any points about DNA structure, if covered, or introduce the rest of the learning episode by looking at the structure of a DNA molecule.

2. DNA in the cell

Give students Worksheet B4.2, which is a cut-and-paste task. If needed, students should use the section on the structure of DNA on page 85 of the Student Book to help them complete their labelling.

3. DNA helix model

If students are having difficulty understanding the idea of a double helix, get them to cut a ladder from an A4 sheet of paper by folding it in half lengthwise and cutting out equal-shaped rectangles, leaving about 1 cm between edges and cuts. They should open out the ladder, and add pairs of bases on the 'rungs' using the correct complementary pairs (A/T and C/G). They should then twist the ladder between top and bottom to produce the double helix shape.

4. Consolidation

Ask students to write down three key points of the learning episode for a student who missed the lesson. Take examples from around the class, and ask the class to refine the points to cover the key points as clearly and succinctly as possible.

Answers

Supplement Page 85

1. It is formed from two strands twisted together into a helical shape.
2. Base A always pairs with base T, and base C always pairs with base G.

Supplement Worksheet B4.2 DNA in the cell

Cut out the diagrams below. Identify each structure and arrange them in order of their real size, starting with the largest. Paste them into your workbook in this order.

Link the diagrams by indicating which part of the previous diagram the following diagram represents, such as by circling the part it shows and adding an arrow to link them.

Add the following labels at the correct points on the diagrams. You may need to use some of the labels more than once: base, chromosome, DNA, nucleus.

Add any other notes to your diagram to help you remember all the details of the structure of DNA.

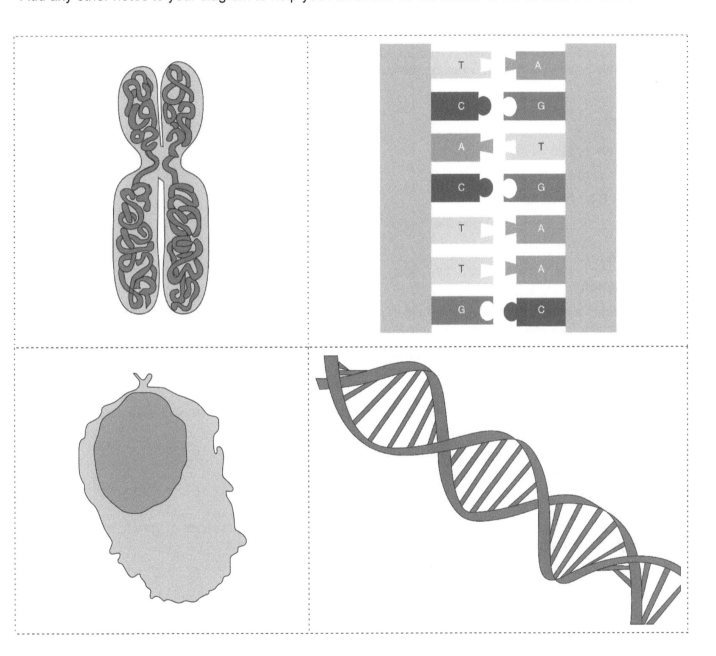

Learning episode B4.3 Food tests

Learning objectives

- Describe the use of:
  - (a) iodine solution test for starch
  - (b) Benedict's solution test for reducing sugars
  - (c) biuret test for proteins
  - (d) ethanol emulsion test for fats and oils
  - (e) DCPIP test for vitamin C

Common misconceptions

Vitamins are introduced in Section 7: *Human nutrition,* in relation to diet. For the current section, students only need to know that vitamins are substances found in some foods, and are essential for health.

Resources

Student Book pages 82–84

B4.3 Technician's notes

Worksheet B4.3 Food tests

Resources for a class practical (see Technician's notes)

Approach

1. Introduction

Show students a selection of nutrition labels from a variety of foods. (These can be found on the back of food packaging. Examples can also be found on the web pages of food manufacturers.)

Ask students to look at the nutrition information and to note the amount of carbohydrate, protein, total fat and vitamin C in each food. Remind them to record only the value per 100 g, so that they can make fair comparisons of the data.

They should then order the foods for each nutrient, starting with the largest amount. Ask questions to check their results, such as:

- Which food contains the largest amount of carbohydrate?
- Which food contains the least fat?

2. Food tests

Worksheet B4.3 guides students in the use of tests used to detect the presence of starch, reducing sugars (glucose), protein, fats and oils, and vitamin C in a range of foods.

Check that students have drawn up a suitable table for their results before they carry out the tests

In the evaluation question, it is suggested that Benedict's test is used to produce a quantitative test.

SAFETY INFORMATION: Safety goggles rather than safety spectacles should be worn because of the corrosive nature of potassium hydroxide solution. Students should be warned to take care when using it and wash off any spillages immediately.

There should not be any naked flames when ethanol is being used.

Remind students to be careful using the hot water and that foods must not be tasted in a science laboratory. Avoid foods that produce known allergies (such as peanuts). Take care when using potassium hydroxide solution. Wash off any spillages immediately.

3. Consolidation

Ask students to produce five quick quiz questions and answers based on what they have learned in this learning episode. They could then try out their questions on a neighbour. Check to see whether there are any recurring errors or weaknesses in knowledge or understanding, so that these can be addressed.

Technician's notes

Be sure to check the latest safety notes on these resources before proceeding.

The following resources are needed for the class practical B4.3, per group:

samples of foods for testing
Benedict's solution
test tubes, measuring pipettes, heatproof tongs or test tube holder and small spoons or spatulas, 1 for each food sample
mortar and pestle
distilled water
white spotting tile
iodine solution
hot water from a kettle that has just boiled
beaker and insulation for beaker
5% potassium hydroxide solution
1% copper sulfate solution
absolute ethanol or industrial denatured ethanol (IDA)
DCPIP dye solution
eye protection (goggles, not safety spectacles)

Include a range of foods, some of which contain larger proportions of some food groups than others, such as biscuits, cake, plain crisps for carbohydrates, and yoghurt, cheese, cooking oil for fats. Dried pulses that have been soaked in water for 24 hours are suitable for proteins. If not already liquid, foods need to be easily broken down to a paste with a mortar and pestle.

You can use a water bath at 100°C, but it is safer to use water that has just boiled and put it in an insulated beaker.

Labels on all bottles should comply with the manufacturer's instructions.

Answers

Page 84

1. a) i) An orange-red precipitate would form because glucose is a reducing sugar.
    ii) The solution wouldn't change colour as there is no starch present.
    b) i) There would be no change in colour because sucrose and the starch in wheat flour are not reducing sugars.
    ii) The solution would turn blue-black because of the starch in flour.
2. Crush the seeds using a mortar and pestle, then:
    a) mix part with ethanol, pour off the liquid and add to water – if the mixture turns cloudy, then fat is present
    b) mix part with water to form a solution, add a few drops of biuret solution – if protein is present, a blue ring forms at the surface, which disappears to form a purple solution
    c) mix part with water to form a solution, add a few drops of this solution to DCPIP – if the blue colour of the dye disappears then vitamin C is present.

Worksheet B4.3 Food tests

In this activity you will use the tests for starch, reducing sugars (glucose), proteins, fats or oils, and vitamin C on a range of foods to see which contain these food groups.

Apparatus

food samples for testing

Benedict's solution

measuring pipettes

test tubes

mortar and pestle

distilled water

white spotting tile

iodine solution

hot water from a kettle, beaker and insulation

heatproof test tube holder

potassium hydroxide solution

copper sulfate solution

ethanol

small spoons or spatulas

safety goggles

SAFETY INFORMATION

Wear safety goggles.

Be careful using the hot water.

Do not taste foods in a science laboratory.

Avoid foods that produce known allergies (such as peanuts).

Take care when using potassium hydroxide solution. Wash off any spillages immediately.

Method

Preparing the food samples

1. Put on safety goggles.
2. Use the mortar and pestle to grind up a small amount (about a teaspoonful) of the food, adding drops of distilled water to help produce a thick paste.
3. Place a teaspoonful of the paste into a test tube, then add 4 cm^3 of distilled water and shake or stir to disperse the paste.
4. Prepare all the solid food samples this way. Liquid food samples do not need preparation.

Starch test

5. Place a drop of iodine solution in each of two dimples in a spotting tile.
6. Add a drop or two of a food solution to one dimple.
7. If starch is present in the food solution, the yellow iodine solution will turn blue-black. (Comparing the food dimple with the iodine-only dimple will help you see even a very slight change in colour.)

Page 1 of 2

Benedict's test for glucose (reducing sugar)

8. Measure 1 cm^3 of the food extract into a clean test tube.
9. Add 1 cm^3 of Benedict's solution and shake to mix the solutions.
10. Using a test tube holder to hold the test tube, place the tube into a boiling water bath and gently shake the solution in the tube to mix it until it reaches the temperature of the water bath.
11. If glucose (a reducing sugar) is present, the blue Benedict's solution will turn green, then yellow, and then orange. If there is a large amount of glucose in the solution, a red-brown precipitate may form.

Biuret test for protein

12. Measure 1 cm^3 of the food extract into a clean test tube.
13. Measure 1 cm^3 of potassium hydroxide solution to the test tube and mix the solutions.
14. Add 2 drops of copper sulfate solution and mix. If protein is present in the food, a mauve or purple colour will develop slowly.

Emulsion test for fats and oils

15. Measure 2 cm^3 ethanol into a clean test tube.
16. Add 1–2 cm^3 of the food extract to the tube. Cover the end of the tube and shake it vigorously to dissolve the food in the ethanol. Allow the solid particles to settle.
17. Pour the ethanol into another clean test tube that is half-full of water, taking care to leave any solid particles in the first test tube. If fats or oils are present, the water will go cloudy white.

DCPIP test for vitamin C

18. Measure 1 cm^3 of DCPIP dye solution into a test tube.
19. Measure 1 cm^3 of the food extract into the dye solution and mix well.
20. If there is vitamin C in the food sample, the blue colour of the dye will disappear.

Handling experimental observations and data

21. Draw up a table to record the results of the tests on each of the food samples you analysed, and complete it by adding your results.
22. Use your table to identify which foods contain each of the main food groups.
23. Identify any patterns in the results, such as whether all the samples that contain glucose come from plant sources or from animal sources.

Planning and evaluating investigations

24. Your results are qualitative, which means they only identify the presence or absence of particular food molecules. Suggest how the method could be adapted to produce a quantitative test, where you could compare the amount of a particular food molecule in different foods.
25. You should also consider whether the tests are *specific* to those food groups (for example, will the test for glucose give positive results with other carbohydrates?).

Learning episode B4.4 Consolidation and summary

Learning aims

- Review the learning points of the topic summarised in the end of topic checklist
- Test understanding of the topic content by answering the end of topic questions

Resources

Student Book pages 86–87

Approach

Introduce the learning episode

Ask students to work with a partner to make a list of key words from this topic. They could then work together to produce a spider diagram showing how the different concepts are linked. They could compare their list with the list of key terms given on page 86 of the Student Book. Discuss the checklist on page 86 and use questioning to see how much of the content they are comfortable with.

Students could make flashcards of the key content and then use the flashcards to quiz each other on the information.

Develop the learning episode

Ask students to work individually through the end of topic questions on page 87 of the Student Book without looking at the text. As they work, walk around the classroom observing their answers and asking questions as necessary to find out which questions are causing difficulties.

Finish the learning episode

After a set period, ask the students to stop working and discuss any areas of difficulty you observed as you walked round the class. Students should complete any unanswered questions for homework, but you should stress that they should try to answer the questions without looking at the text, so that they can see how much they have remembered.

Answers

End of topic questions mark scheme

The marks available for a question can indicate the level of detail you need to provide in your answer.

Question	Correct answer	Marks
1 a)	C	1 mark
1 b)	To make proteins and other compounds needed in the body.	1 mark
2 a)	The tests before the experiment show that the bread contained starch but not glucose (reducing sugar). The tests after the experiment show that the solution contained glucose (reducing sugar) but not starch.	1 mark 1 mark
2 b)	Substance A is an enzyme / digestive chemical which digests / breaks down starch. Starch is a large carbohydrate made up of many molecules of glucose, a reducing sugar. The starch in the bread has been broken into reducing sugars.	1 mark 1 mark 1 mark 1 mark
3	Increases surface area, increasing rate of reaction / colour change	1 mark 1 mark
Supplement 4 a)	i) A twisted ladder shape, formed from two linked strands. ii) Each DNA strand is made of a sequence of bases, each of which joins with a base on the parallel strand.	1 mark 1 mark 1 mark 1 mark
Supplement 4 b)	TTACGTCGA Base A always pairs with base T, and base C always pairs with base G.	1 mark 1 mark
	Total:	16 marks

Introduction

This section looks in greater detail at the structure and function of enzymes, and their importance. Students also investigate the effect of temperature and pH on enzyme activity.

Links to other topics

Topics	Essential background knowledge	Useful links
1 Characteristics and classification of living organisms	1.1 Characteristics of living organisms	
3 Movement into and out of cells	3.1 Diffusion	
4 Biological molecules	4.1 Biological molecules	
6 Plant nutrition		6.1 Photosynthesis
7 Human nutrition		7.1 Diet 7.2 Digestive system 7.3 Physical digestion 7.4 Chemical digestion 7.5 Absorption
12 Respiration		12.1 Respiration 12.2 Aerobic respiration 12.3 Anaerobic respiration
14 Coordination and response		14.4 Homeostasis
16 Reproduction		16.4 Sexual reproduction in humans
17 Inheritance		17.1 Chromosomes, genes and proteins
21 Biotechnology and genetic modification		21.1 Biotechnology and genetic modification 21.2 Biotechnology 21.3 Genetic modification

Topic overview

B5.1	**Enzyme structure**
	This learning episode looks at the role of enzymes in sustaining life. Students will learn about the relationship between the shape of the enzyme's active site and the complementary shape of the substrate.
	Supplement Students will also learn about enzyme-substrate complexes and enzyme specificity.
B5.2	**Enzyme activity**
	This learning episode gives students several practical opportunities to explore the effects of temperature and pH on the rate of enzyme-catalysed reactions.
	Supplement The learning episode continues with a discussion of the 'lock and key model' to help explain the effect of temperature and pH on the rate of enzyme-catalysed reactions.
B5.3	**Consolidation and summary**
	This learning episode quickly recaps on the ideas encountered in the section. Students can answer the end of topic questions in the Student Book.

Careers links

These are some scientific careers that focus on this area of biology but careers in many other fields use the knowledge and skills gained studying science. Formulation biochemists work in companies to develop laundry products that use enzymes to clean clothes with less water and at lower temperatures.

Starting points

The Student Book section opener puts the ideas in the section into context and sets the scene. It also allows students to acknowledge and value their prior learning, and provides a benchmark against which future learning can be compared.

- The questions provide a structure for introducing the section and can be used in a number of different ways:
- You could ask students to consider the questions as an introductory homework task.
- You could put students into groups to share their own ideas and understanding and then to report back to the whole class.
- Students could be given access to the Internet, preferably with a tight timescale, to find out the information required.

You could then use a spider chart or other form of wall chart to summarise everybody's ideas.

Recording these initial ideas allows you to retain them for reference as the individual topics are developed. In this way, your students' progress in learning can be readily acknowledged.

Learning episode B5.1 Enzyme structure

Learning objectives

- Describe a catalyst as a substance that increases the rate of a chemical reaction and is not changed by the reaction
- Describe enzymes as proteins that are involved in all metabolic reactions, where they function as biological catalysts
- Describe why enzymes are important in all living organisms in terms of a reaction rate necessary to sustain life
- Describe enzyme action with reference to the shape of the active site of an enzyme being complementary to its substrate and the formation of products
- Supplement Explain enzyme action with reference to: active site, enzyme-substrate complex, substrate and product
- Supplement Explain the specificity of enzymes in terms of the complementary shape and fit of the active site with the substrate

Common misconceptions

It is important to reinforce the fact that enzymes are proteins, and that all factors that affect proteins will also affect enzymes, and vice versa. This can be reinforced in Section B17: *Inheritance*.

Resources

Student Book pages 90–92

Worksheet B5.1 Know, want to know, learned

Approach

1. Introducing catalysts

Ask students whether they understand the word *catalyst*, and if so, to explain where they have heard it before and what it means. They may know it from catalytic converters in vehicle exhausts, or from their work in chemistry.

Explain that enzymes are catalysts found in living cells. Ask students to suggest what enzymes might do. It does not matter whether they answer correctly or not – this learning episode is to help you identify the limits of their knowledge and any misunderstandings they may have developed.

Tell students that humans have around 75 000 different enzymes in our bodies, and that each enzyme controls a different reaction. Give them a few minutes to work in pairs or small groups to write down as many reactions in the human body as they can think of. (You may need to remind them that a reaction changes reactants into products, so enzymes are not involved in processes such as diffusion and osmosis.)

Take examples from around the class. Make sure these include reactions that students should know from earlier work, such as in respiration, photosynthesis and the digestion of food molecules.

2. Enzyme structure

Worksheet B5.1 is a KWL (Know, Want to know, Learned) grid that can be used to help support learning on the structure of enzymes. Some examples of what students need to know are included in the 'Know' column. Either students could decide what else they need to know using the learning objectives at the start of the section, or these could be added from class discussion. (Note that objectives based on practical work cannot be completed until after Learning episode B5.2.)

Students should first complete the left-hand column and then identify in the middle column what else they need to find out in order to complete the knowledge more securely. If needed, check at this point that students have covered the main points. Note that students must include the term *active site* in their learning.

Students should then complete the right-hand column using the information on pages 92–94 of the Student Book, or information from other sources such as textbooks or the internet. Check that they are accessing information at a suitable level, as there is much content on this at A level and above, which could be confusing.

3. Consolidation

Ask students to prepare questions for which the following are the answers:

1. enzyme
2. catalyst
3. substrate
4. active site
5. this makes enzymes essential for life
6. `Supplement` enzyme-substrate complex
7. `Supplement` this makes enzymes specific to their substrate

They could then test their questions out on a neighbour to help identify any weaknesses in understanding.

For an extra challenge, ask students to write two or more questions per answer.

Answers

Page 91

1. A substance that increases the rate of reaction but remains unchanged at the end of the reaction.
2. A substance that is found in living organisms that acts as a catalyst.
3. Without enzymes, the metabolic reactions of a cell would happen too slowly for life processes to continue.
4. A substrate is a molecule that an enzyme joins with at the start of a reaction. Substrate molecules are changed to product molecules during a reaction.

Page 92

1. When the substrate is joined to the active site this makes it easier for the bonds inside the substrate to be rearranged to form the products.
2. The part of an enzyme into which a substrate fits closely during a reaction.
3. `Supplement` Only a substrate with a shape that is complementary to the shape of the active site can fit it into it. So an enzyme can only work with a particular shape of substrate.

Worksheet B5.1 Know, want to know, learned

This is a KWL grid that organises learning into what you already know, what more you need to know and what you then learn.

- Complete the left-hand column with what you already know on each of the headings given.
- Compare what you know with the learning objectives on page 92 of the Student Book, and complete the middle column with what you still need to learn.
- Carry out research using your Student Book, other books or the internet, to help you complete the right-hand column. Check that you have learned all that is essential by looking at the learning objectives on page 92 of the Student Book.

What I *know* about this	What I *want to know* about this	What I have *learned* about this
enzyme as biological catalyst		
enzymes / proteins / shape / active site		
enzymes in reactions / substrate / products / complementary shapes		
importance of enzymes for life		

Learning episode B5.2 Enzyme activity

Learning objectives

- Investigate and describe the effect of changes in temperature and pH on enzyme activity with reference to optimum temperature and denaturation
- Supplement Explain the effect of changes in temperature on enzyme activity in terms of kinetic energy, shape and fit, frequency of effective collisions and denaturation
- Supplement Explain the effect of changes in pH on enzyme activity in terms of shape and fit and denaturation

Common misconceptions

Many students assume that all human enzymes work best at 'normal' human body temperature. Although this is the case for enzymes that work in the 'core' organs, such as the heart, lungs and digestive system, those that work in cells in the limbs and extremities may have lower optimum temperatures. Enzymes from other organisms may have very different optimum temperatures.

Some students may believe that enzymes are 'killed' by high temperatures. As enzymes are not living things, they cannot be killed – students should learn the correct term, *denaturation*, to explain why some enzymes no longer work at high temperatures. Some students may have the incorrect idea that enzymes can also be denatured by low temperatures – although their activity may be reduced, this is not a permanent change.

Resources

Student Book pages 92–97

Worksheet B5.2a Gelatin, enzymes and temperature

B5.2a Technician's notes

Worksheet B5.2b Trypsin, milk and temperature

B5.2b Technician's notes

Worksheet B5.2c The effect of pH on amylase

B5.2c Technician's notes

Resources for a class practical (see Technician's notes)

Approach

1. Introduction

Write the following words on the board, and ask students to create sentences that use them. Ideally they should use more than one of the words in the same sentence.

- enzyme
- substrate
- complementary
- protein shape
- amino acids

Students could compare their sentences with a neighbour and select the ones combining most words into the same sentence correctly. Take examples from around the class, to try to elicit recall of how proteins are made of amino acids, and introduce the idea that amino acid sequence is related to protein shape (they will come back to this idea in Section B17: *Inheritance*) and how this relates to enzymes.

Briefly introduce the idea that the attraction or repulsion of amino acids within a protein/enzyme can be affected by external factors. Ask students what effect these changes may have on the shape of the active site and the way an enzyme works.

2. Enzyme activity

Ask students to take notes from pages 92–95 of the Student Book on the effect of temperature and pH on enzyme action. Their notes should include the fact that enzymes speed up reactions and that they are affected by temperature and pH.

Students should include references to the *active site* and *denaturation* in their notes.

Supplement Students should include references to *kinetic energy*, *frequency of effective collisions* and *changes in shape* in their notes.

To test students' understanding, ask them to sketch a graph of reaction rate for an enzyme that digests food substances in the stomach, where the pH is around 2, and another for an enzyme that digests food substances in the small intestine, where the pH is near to 8.

3. Investigating enzyme activity

Students can investigate the effect of temperature on the rate of an enzyme-controlled reaction using Worksheet B5.2a or B5.2b, depending on the resources available. It may be difficult to get hold of photographic film for Worksheet B5.2a. Worksheet B5.2b provides an alternative if photographic film is not available (see Technician's notes).

The effect of pH on the rate of an enzyme-controlled reaction can be investigated using Worksheet B5.2c. Note that this practical includes the use of the starch test, which is covered in learning episode B4.3 specifically in relation to food tests. If students have not already used this test, describe how they should carry it out and explain what a change of colour, or lack of it, means in terms of the presence or absence of starch.

If necessary discuss with students how they are going to record their results in a table and what the most suitable graph would be for them to draw. For Worksheet B5.2a, a suitable table would have temperature in one column and time taken for film to clear in another. A suitable graph would show temperature on the *x*-axis and time on the *y*-axis.

For Worksheet B5.2c, you may wish students to carry out a planning exercise for the investigation before giving them the worksheet.

SAFETY INFORMATION: Students should wear eye protection. Remind students to be careful when using the hottest water bath. Care should be taken with enzyme powders and concentrates (follow the precautions described by the supplier).

4. Consolidation: key points

Give students two minutes to write down the four most important points they think they have learned about enzymes. They should then spend another minute discussing their points with a partner and selecting the four best points from both sets.

Take selections from around the class and encourage discussion to produce a class list of key points to remember about enzymes.

Technician's notes

Be sure to check the latest safety notes on these resources before proceeding.

Enzymes purchased from suppliers may not produce the expected trend, or may not work under the conditions given. Please note the following general comments:

- Many purchased enzymes may not be affected by temperatures as high as 50 °C. The enzymes will have been developed for use in biological washing powders, and can still be fully functional up to 95 °C. Please consult suppliers' information before purchase.
- Many enzymes (particularly amylases) are very strongly affected by buffering chemicals. This is very frustrating for all concerned, particularly if the inhibition does not result in an end point. You should

trial and perhaps repurchase enzymes for pH effects. Alternatively, the pH could be very carefully controlled and monitored during the experiments, removing the need for buffer chemicals entirely.

- Care should be taken with enzyme powders and concentrates (follow the precautions described by the supplier).
- Class practical B5.2a may not work above 30–40°C because some films may melt. Technicians should trial this practical before the class to test the film that will be used.

The following resources are needed for class practical B5.2a, per group:

5 test tubes
lengths of exposed and developed 35 mm black-and-white film
scissors
stopwatch or clock
0.1% protease solution, freshly prepared
eye protection
5 water baths set at e.g. 10 °C, 15 °C, 20 °C, 25 °C and 30 °C – only one set needed as all groups will use the same water baths

This is an investigation into the effect of temperature on an enzyme using exposed and developed black-and-white film. If you do not have any suitable film, Worksheet B5.2b can be used instead.

Gelatin is digested by a wide range of protease enzymes. For the best results in the time available, it may be worth carrying out some test runs with different enzymes to see which performs best.

Some people are allergic to enzymes. If you suspect you are allergic, take great care when preparing the solution for students and wear disposable gloves and eye protection. Use a fume cupboard if preparing solutions from powdered enzymes. At the low concentration used in this experiment, students with an allergy to the enzyme are less likely to have a severe reaction, but gloves and eye protection are still recommended.

The following resources are needed for class practical B5.2b, per group:

5 test tubes
10% milk solution
stopwatch or clock
2% trypsin solution freshly prepared
eye protection
5 water baths set at 10 °C, 20 °C, 30 °C, 40 °C and 50 °C – only one set needed as all groups will use the same water baths

Please note that trypsin may not work well if it is made up with acidic distilled water (as is often the case). You may need to adjust the pH using alkali (such as sodium carbonate) to obtain an alkaline pH. The milk and trypsin should both be adjusted to pH 9 before starting, so that the pH optimum of the trypsin is met.

The investigation into the effect of temperature on an enzyme uses trypsin and milk. The milk can be fresh pasteurised milk, or made up into solution with water from powdered milk. Skimmed milk should be used, as the fat in milk will prevent the solution from clarifying.

The worksheet suggests a test run to identify the end point as effectively as possible in order to generate reliable results.

Students should find that trypsin has an optimum temperature at around 37 °C, similar to the core temperature of the human body.

Some people are allergic to enzymes. If you suspect you are allergic, take great care when preparing the solution for students and wear disposable gloves and eye protection. Use a fume cupboard if preparing solutions from powdered enzymes. At the low concentration used in this experiment, students with an allergy to the enzyme are less likely to have a severe reaction, but gloves and eye protection are still recommended.

The time required for this practical can be reduced by giving groups only one or two temperatures from the five, and then collating class results. Additional discussion time will be needed to cover the reliability of shared results.

The following resources are needed for class practical B5.2c, per group:

5 test tubes and rack
5 different pH solutions
1% amylase solution
10% starch solution
2 5 cm^3 syringes or pipettes
iodine solution in dropper bottle (The iodine solution used for testing should be no more than 0.01 M)
stopwatch or clock
marker pen
spotting tile
disposable gloves
eye protection

This practical investigates the effect of pH on amylase using the iodine test for starch to find the point at which all starch is digested. Different amylases have a different optimum pH. Amylase from the pancreas has an optimum between 6.7 and 7.0, but that from yeast has an optimum between 4.6 and 5.2. You may substitute a solution of pancreatin for amylase to give a higher result, more appropriate for the human body. The enzyme and starch solutions should be prepared fresh for the lesson.

The time required for this practical can be reduced by giving groups only one or two pH values from the six, and then collating class results. Additional discussion time will be needed to cover the reliability of shared results.

Different pH/buffer solutions can be made using disodium hydrogenphosphate and citric acid as follows:

pH	Volume of 0.2 mol/dm^3 Na$_2$HPO$_4$	Volume of 0.1 mol/dm^3 citric acid
3	20.55	79.45
4	38.55	61.45
5	51.50	48.50
6	63.15	36.85
7	82.35	17.65
8	97.25	2.75

Alternatively, some suppliers provide separate buffer solutions, or tablets, at the different pH values.

Some people are allergic to enzymes. If you suspect you are allergic, take great care when preparing the solution for students and wear disposable gloves and eye protection. Use a fume cupboard if preparing solutions from powdered enzymes. At the low concentration used in this experiment, students with an allergy to the enzyme are less likely to have a severe reaction, but gloves and eye protection are still recommended.

Answers

Pages 94–95 Developing Practical Skills

1. Similar strips of film would be placed in separate tubes, with the protease, and each tube placed immediately in a separate water bath at each of the temperatures shown in the table. The stopwatch would time from the moment each tube is placed in the water bath, and the tubes would be checked frequently to find the time at which the film became completely clear.
2. Graph with temperature on x-axis, time on y-axis; points clearly marked and an appropriate line drawn.
3. The rate of reaction increases up to an optimum temperature of about 30 °C, and then decreases again.
4. The enzyme has an optimum temperature, below which and above which the enzyme works more slowly. At higher temperatures, this is because the enzyme is denatured by the heat.
5. Test at more temperatures between 20 and 40 °C.

Page 97

1. As temperature increases, the rate of the reaction will increase, up to a maximum point (the optimum), after which it decreases rapidly as the enzyme is denatured.
2. The optimum pH for pepsin is around pH 2, which is very acidic, like the contents of the stomach. The optimum for trypsin is around pH 8, which is more alkaline, like the contents of the small intestine. Each enzyme has an optimum pH that matches the environment in which it works, so that it works most efficiently there.

Page 97

1. Supplement a) The cooler molecules are, the less energy they have and so the slower they move. So the longer it takes for the enzymes and substrate molecules to bump into each other and the substrate to fit into the active site. This means that the cooler the temperature, the slower the rate of reaction.
   Supplement b) As temperature increases the atoms in the enzyme vibrate more as they have more kinetic energy. This changes the shape of the active site, making it more difficult for the substrate to fit into the active site and so slowing down the rate of reaction. Eventually, the atoms vibrate so much that the shape of the active site is destroyed and the enzyme is denatured.
2. Supplement At a pH above and below the optimum of pH 2, the shape of the active site is changed as the interactions between the amino acids in the enzyme are affected by the pH and change. This makes it more difficult for the substrate to fit into the active site, so the rate of reaction slows down.

Worksheet B5.2a Gelatin, enzymes and temperature

Developed black-and-white negative film consists of a celluloid backing covered with a layer of gelatin. Where the film has been exposed, tiny silver particles in the gelatin turn black. Gelatin is composed mainly of the protein collagen, and is easily digested by enzymes called proteases.

This practical may not work over 30–40°C due to melting.

Apparatus

5 test tubes

5 water baths set at e.g. 10 °C, 15 °C, 20 °C, 25 °C and 30 °C

lengths of exposed and developed 35 mm black-and-white film

scissors

stopwatch or clock

protease solution

eye protection

SAFETY INFORMATION
Wear eye protection.
Be careful when using the hottest water bath.
Take care when handling the protease solution and clean up any spills immediately.

Method

1. Put on eye protection.
2. Cut the film into strips and place one strip in each test tube.
3. Add 10 cm^3 of protease solution to each tube, and place one tube in each water bath at the same time. Start the stopwatch or clock.
4. Every minute, check the films to see whether they have cleared. When the gelatin has been digested by the enzyme, the silver grains will fall away from the celluloid backing to leave transparent film. Note down the time taken for each piece of film to clear.
5. Record your results in a suitable table.

Handling experimental observations and data

6. Use your results to draw a suitable graph.
7. Draw a conclusion for the effect of temperature on enzymes using your graph.
8. Explain your conclusion using your scientific knowledge about enzymes.

Planning and evaluating investigations

9. Describe any problems you had with this investigation. How do you think they affected your results?
10. Suggest how the method could be adjusted to avoid these problems.

Worksheet B5.2b Trypsin, milk and temperature

Milk's white colour is the result of a protein called casein. Trypsin is a protease enzyme that digests (breaks down) casein. When trypsin is added to a solution of milk, the white colour is gradually lost as the casein is digested, and the solution becomes translucent.

You can test the effect of temperature on the activity of trypsin by heating it to different temperatures in water baths.

This experiment's end point can be difficult to work out. It is helpful to do a test run at room temperature, to decide an end point and to estimate the time needed. One way to measure the translucency of the solution is to carry out the reaction in a flask, so that it can be stood on a piece of white paper with a large mark on it. You can look down through the flask and judge the clarity of the black mark by eye.

Apparatus

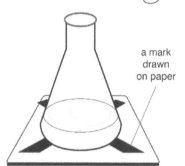

a mark drawn on paper

milk solution (10%)

trypsin solution (2%)

5 small beakers (or test tubes)

stopwatch or clock

eye protection

measuring pipettes

5 water baths set at 10 °C, 20 °C, 30 °C, 40 °C and 50 °C

SAFETY INFORMATION

Wear eye protection.

Be careful when using the hottest water bath.

Method

1. Put on eye protection and gloves.
2. Measure 10 cm^3 milk solution into each small beaker.
3. Add 10 cm^3 of trypsin solution to one beaker, and place the beaker into one of the water baths. Start the stopwatch or clock.
4. Every two minutes check the translucency of the solution in the beaker. When the solution has reached the end point, record the time.
5. Repeat steps 2 and 3 for each of the other water baths.
6. Record your results in a suitable table.

Handling experimental observations and data

7. Use your results to draw a suitable graph.
8. Draw a conclusion for the effect of temperature on enzymes using your graph.
9. Explain your conclusion using your scientific knowledge about enzymes.

Planning and evaluating investigations

10. Describe any problems you had with this investigation. How do you think they affected your results?
11. Suggest how the method could be adjusted to avoid these problems.
12. Explain how you could change the method to find the optimum temperature for this enzyme.

Worksheet B5.2c The effect of pH on amylase

Amylase is the enzyme that digests starch to simple sugars. Amylase has an optimum pH, and this can be investigated by seeing how long it takes a solution of the enzyme to digest a solution of starch. The end point can be measured by testing samples of the starch/enzyme solution at regular intervals with iodine solution.

Apparatus

5 test tubes and rack	5 different pH solutions
small measuring syringes or pipettes	iodine solution in dropper bottle (0.01 M)
starch solution	eye protection
spotting tile	stopwatch or clock
amylase solution	marker pen

SAFETY INFORMATION
Wear eye protection.

Method

1. Put on eye protection.
2. Use the pen to mark each test tube with one of the pH values of the solutions, so that you have one tube for each solution, and place the tubes in the rack.
3. Use the dropper bottle to add a drop of iodine to each dimple in the spotting tile.
4. Measure 2 cm^3 amylase solution into one of the test tubes.
5. Using a clean syringe or pipette, add 1 cm^3 of the appropriate buffer solution to the same tube and swirl gently to mix the solutions.
6. Using a clean syringe or pipette, add 2 cm^3 of the starch solution into the same tube. Start the stopwatch or clock.
7. After ten seconds remove a little solution from the test tube and place one drop into one dimple of the spotting tile and squirt the rest back into the test tube. If there is still starch in the solution the iodine solution will turn blue-black. If all the starch has gone, the iodine will remain yellow-orange.
8. Repeat step 6 every ten seconds until the iodine solution remains yellow-orange. Record this total time and the pH of the solution.
9. Repeat steps 3–7 for each of the other pH solutions, recording the results each time.
10. Draw a suitable table to display your results.

Handling experimental observations and data

11. Use your table to draw a suitable graph of your results.
12. Draw a conclusion for the effect of pH on enzymes using your graph.
13. Explain your conclusion using your scientific knowledge about enzymes.

Planning and evaluating investigations

14. Describe any problems you had with this investigation. How do you think they affected your results?
15. Suggest how the method could be adjusted to avoid these problems.
16. Suggest how you could change the method to find the optimum pH for this enzyme.

Learning episode B5.3 Consolidation and summary

Learning aims

- Review the learning points of the topic summarised in the end of topic checklist
- Test understanding of the topic content by answering the end of topic questions

Resources

Student Book pages 98–99

Approach

Introduce the learning episode

Ask students to work with a partner to make a list of key words from this topic. They could then work together to produce a spider diagram showing how the different concepts are linked. They could compare their list with the list of key terms given on page 98 of the Student Book. Discuss the checklist on page 98 and use questioning to see how much of the content they are comfortable with.

Students could make flashcards of the key content and then use the flashcards to quiz each other on the information.

Develop the learning episode

Ask students to work individually through the end of topic questions on page 99 of the Student Book without looking at the text. As they work, walk around the classroom observing their answers and asking questions as necessary to find out which questions are causing difficulties.

Finish the learning episode

After a set period, ask the students to stop working and discuss any areas of difficulty you observed as you walked round the class. Students should complete any unanswered questions for homework, but you should stress that they should try to answer the questions without looking at the text, so that they can see how much they have remembered.

Answers

End of topic questions mark scheme

The marks available for a question can indicate the level of detail you need to provide in your answer.

Question	Correct answer	Marks
1	A	1 mark
2	There are many reactions in a human body, each with a different substrate,	1 mark
	each substrate needs a different enzyme	1 mark
3	Using an enzyme of choice, explanation should include:	
	• how to measure that the substrate is still present (e.g. using starch test for indicating when starch is broken down to glucose by the enzyme amylase)	1 mark
	• setting up identical tubes of enzyme + substrate at different temperatures	1 mark
	• testing a sample from each tube at regular intervals to see whether the substrate is still present	1 mark
	• stating that the tube in which substrate is broken down most rapidly is the one closest to the optimum temperature for the enzyme.	1 mark
4	The sketch should be similar to the graph on page 95 of the Student Book, in the section: *Enzymes and temperature*.	1 mark
	The optimum temperature (highest point of curve) should be marked at around 37 °C.	1 mark
5 a)	Respiration	1 mark
5 b)	Breathed out through the lungs.	1 mark
5 c)	It would increase	1 mark
	because muscle cells need more energy and so respiration would be happening faster.	1 mark
5 d)	It would make the conditions more acidic.	1 mark
5 e)	The pH change could affect the enzymes, reducing the rate of many cell reactions.	1 mark
	If reactions that are involved in the key life processes slow down, this could damage the cell/organism.	1 mark
Supplement 6	Rate of reaction increases up to the optimum as the molecules move faster and bump into each other more often, forming enzyme-substrate complexes faster.	1 mark 1 mark
	Above the optimum temperature, the atoms in the active site vibrate so much that they change the shape of the active site,	1 mark
	making it more difficult for the enzyme-substrate complex to form, slowing down the rate of reaction,	1 mark
	until the point at which the active site is permanently change and the enzyme is denatured.	1 mark
	Total:	21 marks

Introduction

This section covers plant nutrition in terms of photosynthesis, leaf structure and the mineral requirements of plants for healthy growth.

Links to other topics

Topics	Essential background knowledge	Useful links
1 Characteristics and classification of living organisms	1.1 Characteristics of living organisms 1.3 Features of organisms	
2 Organisation of the organism	2.1 Cell structure and size of specimens	
3 Movement into and out of cells	3.1 Diffusion 3.2 Osmosis 3.3 Active transport	
4 Biological molecules	4.1 Biological molecules	
5 Enzymes	5.1 Enzymes	
8 Transport in plants		8.1 Xylem and phloem 8.2 Water uptake 8.3 Transpiration 8.4 Translocation
12 Respiration		12.2 Aerobic respiration
14 Homeostasis		14.5 Tropic responses
19 Organisms and their environment		19.1 Energy flow 19.2 Food chains and food webs 19.3 Nutrient cycles
20 Human influences on ecosystems		20.1 Food supply 20.2 Habitat destruction 20.3 Pollution 20.4 Conservation
21 Biotechnology and genetic modification		21.2 Biotechnology 21.3 Genetic modification

125

Topic overview

B6.1	**Photosynthesis**
	This learning episode looks at the process of photosynthesis, including the factors that affect it. It provides many opportunities for practical work. You may have to select from the suggestions offered, in order to fit the time and resources available. Some might also be presented as class demonstrations.
	Supplement Students also learn about the role of limiting factors in photosynthesis.
B6.2	**Mineral requirements of plants**
	This learning episode looks at why plants need minerals for healthy growth. It offers a practical to investigate the effect of nitrate and magnesium ions on plant growth. This practical needs at least a month to produce significant results.
B6.3	**Leaf structure**
	In this learning episode, students learn about the structure of the leaf and how it is adapted for photosynthesis.
B6.4	**Consolidation and summary**
	This learning episode quickly recaps on the ideas encountered in the section. Students can answer the end of topic questions in the Student Book.

Careers links

These are some scientific careers that focus on this area of biology but careers in many other fields use the knowledge and skills gained studying science. **Agronomists** study the factors that affect crop growth, and work with farmers to improve plant yields. **Floriculturists** grow flowering and ornamental plants and produce cut flowers for decoration. Both of these jobs need to understand the conditions that promote healthy growth of plants.

Starting points

The Student Book section opener puts the ideas in the section into context and sets the scene. It also allows students to acknowledge and value their prior learning, and provides a benchmark against which future learning can be compared.

The questions provide a structure for introducing the section and can be used in a number of different ways:

- You could ask students to consider the questions as an introductory homework task.
- You could put students into groups to share their own ideas and understanding and then to report back to the whole class.
- Students could be given access to the Internet, preferably with a tight timescale, to find out the information required.

You could then use a spider chart or other form of wall chart to summarise everybody's ideas.

Recording these initial ideas allows you to retain them for reference as the individual topics are developed. In this way, your students' progress in learning can be readily acknowledged.

Learning episode B6.1 Photosynthesis

Learning objectives

- Describe photosynthesis as the process by which plants synthesise carbohydrates from raw materials using energy from light
- State the word equation for photosynthesis as:
    carbon dioxide + water → glucose + oxygen
  in the presence of light and chlorophyll
- State that chlorophyll is a green pigment that is found in chloroplasts
- State that chlorophyll transfers energy from light into energy in chemicals, for the synthesis of carbohydrates
- Supplement State the balanced chemical equation for photosynthesis as:
  $6CO_2 + 6H_2O \rightarrow C_6H_{12}O_6 + 6O_2$
- Outline the subsequent use and storage of the carbohydrates made in photosynthesis, limited to:
    (a) starch as an energy store
    (b) cellulose to build cell walls
    (c) glucose used in respiration to provide energy
    (d) sucrose for transport in the phloem
    (e) nectar to attract insects for pollination
- Investigate the need for chlorophyll, light and carbon dioxide for photosynthesis, using appropriate controls
- Investigate and describe the effects of varying light intensity, carbon dioxide concentration and temperature on the rate of photosynthesis
- Investigate and describe the effect of light and dark conditions on gas exchange in an aquatic plant using hydrogencarbonate indicator solution
- Supplement Identify and explain the limiting factors of photosynthesis in different environmental conditions

Common misconceptions

Some students may not recognise that light is essential for plant growth, but believe that sunlight helps plants to grow by keeping them warm. This is often reinforced by the idea that plants need 'plant food' in a similar way to animals needing food to provide energy and nutrients. In this first learning episode, it is crucial that students develop an understanding of the roles of light and photosynthesis in plant growth. This will be reinforced in the next learning episode on mineral requirements.

Resources

Student Book pages 103–112

Worksheet B6.1a Testing leaves for starch

B6.1a Technician's notes

Worksheet B6.1b Chlorophyll and photosynthesis

B6.1b Technician's notes

Worksheet B6.1c Carbon dioxide and photosynthesis

B6.1c Technician's notes

Worksheet B6.1d Light and the rate of photosynthesis

B6.1d Technician's notes

Resources for class practical (see Technician's notes)

Approach

1. Introduction to photosynthesis

Show students a potted plant and give them three minutes to jot down everything they can think of that relates to how the plant gets its food. Take examples from around the class, focusing on the answers that relate to photosynthesis.

Note any misconceptions that you will need to tackle during the work on this section.

Use the Student Book to introduce the word equation (and for `Supplement` content, the balanced chemical equation) for photosynthesis. Students should make suggestions for which factors might affect photosynthesis. Challenge students to use the equation for photosynthesis to justify their suggestions.

2. Investigating factors needed for photosynthesis

There are several worksheets for this part of the section.

Worksheet B6.1a explains how to prepare and test a leaf for starch. This method is needed for Worksheets B6.1b and B6.1c, so students should be given the opportunity to do this. If you keep one plant in the dark and test a leaf from that at the end of the students' work, you will be able to show the 'need for light'.
Worksheet B6.1b investigates the need for chlorophyll.
Worksheet B6.1c investigates the need for carbon dioxide. Note that this might be better done as a demonstration, due to the apparatus used (see Technician's notes). You could also introduce the concept of a control in this investigation.

If time is limited, different groups of students could carry out different practicals at the same time, and then present their findings to the rest of the class.

SAFETY INFORMATION: Do not use open flames (such as that on a Bunsen burner) when heating ethanol. It is highly flammable. Students should wear eye protection when using ethanol or iodine solution. Remind students to take care to avoid scalding with very hot water. Students should rinse hands thoroughly at the end of the practical.

The soda lime should not be handled by the students.

You should supervise the pouring of very hot water from kettles to minimise the risk of scalding.

3. Investigating the rate of photosynthesis

Worksheet B6.1d provides a method using a water plant to give quantitative data on the effect of light intensity on the rate of photosynthesis. The method assumes that the bubbles produced are all oxygen and of the same size. A more accurate method is to collect and measure the volume of oxygen produced over a fixed time, but this will take longer to produce measurable results.

The method could be adapted by adding different concentrations of sodium hydrogencarbonate to the water to produce quantitative data of the effect of carbon dioxide concentration on the rate of photosynthesis, or by changing the temperature of the water to investigate the effect of temperature on the rate of photosynthesis.

Check that students have drawn up suitable tables for their results before they carry out the tests.

SAFETY INFORMATION: Cover any cuts with a waterproof plaster. Students should wash their hands thoroughly after the practical work.

The 'Developing practical skills' box on pages 111–112 of the Student Book describe an alternative method, using hydrogencarbonate indicator, of investigating the effect of light and dark conditions on gas exchange in an aquatic plant.

`Supplement` Use the graphs on page 113 of the Student Book to introduce the concept of limiting factors in photosynthesis. Ask students to explain the effects of changing light intensity, carbon dioxide concentration and temperature on the rate of photosynthesis. To help explain the latter, refer students back to their knowledge of the effect of temperature on enzymes from Section B5: *Enzymes,* by telling them that the reactions of photosynthesis are controlled by enzymes.

To provide a 'real-life' application, ask students to annotate a sketch of a glasshouse to explain how conditions could be adjusted to maximise the rate of growth of plants inside it.

4. Use of carbohydrates

Ask students to sketch a diagram of a whole plant, including leaves, stem and roots. They should annotate their sketch to show how the carbohydrates made during photosynthesis are used by the plant. They could use the information on page 105 of the Student Book to supplement their own ideas.

5. Consolidation: question tag

Ask students to jot down three key facts that they think the class should have learned in this learning episode on photosynthesis. They should then write a question for each of the facts.

Take examples from around the class, and select another student to answer it. If they answer it successfully, they should then read out one of their own questions.

Technician's notes

Be sure to check the latest safety notes on these resources before proceeding.

The method described on Worksheet for B6.1a is also used in Worksheets B6.1b and B6.1c, so the following applies to all three investigations.

The following resources are needed for the class practical B6.1a, per group:

Industrial denatured ethanol (IDA)
kettle of boiling water
water bath at 90 °C with boiling tube rack
marker pen
iodine solution in dropper bottle (a 0.1 M solution should be strong enough but you may need to use up to 0.5 M solution; you may need to carry out trials to check that the iodine solution is strong enough)
large beaker
leaf (best taken from the plant at the start of the practical so that it is fresh)
forceps
boiling tube
glass rod
Petri dish
white tile
eye protection

Please note that the plants need a large amount of daylight to produce starch. If this practical is being carried out in winter, the leaves will need to have several hours of strong, safe illumination to make sure that they have produced some starch.

IDA is highly flammable. All naked flames must be put out. IDA is also harmful (because of the methanol it contains), so eye protection should be worn.

The following resources are needed for the class practical B6.1b, per group, in addition to the apparatus for B6.1a:

destarched plant, either with variegated or green leaves
metal foil and paper clip (if green leaves) – students will need a sufficiently long strip to cover both sides of the leaf (if metal foil is not available, black card can be used instead)

The plants for use in this practical will need to be destarched, by placing them in the dark for about 48 hours. Test a leaf to check that the plants are fully destarched, otherwise this will affect the results.

The following resources are needed for the class practical B6.1c per group, in addition to the apparatus for B6.1a:

2 bell jars
2 destarched plants
2 plastic trays
alternatively glass sheets – to create a perfect seal with the bell jar using the jelly/grease
petroleum jelly or silicone grease
small beaker of soda lime
small beaker containing 10–20 marble chips
dilute hydrochloric acid
safety goggles (not safety spectacles)

The plants for use in this practical will need to be destarched, by placing them in the dark for about 48 hours. Test a leaf to check that the plants are fully destarched, otherwise this will affect the results.

Soda lime is corrosive – please consult safety guidance on how to handle it.

It would be best to have the soda lime beaker already under the bell jar, for about 24 hours. Students can then be supervised adding the plant. This will allow the atmospheric CO_2 to be removed before the plant is added. Otherwise the plant will carry out some initial photosynthesis.

The following resources are needed for the class practical B6.1d, per group:

pondweed, for example *Elodea*
paper clip
boiling tube containing pond water
250 cm^3 beaker of water
metre rule
lamp – ideally low-voltage bulb or LED light
stopwatch or clock

Prepare the pondweed before the lesson by cutting off 10 cm lengths, one for each group. Check that there are no small invertebrates on the plants. Dispose of plant material safely to ensure non-native plants cannot contaminate the environment.

You will need to dim the lights in the classroom for this practical. Students should be reminded how to prepare for this (e.g. removing all bags, etc. from the floor) and how to behave appropriately in a darkened classroom.

If available, you could use a light meter, or sensor, to measure light intensity directly, instead of recording the distance between the light and the pondweed.

The voltage of the bulb should not exceed the safety limits of the bench lamp. Low-energy bulbs should have some sort of protection so that they do not protrude beyond the protective metal of the bench lamp; or they can be protected by a plastic or glass screen.

Low-energy bulbs contain mercury and so can be hazardous if the bulb is broken.

Be vigilant during this practical because of water being used close to electrical equipment.

LED lights emit a very narrow range of wavelengths of light, which might not be absorbed by the plant. Halogen light sources are fine for photosynthesis but they do get hot. A large beaker of tap water placed in front of the light source will absorb the heat.

Answers

Page 104

1. carbon dioxide + water $\xrightarrow[\text{light energy}]{\text{chlorophyll}}$ glucose + oxygen

2. Light provides the energy needed for photosynthesis.

3. Supplement a) $6CO_2 + 6H_2O \xrightarrow[\text{light energy}]{\text{chlorophyll}} C_6H_{12}O_6 + 6O_2$

   b) Labels should show: 'CO$_2$ from air', 'H$_2$O from soil water', 'C$_6$H$_{12}$O$_6$ used in cells for respiration or converted to other chemicals for use in cells', 'O$_2$ released into air if not needed in respiration'.

4. Supplement Most organisms other than plants get their energy in chemical form from the food that they eat. That energy was originally transferred from light to energy in chemicals during photosynthesis in a plant cell and then transferred along the food chain.

Page 106

1. Test the leaf of a variegated plant for starch. Starch is only produced in the green parts of the leaf, where there is chlorophyll, so only the green parts of the leaf photosynthesise. The pale parts of the leaf, without chlorophyll, act as the control.

2. Heat in a water bath, keeping the ethanol away from open flames such as from a Bunsen burner, because ethanol gives off flammable fumes.

3. Place one destarched plant in an atmosphere with no (or limited) carbon dioxide (due to the presence of potassium hydroxide) and one in an atmosphere high in carbon dioxide (for example, due to carbon dioxide given off in a reaction between marble chips and dilute acid). Shine light on the plants. After several hours, test one leaf from each plant. Only the leaf in high carbon dioxide will have produced significant amounts of starch as a result of photosynthesis.

Pages 107–108 Developing Practical Skills

1. a) Oxygen is a waste product of photosynthesis, so the rate of production of oxygen is a good estimate of the rate of photosynthesis.
   b) With a glowing splint – because it is oxygen, the splint should relight.

2. a) Line graph should be shown, with distance on x-axis and number of bubbles on y-axis; line of best fit drawn through the points.

   b) The graph shows that as distance increases, the rate of bubble production decreases. This is because light energy is needed for the process of photosynthesis – as light decreases so the rate of photosynthesis decreases.

3. a) A filament light bulb releases heat energy as well as light energy, so temperature is another factor that might have had an effect on the rate of photosynthesis.
   b) Use a low-energy bulb that releases little heat, or place a transparent heat barrier (such as a thin water tank) between the lamp and the plant.

Page 109 Developing Practical Skills

1. Plan should include:
   - piece of *Elodea* (pondweed) or similar plant in beaker of pond water with added hydrogencarbonate indicator

- at least two sets of identical apparatus, for use in different light regimes, or the same apparatus used in different light regimes starting with fresh pond water and indicator each time
- as soon as apparatus is set up, placed in appropriate light regime, including at minimum bright light and dark (such as in dark cupboard)
- apparatus left for sufficient time (20 minutes) without disturbance, to allow for photosynthesis and respiration to have an effect on indicator.

2. a) The colour of the indicator turned from red-orange to purple.
   b) The colour of the indicator turned from red-orange to yellow.
3. The solution was becoming more acidic because carbon dioxide concentration in the water was increasing. This suggests that the plant was releasing carbon dioxide into the water from respiration.
4. The solution was becoming more alkaline because carbon dioxide was being taken out of the water by the plant for photosynthesis.
5. Respiration is also happening while photosynthesis is taking place, but no carbon dioxide is added to the solution because the cells use it for photosynthesis.
6. It would mean no carbon dioxide being added or removed from the solution – the plant is at its compensation point.
7. To find the compensation point, you need to have several repeats of the apparatus, and be able to control the light intensity. Then you could find the intensity at which there was no change in colour of the indicator from red-orange, which means the solution remains at neutral pH and there is no net uptake or release of carbon dioxide.

Page 110

1. Carbon dioxide is soluble and acidic, so when more gas is being produced, such as during respiration, the solution becomes more acidic. When carbon dioxide is removed from the solution, such as during photosynthesis, the solution becomes less acidic.

Page 112

1. Supplement A limiting factor is the factor that is limiting the rate of a reaction because it is the one in short supply
   at that particular time.
2. a) As light intensity increases, so rate of photosynthesis increases.
   b) As carbon dioxide concentration increases, so rate of photosynthesis increases.
   c) As temperature increases, the rate of photosynthesis increases up to a maximum, after which it decreases rapidly.
3. Supplement a)As light intensity increases, more energy is supplied to drive the process of photosynthesis.
   Supplement b) As carbon dioxide concentration increases, so there is more reactant for the process.
   Supplement c) As temperature increases, up to the maximum the particles in the reaction including enzymes are moving faster and collide into each other more. Above the maximum the rate of photosynthesis decreases because the enzymes that control the process start to become denatured.

Worksheet B6.1a Testing leaves for starch

Photosynthesis does not form starch directly. However, it does form simple sugars, which are quickly converted to starch in other reactions. So the production of starch in a leaf can be used as evidence of photosynthesis.

The starch test uses iodine solution, which changes from orange-yellow to blue-black in the presence of starch. Before testing a leaf for starch, you must remove the green colour of chlorophyll and the waxy cuticle, so that the iodine solution can penetrate the cells of the leaf. This practical provides the method for preparing and testing leaves for starch, which you can use in the following practicals.

Apparatus

ethanol	kettle of boiling water	leaf
water bath at 80 °C with boiling-tube rack	marker pen	forceps
iodine solution in dropper bottle	large beaker	boiling tube
glass rod	Petri dish	white tile
eye protection		

SAFETY INFORMATION

Do not use open flames (such as a Bunsen burner) when heating ethanol. It is highly flammable.

Wear eye protection.

Take care to avoid scalding with very hot water.

Rinse hands thoroughly at the end of the practical.

Method

1. Put on eye protection.
2. Pour boiling water into a large beaker.
3. Pick up the leaf, by the stalk, with the forceps and hold it in the water for about a minute.
4. Remove the leaf from the water with the forceps and place it in a boiling tube. Push the leaf to the bottom of the tube with the glass rod. Label the tube with your initials.
5. Pour enough ethanol into the tube to cover the leaf. Place the tube into the tube rack in the water bath. The ethanol will boil and remove the chlorophyll.
6. Check the leaf after a few minutes. If there is still green chlorophyll in the leaf, replace the tube for a few more minutes.
7. When the leaf has lost all of its green colour, use forceps to remove the leaf from the tube and rinse it in cold water. This will also soften the leaf so it does not become brittle.
8. Place the leaf on a white tile and add a few drops of iodine solution to cover the leaf.
9. Where there is any starch in the leaf, a blue-black colour will form. Record your results as a drawing.

Handling experimental observations and data

10. Explain fully why a leaf turns blue-black after using this method. What does this evidence show?
11. A plant is left in a dark cupboard for two days. What colour would you expect to see after one of its leaves is tested with iodine solution: yellow-orange or blue-black? Explain your answer.
12. Suggest why a plant should be left in a dark cupboard for two days before carrying out experiments on photosynthesis.

Worksheet B6.1b Chlorophyll and photosynthesis

You can test the need for chlorophyll in the process of photosynthesis using the method of testing leaves for starch described in Worksheet B6.1a.

You will start with a plant that has been left in a cupboard for two days (to become destarched). You may be given a variegated plant (one that has leaves that are a mix of white and green). The white parts of the leaf contain cells that do not make chlorophyll.

Apparatus

destarched plant, either with variegated or green leaves

metal foil and paper clip (if green leaves)

apparatus from Worksheet B6.1a

SAFETY INFORMATION

Do not use open flames (such as a Bunsen burner) when heating ethanol. It is highly flammable.

Wear eye protection when using ethanol or iodine solution.

Take care to avoid scalding with very hot water.

Rinse hands thoroughly at the end of the practical.

Method

1. Put on eye protection.
2. If you are using a plant with variegated leaves, choose one leaf on the plant and draw the green/white markings as carefully as you can. Make sure you can identify the leaf again later.
3. If you are using a plant with leaves that are completely green, carefully fold a piece of metal foil around part of one leaf and hold it in place with a paper clip. Avoid damaging the leaf and plant during handling. Draw the leaf with the foil, to show which areas are covered and which can receive light.
4. Place the plant in bright light for about 24 hours.
5. Remove the leaf from the plant and test for starch using the method on Worksheet B6.1a.
6. Record your results as a drawing.

Handling experimental observations and data

7. Describe your results.
8. Explain your results using your knowledge of photosynthesis.
9. Explain the importance of chlorophyll for plants.

Worksheet B6.1c Carbon dioxide and photosynthesis

Soda lime is a chemical that absorbs carbon dioxide by reacting with it. It can be used to remove carbon dioxide from the air. Marble chips are a form of calcium carbonate, and react with dilute acid to release carbon dioxide. You can use these chemicals to investigate the role of carbon dioxide in photosynthesis.

Apparatus

2 bell jars

2 destarched plants

2 plastic trays

petroleum jelly or silicone grease

small beaker of soda lime (handled by your teacher)

small beaker containing 10–20 marble chips

dilute hydrochloric acid

apparatus from Worksheet B6.1a

SAFETY INFORMATION

The soda lime should only be handled by the teacher.

Do not use open flames (such as a Bunsen burner) when heating ethanol. It is highly flammable.

Wear eye protection when using ethanol or iodine solution.

Take care to avoid scalding with very hot water.

Rinse hands thoroughly at the end of the practical.

Method

1. Put on eye protection – these should be safety goggles.
2. With the teacher's help, set up one plant on a plastic tray with the beaker of soda lime beside it.
3. Rub a layer of petroleum jelly or silicone grease around the edge of the bell jar and carefully place the jar over the plant and beaker onto the tray. Press the bell jar down gently so that the jelly/grease forms a good seal between the jar and tray.
4. Pour sufficient acid into the other beaker to just cover the marble chips. Set up the other bell jar with the other plant and the beaker of marble chips and acid on another tray, sealed in the same way with jelly/grease.
5. Leave the plants for about 24 hours. Then test one leaf from each plant for starch using the method on Worksheet B6.1a. Record your results.

Handling experimental observations and data

6. Describe the results from the plant with carbon dioxide (marble chips) and the plant without carbon dioxide (soda lime).
7. Explain any differences in the leaves from the two plants.
8. Use your results to explain the importance of carbon dioxide for photosynthesis.

Worksheet B6.1d Light and the rate of photosynthesis

To measure changes in the rate of photosynthesis in relation to changes in one of the conditions that affects it, we need a different method of measuring photosynthesis from the methods used in earlier practicals. Photosynthesis produces oxygen gas as a waste product. The rate of oxygen production, as bubbles from a water plant, can be used as a measure of the rate of photosynthesis.

In this practical you will investigate the effect of different light intensities on the rate of photosynthesis. The work will need to be carried out in a darkened room.

Apparatus

pondweed

boiling tube containing pond water

metre rule

stopwatch or clock

paper clip

beaker of water

lamp

SAFETY INFORMATION
Cover any cuts with a waterproof plaster.
Wash hands thoroughly after the practical work.

Method

1. Attach a paper clip to the tip of a fresh sprig of pondweed. Then place the pondweed in a boiling tube of water with the cut stem uppermost. Make sure there is at least 1 cm of water above the end of the pondweed.
2. Place the boiling tube inside a water-filled beaker. This beaker will act as a heat shield.
3. Set up the lamp. Use the ruler to make sure that it is at a set distance away from the pondweed. Record this distance.
4. Look carefully at the cut end of the pondweed. You should be able to see a stream of oxygen bubbles.
5. Leave the pondweed for two minutes, to allow it to adjust to the conditions. Then count the number of bubbles produced in one minute. Record your results.
6. Repeat the count two more times and record those results also.
7. Move the lamp to a different distance from the pondweed. Record this distance and repeat steps 5 and 6.
8. Repeat step 7 for several different distances.

Handling experimental observations and data

9. For each distance, calculate the average number of bubbles produced in one minute.
10. Use your averaged results to draw a graph of number of bubbles produced against distance of light from the pondweed.
11. Describe the shape of your graph.
12. Explain the shape of your graph using your knowledge of photosynthesis.

Planning and evaluating investigations

13. Describe any problems you had with this investigation. How do you think they affected your results?
14. Suggest how the method could be adjusted to avoid these problems.

Learning episode B6.2 Mineral requirements of plants

Learning objective

- Explain the importance of:
  (a) nitrate ions for making amino acids
  (b) magnesium ions for making chlorophyll

Common misconceptions

Students may believe that plants need mineral nutrients in order to grow, because these are often supplied as 'plant food', which students confuse with animal food. It is essential for students to realise that plants produce their own food in the form of simple carbohydrates from photosynthesis, and to understand that the additional minerals are nutrients needed for converting carbohydrates to other biological molecules within cells.

Resources

Student Book pages 104–105

Worksheet B6.2 Plants and mineral ions

B6.2 Technician's notes

Resources for class practical (see Technician's notes)

Approach

1. Introduction

Ask students to think of all the different substances in a plant. If necessary, remind them of their work on biological molecules in Section 4, from which they should be able to identify *carbohydrates*, *fats* and *proteins*.

Supplement Students covering this content should also identify nucleic acids, such as DNA. From these, students should then identify the chemical elements that are found in plants, for which they should be able to give *carbon*, *hydrogen*, *oxygen*, *nitrogen* and *sulfur*.

Then ask students where they get these elements from, in order to make new cells and grow. From their work in Learning episode 6.1, they should be able to identify photosynthesis as the source of carbon, hydrogen and oxygen (from carbon dioxide and water). Some students may recognise that nitrogen and sulfur enter through the plant's roots. If not, ask how substances enter plants. They should remember from earlier work that water enters through the roots, and that therefore any soluble substances may also be able to enter through that route.

2. Plants need mineral ions

Use the Student Book to introduce the idea that plants need a range of mineral ions to make a range of chemicals and that, without these, growth will be affected.

Worksheet B6.2 provides a simple method for investigating the need of seedlings for magnesium ions and nitrate ions using duckweed plants. This will need 4–8 weeks to produce measurable results. Other methods are available on the internet, which may take less time but are not as easy to set up.

SAFETY INFORMATION: Students should wash their hands thoroughly after handling plants. Dispose of plant material safely to ensure non-native plants cannot contaminate the environment.

Technician's notes

Be sure to check the latest safety notes on these resources before proceeding.

The following resources are needed for the class practical B6.2, per group:

duckweed plants
150 cm^3 culture solution containing all nutrients
150 cm^3 culture solution deficient in nitrate ions
150 cm^3 culture solution deficient in magnesium ions
three 250 cm^3 beakers or glass jars
plastic film (clingfilm)
access to light
marker pen

Duckweed (*Lemna*), or any other small floating aquatic plant, is suitable for this investigation. Be aware that duckweed is not available from ponds over winter. Select healthy plants for the experiment.

Culture solutions can be bought ready prepared from some suppliers. Alternatively, the culture solutions can be prepared using suitable recipes.

There are safety issues if the plants die and start to decompose. You should be alert to this possibility and dispose of decomposing cultures as soon as you notice this. Dispose of plant material safely to ensure non-native plants cannot contaminate the environment.

Answers

Page 105

1. Plants make their own foods and need to convert the carbohydrates made by photosynthesis into other substances, such as proteins, which contain additional elements.
2. a) Nitrate ions are needed to make amino acids and proteins.
   b) Magnesium ions are needed to make chlorophyll, which is the green substance in plants.

Worksheet B6.2 Plants and mineral ions

Plants make simple carbohydrates using photosynthesis, but they need mineral ions absorbed from the environment to convert these into all the other substances they need. This practical investigates the need for nitrate ions and magnesium ions in duckweed plants.

Duckweed plants are simple plants that float on the surface of water. When they grow large enough they split off a leaf which becomes a new plant. Rate of growth can be measured by estimating the area of the solution covered by the plants.

Method adapted from The Nuffield Foundation and Royal Society of Biology's Practical Biology website: www.practicalbiology.org

Apparatus

duckweed plants

3 beakers

culture solution containing all mineral ions

plastic film culture solution deficient in nitrate ions culture

access to light solution deficient in magnesium ions marker pen

SAFETY INFORMATION
Wash hands thoroughly after handling plants.
Do not remove the covering from the beakers.
Wash hands thoroughly after the practical work.

Method

1. Label each of the beakers with the pen as follows:
   - complete culture solution
   - minus nitrate
   - minus magnesium
2. Half fill each beaker with the appropriate solution. Beakers should be no more than half full.
3. Select five healthy plants and float them on the top of the solution in one of the beakers. Repeat this for the other two beakers.
4. Cover the top of each beaker with plastic film, and use a sharp point, such as that of a pencil, to make a few air holes in the plastic.
5. Leave the beakers in a bright place for 4–8 weeks.
6. Estimate the coverage of the surface of the solution in each beaker with duckweed.
7. Note any differences between the plants in the three beakers, such as colour or length of root.
8. Record all your results in a suitable table.

Handling experimental observations and data

9. Describe any differences between the plants grown in the three solutions.
10. Explain any differences between the plants using your knowledge of plant mineral ions.

Learning episode B6.3 Leaf structure

Learning objectives

- State that most leaves have a large surface area and are thin, and explain how these features are adaptations for photosynthesis
- Identify in diagrams and images the following structures in the leaf of a dicotyledonous plant: chloroplasts, cuticle, guard cells and stomata, upper and lower epidermis, palisade mesophyll, spongy mesophyll, air spaces, vascular bundles, xylem and phloem
- Explain how the structures listed above adapt leaves for photosynthesis

Resources

Student Book pages 112–115

B6.3 Technician's notes

Resources for class practical (see Technician's notes)

Approach

1. Introduction

Ask students to imagine that they are designing a solar panel that collects energy from light for producing electricity. Give them a few minutes to consider what features the panel would need to make it efficient at collecting light. Ask students to work with one or more neighbours to gather together their ideas and select the best. Take examples to share with the class. Expect students to suggest a large surface area to gather light, chemical, or other system that captures the energy from light. Some students should also identify a very thin panel, to minimise distance through which the light has to travel before it is captured. Using a potted plant to demonstrate, explain that leaf structure has evolved over millions of years to be as efficient as possible at capturing energy from light – and for obtaining the other requirements for photosynthesis: carbon dioxide and water (see Technician's notes).

2. Structure of a leaf

If available, ask students to look at prepared slides of transverse sections through a leaf (see Technician's notes). Alternatively, use a digital microscope to display such a slide, or display an example from the internet. Ask students to identify the different tissues in the leaf, and describe how they are adapted to support photosynthesis. This could be done by annotating a sketch of a leaf section. Alternatively, ask students to draw a diagram of the micrograph of a plant leaf in Fig. 6.14, on page 114 in the Student Book, and annotating it to show the different cells and tissues.

Note that the functions of xylem and phloem are covered in more detail in Section 8: *Transport in plants,* but it is helpful to identify these tissues in the leaf during this learning episode.

Students should add notes to their diagram explaining how the structure of the leaf is adapted for photosynthesis. Prompt students to make links with the work they did on diffusion in Section 3: *Movement into and out of cells*, and the factors that affect the rate at which diffusion occurs.

3. Consolidation

Ask students to write ten crossword clues for a crossword on photosynthesis. They could then test out their clues on a neighbour to identify any areas of learning that need further work.

Technician's notes

Be sure to check the latest safety notes on these resources before proceeding.

The following resources are needed for the demonstrations in the introduction and task 2:

potted plant
prepared slides of transverse sections through a leaf and microscope *or* digital microscope

Answers

Page 115

1. Any four from: epidermis, spongy mesophyll, palisade mesophyll, phloem, xylem.
2. Thin broad leaves, chlorophyll/chloroplasts in cells, veins containing xylem tissue that transports water and mineral ions to the leaves and phloem tissue that takes products of photosynthesis to other parts of the plant, transparent epidermal cells, palisade cells tightly packed in a single layer near top of leaf, stomata to allow gases into and out of leaf, spongy mesophyll layer with large internal surface.
3. A large surface area helps to maximise the rate of diffusion, in this case diffusion of carbon dioxide into cells for photosynthesis and oxygen out of cells so that it can be released into the air.
4. It allows as much light as possible to pass through the epidermal cells to reach the palisade (and spongy mesophyll) cells below, where there are chloroplasts.

Learning episode B6.4 Consolidation and summary

Learning aims

- Review the learning points of the topic summarised in the end of topic checklist
- Test understanding of the topic content by answering the end of topic questions

Resources

Student Book pages 116–119

Approach

Introduce the learning episode

Ask students to work with a partner to make a list of key words from this topic. They could then work together to produce a spider diagram showing how the different concepts are linked. They could compare their list with the list of key terms given on page 116 of the Student Book. Discuss the checklist on pages 116–117 and use questioning to see how much of the content they are comfortable with.

Students could make flashcards of the key content and then use the flashcards to quiz each other on the information.

Develop the learning episode

Ask students to work individually through the end of topic questions on pages 118–119 of the Student Book without looking at the text. As they work, walk around the classroom observing their answers and asking questions as necessary to find out which questions are causing difficulties.

Finish the learning episode

After a set period, ask the students to stop working and discuss any areas of difficulty you observed as you walked round the class. Students should complete any unanswered questions for homework, but you should stress that they should try to answer the questions without looking at the text, so that they can see how much they have remembered.

Answers

End of topic questions mark scheme

The marks available for a question can indicate the level of detail you need to provide in your answer.

Question	Correct answer	Marks
1	C	1 mark
2 a)	When the light was switched on, the proportion of oxygen dissolved in the water increased.	1 mark
	This is because more oxygen was released by the plant as the result of photosynthesis than was used in respiration.	1 mark
2 b)	When the light was switched off, the proportion of dissolved oxygen in the water decreased.	1 mark
	This is because no oxygen was being released by the plant from photosynthesis, and the oxygen that was available was being used in respiration.	1 mark
3	Plants need nitrate ions to make proteins for growth and chlorophyll, and they need magnesium ions to make chlorophyll.	1 mark
	If plants do not get enough of these minerals they will grow more slowly / may show deficiency symptoms, so they are added to make sure they grow well.	1 mark

4	Any six from: • broad / large surface area for collecting light • thin so short diffusion pathway • transparent cuticle / epidermis • many chloroplasts near upper surface / in palisade mesophyll to collect light • air spaces in spongy mesophyll for gas exchange • stomata for diffusion of gases in / out of leaf • xylem for bringing water • phloem for taking away sugars	1 mark each to a max of 6 marks
5 a)	The detail of the shape of the line on the graph is unimportant, as long as it shows an increase in rate from the baseline after sunrise, to a maximum midday/early afternoon, and a decrease to the baseline at sunset.	1 mark
Supplement 5 b)	The increase after sunrise to the maximum should be labelled to indicate that light is a limiting factor (and possibly temperature, which will also increase as the sun warms the air and land).	1 mark
	At the maximum, light and temperature should be in excess, so carbon dioxide is most likely to limit the rate of photosynthesis.	1 mark
	As the rate falls again, the controlling (limiting) factor is most likely to be light.	1 mark
6 a)	to continue growth through the night when darkness usually stops photosynthesis	1 mark
	and so continue the production of sugars, which can be used for growth	1 mark
6 b)	Closing them at night prevents the temperature falling as low as it is outside,	1 mark
	which would slow down the rate of photosynthesis and plant growth.	1 mark
	Opening the windows during the day helps to prevent the temperature rising too high in the glasshouse,	1 mark
	which would reduce the rate of photosynthesis and so the rate of plant growth.	1 mark
7	The statement only refers to net production / the balance between gases	1 mark
	taken in from the air and gases given out by the plant as a result of photosynthesis and respiration.	1 mark
	During the day, when there is light, the plant photosynthesises as well as respires.	1 mark
	It releases more oxygen from photosynthesis than it uses in respiration, but all the carbon dioxide from respiration is used in photosynthesis.	1 mark
	At night, when it is dark, there is no photosynthesis but respiration continues. So the plant takes in oxygen and releases carbon dioxide.	1 mark
	Total:	28 marks

Introduction

This section covers human nutrition in terms of diet, and digestion and absorption in the human digestive system.

Links to other topics

Topics	Essential background knowledge	Useful links
1 Characteristics and classification of living organisms	1.1 Characteristics of living organisms 1.3 Features of organisms	
2 Organisation of the organism	2.1 Cell structure and size of specimens	
3 Movement into and out of cells	3.1 Diffusion 3.2 Osmosis 3.3 Active transport	
4 Biological molecules	4.1 Biological molecules	
5 Enzymes	5.1 Enzymes	
9 Transport in animals		9.3 Blood vessels 9.4 Blood
10 Disease and immunity		10.1 Disease and immunity
12 Respiration		12.1 Respiration 12.2 Aerobic respiration 12.3 Anaerobic respiration
13 Excretion in humans		13.1 Excretion in humans
15 Drugs		15.1 Drugs
19 Organisms and their environment		19.1 Energy flow 19.2 Food chains and food webs 19.3 Nutrient cycles 19.4 Populations
20 Human influences on ecosystems		20.1 Food supply

Topic overview

B7.1	Diet
	This learning episode looks at what is meant by a balanced diet, and identifies the sources of some nutrients and some of the effects of nutrient deficiencies.
B7.2	Digestive system
	In this learning episode, students study the structure of the digestive system and are introduced to the processes that take place there.
B7.3	Physical digestion
	In this learning episode, students study the structure and function of teeth. Supplement Students are introduced to the role of bile in digestion.

B7.4	**Chemical digestion**
	This learning episode builds on the work in Section 5: *Enzymes*, by looking at digestive enzymes, as well as the role of stomach acid. Some of the practical work from Section 5 could be used to support this learning.
	Supplement Students learn more about the role of bile in the digestive system, and go into greater detail about how some digestive enzymes work.
B7.5	**Absorption**
	This learning episode covers the absorption of digested food molecules and water from the alimentary canal.
	Supplement Students apply their learning on diffusion from Section 3: *Movement in and out of cells* to explain how the small intestine lining is adapted for absorption. They also learn about the roles of the capillaries and lacteals in villi.
B7.6	**Consolidation and summary**
	This learning episode quickly recaps on the ideas encountered in the section. Students can answer the end of topic questions in the Student Book.

Careers links

These are some scientific careers that focus on this area of biology but careers in many other fields use the knowledge and skills gained studying science. **Dietitians** are health professionals who diagnose and treat people with nutrition problems. Some dietitians specialise in helping patients after illness or treatment for cancer. **Dental technicians** make dentures to replace broken or diseased teeth to allow people to chew and eat food properly.

Starting points

The Student Book section opener puts the ideas in the section into context and sets the scene. It also allows students to acknowledge and value their prior learning, and provides a benchmark against which future learning can be compared.

1. The questions provide a structure for introducing the section and can be used in a number of different ways:
- You could ask students to consider the questions as an introductory homework task.
- You could put students into groups to share their own ideas and understanding and then to report back to the whole class.
- Students could be given access to the Internet, preferably with a tight timescale, to find out the information required.

You could then use a spider chart or other form of wall chart to summarise everybody's ideas.

Recording these initial ideas allows you to retain them for reference as the individual topics are developed. In this way, your students' progress in learning can be readily acknowledged.

Learning episode B7.1 Diet

Learning objectives

- Describe what is meant by a balanced diet
- State the principal dietary sources and describe the importance of:
  (a) carbohydrates
  (b) fats and oils
  (c) proteins
  (d) vitamins, limited to C and D
  (e) mineral ions, limited to calcium and iron
  (f) fibre (roughage)
  (g) water
- State the causes of scurvy and rickets

Common misconceptions

It is important that students understand that dietary advice is based on an idea of the 'average' person. Such advice is useful in guiding us towards healthy ideals, but everyone is different and has different needs. It is important to distinguish between the needs of the 'average' population and the needs of the individual.

Note that the term *malnutrition* is most commonly associated with deficiency diseases and starvation when people get insufficient food to eat. However, it correctly refers to any situation where the diet is not appropriate for health, and so relates to eating *too much* of some foods as well as too little.

Resources

Student Book pages 123–125

Worksheet B7.1 Government health advice

Approach

1. Introduction

Give students a few minutes to write a definition of the term *balanced diet.* Ask them to compare their definition with that of a neighbour and to work together to discuss what this means. Then ask pairs to work together to improve their definitions.

Finally, take examples from around the class and discuss with students any differences between what they think it means. It is not essential to reach a consensus on what the term means. However, this should raise some of students' misconceptions on the topic – make sure that these are tackled through this learning episode.

2. Food sources

Either provide images of plates of food, or ask students to research some. They should then annotate the images to show which foods are good sources of the main nutrients (carbohydrates, fats, proteins, vitamins C and D, calcium and iron, fibre and water). This could be extended with research on which foods are the best sources of these nutrients.

Students should also add annotations that identify the functions of these nutrients in the body. Pages 123–125 in the Student Book could be used to help them do this.

3. A balanced diet

This part of the learning episode provides opportunities for research work on many diet-related topics that link to what we mean by a balanced diet. For example, students could find out about:

- The association between obesity and diseases such as diabetes and heart disease – how concerned should we be with our weight?

- The huge market in vitamin pills – do we need them?
- Evidence for the role of fibre in preventing bowel cancer, diverticulitis and constipation – why should we eat fibre?
- The increase in cases of rickets in young children in some parts of the world over the past ten years, and its cause – is it preventable?

Be aware of students' sensitivities to ethnic and social differences and to concerns about individual body mass, and treat these with consideration.

Students could carry out their research individually, in pairs or in small groups, according to time and ability. They could use their findings to present a verbal or written report, or create a poster for display.

Alternatively, Worksheet B7.1 provides a set of questions on UK government health information that covers advice on the amount of nutrients and energy we get from our food. The final question might be used as a class activity, to explore what a healthy diet is and how you persuade people to eat healthily.

Before starting on this work, it may help to discuss some of the terminology, such as saturated fats, calories and salt, and why these are focused on when thinking about healthy eating.

4. Scurvy and rickets

Students should research scurvy and rickets, using page 124 of the Student Book and/or the internet. They should make notes to identify the causes and effects of the diseases.

5. Consolidation: healthy eating notes

Ask students to imagine that they are part of a government health advice group. They should work in pairs or small groups to prepare a list of five points that cover how to make sure the population of the country eats healthily.

Take examples from around the class and group the responses according to whether they are advice to individuals, advice to the government department that makes sure sufficient quality food is available for everyone, or advice to farmers who grow the food, and shops and supermarkets who sell the food.

Answers

Page 125

1. Carbohydrates, proteins, and fats/oils
2. Carbohydrates from pasta, rice, potato, bread, wheat flour; proteins from meat, pulses, milk products, nuts; fats/oils from vegetable oils, butter, full-fat milk products, red meat
3. Vitamins, minerals, water, and fibre
4. Vitamins and minerals are needed for maintaining the health of skin, blood, bones, etc. Water is needed as a solvent and to maintain the water potential of cells. Fibre is needed to help digested food to move easily through the alimentary canal.
5. A diet that contains all the groups of food in the correct proportions.

Worksheet B7.1

1. Vitamins, mineral ions, fibre (also carbohydrates, some protein, some oils, water)
2. Fibre is essential for healthy working of peristalsis in the alimentary canal, and prevention of constipation and related problems. Vitamins and minerals are needed for the healthy functioning of the body.
3. Any suitable answers, such as: does not identify which nutrients so cannot tell whether you are ending up with lots of some and not enough of others; does not identify portion size; what if you don't like fruit and/or vegetables, what can you eat instead?
4. They include all the main food groups, not just fruit and vegetables.
5. From a third to a half.
6. Any suitable answer, such as: simple visual image easily translated into a meal on a plate; shows balance of food groups in diet; easy to substitute one food in a group for another.
7. Any suitable answers, such as: what if you are not eating a whole meal?; does not apply well to snacks; difficult to apply to food eaten over whole day.
8. Energy content; main food groups, i.e. fats, proteins and carbohydrates; saturated fats; sugars; fibre; sodium and salt equivalent.
9. Energy content is the amount of energy in the food, so helps you see what energy you need to use to maintain an energy balance. Fats are needed but high saturated fat intake is associated with heart disease. Carbohydrates are needed for energy, but having too much as sugar is associated with a risk of diabetes and obesity. Fibre is important for healthy working of the alimentary canal. Sodium/salt equivalent is given because high levels in a diet may be a problem for people with heart disease.
10. It shows what proportion of the average daily adult intake of each of the risk food groups is in the food.
11. Any suitable answers, such as: lists of numbers are not easy to make sense of; takes time to add up all the information from all the labels on the foods; who will bother?
12. These are the key elements of food that can cause concern as noted in answer to question 9 (cals = energy) – a simplified way of displaying data from table.
13. It shows you how much you will get of each of the food groups that are of concern, so you can balance it with other foods with lower amounts of those during the rest of the day.
14. Any suitable answers, such as: do the people who need to actually bother to look at the labelling?; does everyone understand percentages?; the guideline daily amount is an average, so may not be appropriate advice for everyone
15. Any suitable answers, such as: lots of different advice so confusing; few people take notice of food advice when eating.
16. Any suitable answers that consider trying to get a complex message across simply to different groups of people.

Worksheet B7.1 Government health advice

In some countries, it is recommended that a person should eat five or more portions of fruit and vegetables each day. (Portion size differs for different ages.)

1. Which nutrients are contained in fruit and vegetables? Try to include at least three different groups in your answer.
2. Explain why each of your answers to question 1 are important in the body.
3. What limitations are there with this advice? Try to think of as many limitations as you can.
4. Look at the 'healthy plate' diagrams on page 125 of the Student Book.
5. How does this differ from the '5+-a-day' advice?
6. What proportion of the healthy plate does the '5+-a-day' match?
7. Why is the 'healthy plate' image useful in showing what we should eat?
8. What limitations can you think of with this image in guiding us on what to eat? Try to think of as many limitations as you can.

Food packaging also has advice on healthy eating.

All packaged foods should have a nutrition advice label, like the one below.

Nutrition				
Typical values (as consumed)	per 100 g	per pack	%GDA	your GDA*
Energy	541 kJ	2011	24%	8500 kJ
Protein	4.9 g	18.2 g		
Carbohydrates	20.8 g	77.4 g		
of which sugars	1.5 g	5.6 g	6.2%	90 g
Fat	2.8 g	10.4 g	15%	70 g
of which saturates	2.3 g	8.6 g	43%	20 g
Fibre	2.1 g	7.8 g		
Sodium	0.1 g	0.5 g		
Salt equivalent	0.3 g	1.3 g	22%	6 g
*Recommended guideline daily amount for adults (GDA)				

9. What information does this labelling show?
10. Explain as fully as you can why it shows this information.
11. What is useful about this labelling?
12. What limitations are there with this labelling? Try to think of as many limitations as you can.

Another kind of food labelling uses a 'traffic lights' system, with red for high content, orange for medium content and green for low content. The colours are worked out from the proportion of the recommended daily intake for an average adult, as shown by the percentages on the label.

SERVES 2 – HALF PIZZA PROVIDES				
CALS	SUGAR	FAT	SAT. FAT	SALT
495	9.0 g	18.3 g	9.2 g	2.0 g
25%	10%	26%	46%	33%
MEDIUM	LOW	MEDIUM	HIGH	MEDIUM
OF YOUR GUIDELINE DAILY AMOUNT				

13. Why are these particular groups included in the 'traffic lights' system?
14. How is this system helpful?
15. What limitations are there with this labelling? Try to think of as many limitations as you can.
    Even with all this advice, there is still concern that many people in developed countries are eating too much of the wrong foods, and so risk diet-related illness later in life.
16. Use your answers to suggest why healthy eating may be difficult.
17. How would you prepare advice on healthy eating for everyone? Explain your answer.

Page 2 of 2

Learning episode B7.2 Digestive system

Learning objectives

- Identify in diagrams and images the main organs of the digestive system, limited to:
    - (a) alimentary canal: mouth, oesophagus, stomach, small intestine (duodenum and ileum) and large intestine (colon, rectum, anus)
    - (b) associated organs: salivary glands, pancreas, liver and gall bladder
- Describe the functions of the organs of the digestive system listed above, in relation to:
    - (a) ingestion – the taking of substances, e.g. food and drink, into the body
    - (b) digestion – the breakdown of food
    - (c) absorption – the movement of nutrients from the intestines into the blood
    - (d) assimilation – uptake and use of nutrients by cells
    - (e) egestion – the removal of undigested food from the body as faeces

Common misconceptions

Remind students of the difference between *egestion* (the expelling of undigested food from the alimentary canal) and *excretion* (the expelling of waste substances from metabolism, such as urea in urine).

Some students may think that digestion is the release of energy from food molecules, confusing it with respiration, which takes place in all cells. It is important to clarify that digestion produces the small food molecules that enter the body before being assimilated.

Resources

Student Book pages 125–128

Approach

1. Introduction

Write the following words on the board in alphabetic order: absorption, assimilation, digestion, egestion, ingestion. Ask students to identify the organ system in the body in which all these processes occur (they may not understand the meaning of some of the words but should be able to guess the answer from those they know), and then to put the terms in the order in which they occur, starting at the mouth.

Take suggestions from around the class – the answers may not be exact (for example, digestion covers chewing in the mouth as well as enzymes in the small intestine), but this will provide an opportunity to discuss the meaning of each term.

2. The digestive system

Learning the structure of the digestive system, and which process happens where, is mainly a case of learning the facts. The Student Book pages 125–128 provide what students need to know.

Consider asking students to use the facts in a way that makes learning more fun, e.g. asking them to write the 'story' of the digestion of a meal as it passes through the alimentary canal, to explain the role of each organ in the digestion and assimilation of nutrients from large pieces of food.

3. Consolidation: crossword clues

Give students a few minutes to produce clues for a crossword on the human digestive system, to cover at least ten of the following 'answers': liver, stomach, physical digestion, chemical digestion, assimilation, absorption, egestion, pancreas, oesophagus, enzyme, duodenum, colon, rectum, anus. They should exchange their clues (without the answers) with a neighbour and attempt to answer the ones they have been given.

Ask students to make suggestions on how to improve any clues that they feel are not clear, and to discuss them with the original writer.

Answers

Page 128

1. Sketch should show the following labels correctly attached to organs shown on the diagram:
   • mouth, where food is broken down by physical digestion (chewing) and amylase enzyme starts digestion of starch in food
   • oesophagus moves food from mouth to stomach by peristalsis
   • stomach, where churning mixes food with protease enzymes and acid to start digestion of protein molecules
   • small intestine, where alkaline bile neutralises the acid chyme and enzymes from pancreas complete digestion of proteins, lipids and carbohydrates, and where digested food molecules are absorbed into the body
   • large intestine, where water is absorbed from undigested food
   • rectum, where faeces are held until they are egested through the anus
   • liver, where bile is made and where some food molecules are assimilated
   • gall bladder, where bile is stored until needed
   • pancreas, where proteases, lipases and amylase which pass to the small intestine.
2. Egestion is the removal of undigested food from the alimentary canal – food that has never crossed the intestine wall into the body.
   Excretion is the removal of waste substances that have been produced inside the body.

Page 131 Science in Context: Lactose

Adults are usually lactose-intolerant and drinking milk will cause discomfort.

Learning episode B7.3 Physical digestion

Learning objectives

- Describe physical digestion as the breakdown of food into smaller pieces without chemical change to the food molecules
- State that physical digestion increases the surface area of food for the action of enzymes in chemical digestion
- Identify in diagrams and images the types of human teeth: incisors, canines, premolars and molars
- Describe the structure of human teeth, limited to: enamel, dentine, pulp, nerves, blood vessels and cement, and understand that teeth are embedded in bone and the gums
- Describe the functions of the types of human teeth in physical digestion of food
- Describe the function of the stomach in physical digestion
- Supplement Outline the role of bile in emulsifying fats and oils to increase the surface area for chemical digestion

Common misconceptions

Some students may think that digestion is a purely chemical process involving enzymes. This learning episode should help them to realise that digestion by physical means is a useful first stage.

Resources

Student Book pages 128–130

Worksheet B7.3 Human teeth

Approach

1. Introduction

Ask students the question 'What are teeth for?' and give them about a minute to think up one or more answers to the question. Most students should be able to answer that they are for breaking down large pieces of food. Extend the questioning to ask why teeth are in the mouth and not in another part of the digestive system. Take examples of useful answers, but do not elaborate further.

2. Human teeth

Worksheet B7.3 provides an unlabelled diagram of the human teeth that students can label and annotate to describe the role of the different kinds of human teeth in mechanical digestion (the cutting and chewing of food). Page 129 in the Student Book can help with this. Students should then complete the sentence at the bottom of the worksheet to summarise the role of physical digestion.

Lead students to make the link between surface area and the rate of digestion by enzymes, and thus the value of chewing food thoroughly into small pieces to aid chemical digestion.

3. Structure of a tooth

Ask students to use Fig. 7.7 on page 129 of the Student Book to write a webpage article on the structure of a human tooth and the gums that surround it.

4. Other examples of physical digestion

Remind students of the work they did in learning episode 7.2 about the role of the stomach. Explain that churning food is another example of physical digestion.

Supplement Ask students to look at Fig. 7.8 on page 130 of the Student Book, describing the role of bile in emulsification. Ask them to explain why this is also an example of physical – not chemical – digestion.

5. Consolidation: concept map

Ask students to write the term *physical digestion* in the middle of a sheet of paper and then to add linked words to form a concept map on the section covering what they have learned in this learning episode. After a few minutes, ask students to compare their maps and to identify any improvements they could make.

Answers

Page 130

1. Chemical digestion uses chemicals (enzymes) to help break down large food molecules into smaller ones.
   Physical digestion is the chewing by the teeth to break large pieces of food into smaller ones before swallowing, or the breaking up of large fat droplets into smaller ones by bile – there is no chemical change to the food molecules.
2. Incisors have a flat blade shape for biting off food. Canines have a sharp point for holding food while other teeth chew off pieces. Premolars and molars have grinding surfaces for chewing.
3. Blood vessels supply materials needed for tooth growth. Nerves are used to sense how hard you are biting or if there is any tooth damage/ decay.
4. Supplement Bile helps to emulsify fats in food, breaking them up into much smaller droplets and creating a much larger surface area for lipase enzymes to act on.

Worksheet B7.3 Human teeth

The diagram below shows the arrangement of teeth in the upper and lower human jaws.

Label the diagram to show the names of the different kinds of teeth. Then annotate the diagram to explain the role of the different kinds of teeth in the physical digestion of food.

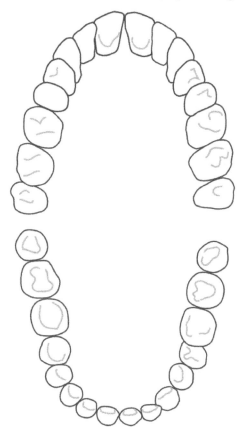

Now complete this sentence:

Physical digestion of food before it leaves the mouth is useful because ...

Learning episode B7.4 Chemical digestion

Learning objectives

- Describe chemical digestion as the breakdown of large insoluble molecules into small soluble molecules
- State the role of chemical digestion in producing small soluble molecules that can be absorbed
- Describe the functions of enzymes as follows:
  (a) amylase breaks down starch to simple reducing sugars
  (b) proteases break down protein to amino acids
  (c) lipase breaks down fats and oils to fatty acids and glycerol
- State where, in the digestive system, amylase, protease and lipase are secreted and where they act
- Describe the functions of hydrochloric acid in gastric juice, limited to killing harmful microorganisms in food and providing an acidic pH for optimum enzyme activity
- Supplement Describe the digestion of starch in the digestive system:
  (a) amylase breaks down starch to maltose
  (b) maltase breaks down maltose to glucose on the membranes of the epithelium lining the small intestine
- Supplement Describe the digestion of protein by proteases in the digestive system:
  (a) pepsin breaks down protein in the acidic conditions of the stomach
  (b) trypsin breaks down protein in the alkaline conditions of the small intestine
- Supplement Explain that bile is an alkaline mixture that neutralises the acidic mixture of food and gastric juices entering the duodenum from the stomach, to provide a suitable pH for enzyme action

Common misconceptions

Some students may not realise that certain enzymes are released in the mouth, to begin the process of the chemical digestion of starch.

Reinforce the fact that enzyme names always end in -ase.

Supplement Be aware of possible confusion between maltase and maltose.

Resources

Student Book pages 130–132

Worksheet B7.4 The digestion and absorption of carbohydrates

B7.4 Technician's notes

Resources for a class practical (see Technician's notes)

Approach

1. Introduction

Write the following words on the board: protease, amylase, lipase. Then give students a minute or so to discuss in pairs or small groups what the words have in common and what they mean. They should associate protease and lipase with two of the major food groups that they learned about in Section 4: Biological molecules, and may remember amylase from the work on food tests. If needed, identify the -ase syllable as the part that means 'enzyme'.

Ask students to identify the smaller, simpler, molecules that each of the food groups is broken down into. Then give students a few minutes to jot down all that they remember about enzymes from Section 5: Enzymes. Take examples from around the class and note down any weaknesses that can be tackled in this learning episode.

2. Enzymes in digestion

Worksheet B7.4 applies the methods from investigating the movement of molecules through visking tubing (dialysis tubing), from Section 3: *Movement into and out of cells* and investigating amylase from Section 5: *Enzymes*, to the digestion of starch and absorption of the products of digestion. As students should be familiar with these methods, this practical sheet is presented as a planning exercise.

If there is time, and students need to practise their investigative skills, allow them to carry out their plans after you have approved them.

Alternatively, students could carry out any of the practicals in Section 5 that they were unable to do earlier.

3. Sites of secretion

Give students a few minutes to look at Table 7.3 on page 131 of the Student Book and to think of any way they can to remember the information. They might do this with a mnemonic, a brief sketch or a cartoon. They should then test each other by asking questions, trying to remember without the assistance of prompts.

`Supplement` Students should include the detail of the digestion of starch and protein on page 132 of the Student Book as part of their learning.

4. Other chemicals in digestive juices

Tell students that the acid produced in the stomach is as strong as the hydrochloric acid that is used in school laboratory experiments. Give them a minute or so to think about why the digestive system does this. Encourage them to make links to their work in Section 5: *Enzymes*, on the effect of pH on enzyme activity, and to identify the optimum pH of gastric enzymes (around pH 2). They should also consider the effect of acid on living cells and identify that acid will kill many microorganisms that we take in with our food.

(If students wonder why the acid does not attack the stomach lining cells, explain that some cells in the lining produce large amounts of mucus (slime), which protects the surface. However, if these cells are affected (e.g. by infection) so that they stop producing mucus, the acid will start to digest holes in the surface, causing painful stomach ulcers.)

`Supplement` Extend the discussion to include bile. Explain that bile is alkaline, and give students a few minutes to consider why it is released into the duodenum. They should be able to identify that enzymes in the duodenum have an optimum pH that is slightly alkaline, and that therefore the acidic chyme released from the stomach into the duodenum must be neutralised. To reinforce this, ask students to compare pepsin and trypsin, where they are secreted and their optimum pH values, using the information in the section *Details of digestion* on page132 of the Student Book.

5. Consolidation

Ask students to summarise the learning episode in three to five key points for a student who was away. Take examples from around the class to make sure all points of learning are covered.

Technician's notes

Be sure to check the latest safety notes on these resources before proceeding.

The following resources are needed for the class practical B7.4, per group:

10% glucose solution and 10% starch solution
15 cm length visking tubing (dialysis tubing) and elastic band, to attach visking tubing to syringe end
sawn-off end of 5 cm³ plastic syringe (the plunger end)
100 cm³ beaker; tap water
iodine solution in dropper bottle
spotting tile
Benedict's solution and water bath at 100 °C and heatproof tongs (or Clinistix®)
boiling tube and teat pipette
eye protection

This is a planning exercise, but students may carry out their planned investigations. If so, the apparatus shown above will be needed. However, students may suggest other pieces of apparatus. If not available, offer suitable alternatives.

Answers

Page 132

1. The digestive enzymes break down food molecules that are too large to cross the wall of the small intestine into smaller ones that can be absorbed across cell membranes and so enter the body. If we did not have enzymes, we would not be able to absorb many nutrients from our food.
2. a) amylase
   b) fatty acids and glycerol
3. a) The acid increases stomach acidity, providing the right conditions for enzymes that digest food in the stomach.
   Supplement b) Bile neutralises the acidity of food from the stomach, providing the right conditions for enzymes that digest food in the small intestine. It also emulsifies fats, providing a larger surface area for lipase enzymes to work on.

Worksheet B7.4 The digestion and absorption of carbohydrates

In Section 5: *Enzymes,* you were introduced to the use of amylase to digest starch, and in Section 3: *Movement into and out of cells*, you used an artificial membrane called visking tubing (dialysis tubing) to model a cell membrane.

1. Plan an investigation, using visking tubing as a model for the human alimentary canal, to investigate the time taken for amylase to digest starch and the products of digestion being absorbed by the blood. You will need to consider:
   - what apparatus you will need
   - the risks associated with the use of any of your suggested apparatus and how you will minimise them
   - how you will set up the apparatus
   - what you will measure
   - how often you will take measurements
   - which variables you will need to control and how you will control them
   - how you will record your results.

2. Write a plan for your investigation. Worksheets B3.1a, B3.2a, B4.3 and B5.2c may help you.

3. When you have completed your plan, give it to your teacher for checking. Your teacher will tell you whether you should now carry out your plan. If you have been given permission to proceed, below is an example of the apparatus you can use.

Apparatus

glucose solution	starch solution
visking tubing (dialysis tubing)	sawn-off end of plastic syringe
elastic band	small beaker
tap water	iodine solution in dropper bottle
spotting tile	boiling tube
pipettes	eye protection
disposable gloves	

Benedict's solution, plus water bath at 100 °C and heatproof tongs (or Clinistix®)

Learning episode B7.5 Absorption

Learning objectives

- State that the small intestine is the region where nutrients are absorbed
- State that most water is absorbed from the small intestine but that some is also absorbed from the colon
- Supplement Explain the significance of villi and microvilli in increasing the internal surface area of the small intestine
- Supplement Describe the structure of a villus
- Supplement Describe the roles of capillaries and lacteals in villi

Common misconceptions

Some students may think that the body only absorbs substances that are good for it, or that it needs. It is important that they remember their work on diffusion from Section 3: *Movement into and out of cells*, and realise that almost every small soluble molecule can diffuse across the lining of the small intestine and so be absorbed into the body.

Check that students learn the correct spelling *absorption*, as some spell it *absorbtion*.

Resources

Student Book pages 133–134

Approach

1. Introduction: quick quiz

Write the following questions on the board, and give students a few minutes to jot down the answers.

- What is *absorption*?
- Where does absorption occur in the alimentary canal?
- What is absorbed in the alimentary canal?
- Why must food be digested before it can be absorbed?

When students have had time to answer, ask them to compare their answers with their neighbour and try to improve them. Pairs of students should then compare their answers with another pair and try to improve again. Then take answers from a spokesperson from each group and compare answers between groups.

2. The intestines

Ask students to read the section *Absorption* on page 133 in the Student Book to identify what nutrients are absorbed, and where, in the alimentary canal. They should take notes and save this information for the consolidation task.

Supplement 3. Villi

Students should sketch a diagram of a villus, and annotate it to show how the small intestine lining is adapted for absorption. Pages 133–134 of the Student Book can help with this. Encourage students to make the link between the adaptations and their effect on the rate of diffusion, which they studied in Section 3: *Movement into and out of cells*.

They should include the capillaries and lacteal on their sketch and annotate them to explain their role in absorption.

4. Consolidation: human nutrition concept map

Allow 15–20 minutes for this task. Arrange students into small groups and give each group a large sheet of paper. The words *Human nutrition* should be written in the middle of the sheet and the following words arranged around the centre, linked to the middle:

- structure of the alimentary canal
- processes in the digestive system
- organs of the digestive system and their functions
- enzymes and other chemicals in digestive juices.

Students should then work in their groups to add to the concept map anything they can remember from this section, and arrange the information to link in the best way. Encourage discussion within each group to help decide what information should go where.

Use the last five minutes of the learning episode to compare concept maps from different groups, to identify any differences and add anything important that is missing.

Answers

Page 134

1. The movement of digested food molecules across the alimentary canal (small intestine) wall into the body.
2. small intestine (ileum) and large intestine (colon)
3. Supplement The larger the surface area, the more rapidly small molecules can be absorbed across the intestine wall into the body.
4. Supplement There are millions of villi on the surface of the intestine wall projecting into the alimentary canal to greatly increase the surface area of the wall. In addition, each villus has many microvilli to increase its surface area.
5. Supplement The capillaries provide a large blood supply to remove absorbed food molecules quickly, so maintaining a high concentration gradient for diffusion; lacteals in the villi carry absorbed lipid molecules away to the rest of the body.

Learning episode B7.6 Consolidation and summary

Learning aims

- Review the learning points of the topic summarised in the end of topic checklist
- Test understanding of the topic content by answering the end of topic questions

Resources

Student Book pages 135–139

Approach

Introduce the learning episode

Ask students to work with a partner to make a list of key words from this topic. They could then work together to produce a spider diagram showing how the different concepts are linked. They could compare their list with the list of key terms given on page 135 of the Student Book. Discuss the checklist on pages 135–136 and use questioning to see how much of the content they are comfortable with.

Students could make flashcards of the key content and then use the flashcards to quiz each other on the information.

Develop the learning episode

Ask students to work individually through the end of topic questions on page 137–139 of the Student Book without looking at the text. As they work, walk around the classroom observing their answers and asking questions as necessary to find out which questions are causing difficulties.

Finish the learning episode

After a set period, ask the students to stop working and discuss any areas of difficulty you observed as you walked round the class. Students should complete any unanswered questions for homework, but you should stress that they should try to answer the questions without looking at the text, so that they can see how much they have remembered.

Answers

End of topic questions mark scheme

The marks available for a question can indicate the level of detail you need to provide in your answer.

Question	Correct answer	Marks
1	C	1 mark
2 a)	Vitamin C for healthy skin and gums and blood vessel linings.	1 mark
	Vitamin D for strong bones and teeth.	1 mark
2 b)	Calcium for strong bones and teeth, also blood clotting.	1 mark
	Iron needed for haemoglobin in red blood cells.	1 mark
2 c)	Water needed in all cells because many reactions take place in watery cytoplasm.	1 mark
2 d)	Fibre keeps undigested food in alimentary canal bulky and soft, helping peristalsis to work efficiently.	1 mark
	A low-fibre diet may increase the risk of bowel cancer.	1 mark
3	Physical digestion of the food in the mouth breaks large food pieces up into smaller pieces, which not only makes them easier to swallow,	1 mark
		1 mark
	but also increases the exposed surface area,	
	making it easier for enzymes to combine with more food molecules more quickly and so speeding up the rate of digestion.	1 mark

4 a)	In the mouth,	1 mark
	where food enters the alimentary canal.	1 mark
4 b)	In the mouth – physical by teeth,	1 mark
	chemical by amylase	1 mark
	in the stomach – physical by churning	1 mark
	chemical by protease enzymes	1 mark
	in the small intestine – amylase, protease and lipase enzymes produced by the pancreas	1 mark
4 c)	In the small intestine,	1 mark
	where digested food molecules can cross the cell membranes of the villi cells and into the blood.	1 mark
	(Also credit absorption of water in the colon)	
5 a)	i) Cereals, banana, bread, honey/jam, beans, sugars in juices and squash, fruit, potatoes/rice, bagel, chocolate.	1/2 mark each max. of 5 marks
	ii) Milk, eggs, tofu, fish/meat, chicken, yoghurt.	
	iii) Milk, olive/sunflower spread, eggs, sauce (possibly), yoghurt.	1/2 mark each max. of 3 marks
	iv) Dairy for vitamin D and calcium; fish oil for vitamin D (if oily fish eaten); vegetables and fruit for vitamin C; green vegetables for iron.	1/2 mark each max. of 2 marks
	v) Wholegrain bread, cereals, vegetables.	1/2 mark each max. of 2 marks
		1/2 mark each max. of 1 marks
5 b)	carbohydrates	1 mark
5 c)	They break down easily to glucose for use in respiration to release energy during training.	1 mark
5 d)	Proteins,	1 mark
	to help with building more muscle.	1 mark
5 e)	It contains a lot of energy.	1 mark
	Most people do not use this amount of energy in one day. The excess energy in the diet would be stored as fat in the body, leading to a risk of obesity and related health problems.	1 mark
6	Emulsifies fats/oils/lipids OR breaks fats/oils/lipids into small droplets,	1 mark
	increasing surface area,	1 mark
	increasing digestion by/activity of enzymes/lipases.	1 mark
	Bile is alkaline,	1 mark
	neutralises acidic mixture entering duodenum from stomach,	1 mark
	providing a suitable/optimum pH for enzyme action.	1 mark
	Total:	37 marks

Introduction

In this section, students will learn about the transport systems in flowering plants, including the processes of water uptake, transpiration and translocation.

Links to other topics

Topics	Essential background knowledge	Useful links
1 Characteristics and classification of living organisms	1.1 Characteristics of living organisms 1.3 Features of organisms	
2 Organisation of the organism	2.1 Cell structure and size of specimens	
3 Movement into and out of cells	3.1 Diffusion 3.2 Osmosis 3.3 Active transport	
4 Biological molecules	4.1 Biological molecules	
6 Plant nutrition	6.1 Photosynthesis 6.2 Leaf structure	
12 Respiration		12.1 Respiration 12.2 Aerobic respiration 12.3 Anaerobic respiration
14 Coordination and response		14.4 Homeostasis 14.5 Tropic responses

Topic overview

B8.1	**Xylem and phloem**
	In this learning episode, students will learn about the tissues involved in transporting materials around a plant.
	Supplement Students relate the structure of xylem vessels to their function.
B8.2	**Water uptake**
	In this learning episode, students look at how water is gained through the root hair cells, and how it moves through the plant.
B8.3	**Transpiration**
	Students learn about the route of water loss through leaves. There is also the opportunity to plan and carry out practical work on factors that affect the rate of transpiration from leaves.
	Supplement Students look at the mechanism of transpiration, and explain why different factors affect transpiration rate.
Supplement B8.4	**Translocation**
	Students learn about translocation, and the roles of sources and sinks in the movement of sucrose and amino acids around a plant.

B8.5	**Consolidation and summary**
	This learning episode provides an opportunity for a quick recap on the ideas encountered in this section. Students can answer the end of topic questions in the Student Book.

Careers links

These are some scientific careers that focus on this area of biology but careers in many other fields use the knowledge and skills gained studying science. **Forestry managers** look after forests and woodland areas and need to ensure that newly planted trees will have sufficient space and water to grow. **Maple tree tappers** carefully remove some of the sap in the phloem of maple trees to produce maple syrup.

Starting points

The Student Book section opener puts the ideas in the section into context and sets the scene. It also allows students to acknowledge and value their prior learning, and provides a benchmark against which future learning can be compared.

- The questions provide a structure for introducing the section and can be used in a number of different ways:
- You could ask students to consider the questions as an introductory homework task.
- You could put students into groups to share their own ideas and understanding and then to report back to the whole class.
- Students could be given access to the Internet, preferably with a tight timescale, to find out the information required.

You could then use a spider chart or other form of wall chart to summarise everybody's ideas.

Recording these initial ideas allows you to retain them for reference as the individual topics are developed. In this way, your students' progress in learning can be readily acknowledged.

Learning episode B8.1 Xylem and phloem

Learning objectives

- State the functions of xylem and phloem:
  (a) xylem – transport of water and mineral ions, and support
  (b) phloem – transport of sucrose and amino acids
- Identify in diagrams and images the position of xylem and phloem as seen in sections of roots, stems and leaves of non-woody dicotyledonous plants
- Supplement Relate the structure of xylem vessels to their function, limited to:
  (a) thick walls with lignin (details of lignification are **not** required)
  (b) no cell contents
  (c) cells joined end to end with no cross walls to form a long continuous tube

Common misconceptions

Check that students learn the correct spelling *absorption*, as some spell it *absorbtion*.

This is a good opportunity to revise diffusion and absorption from earlier sections and check that students have not formed any misconceptions on that work.

Resources

Student Book pages 143–145

B8.1 Technician's notes

Resources for demonstrations and class practicals (see Technician's notes)

Approach

1. Introduction

Remind students of their work on Section 6: *Plant nutrition*, and give them a minute or so to jot down which substances the plant needs from the environment for growth. They should identify what these substances are used for, which provides an opportunity to briefly revise photosynthesis and the formation of proteins and other substances that need mineral ions from the environment. Take examples from around the class and note any weaknesses in understanding that need addressing during work on this section. Ask students which parts of the plants take in the substances they need. They should be able to answer: leaves for taking in carbon dioxide and roots for absorbing water and mineral ions.

2. Xylem and phloem

Students should use prepared slides of transverse sections of plant roots, stems and leaves to help identify the vascular bundles in the different plant organs. Make sure that the slides show non-woody, herbaceous, dicotyledonous plants. (If slides are not available, students could use the photomicrographs in Fig. 8.3 on page 144 of the Student Book.)

Students should draw labelled diagrams from each slide, to show the main features. This is a good opportunity to remind students how to draw plan diagrams, which outline the main areas, and then add detail only to the areas of interest, in this case the xylem and phloem. Remind students that they should use a sharp pencil, only draw what they see, and should label the key features on their diagrams. Fig. 8.2 on page 144 in the Student Book could help students identify and label the tissues correctly on their diagrams.

Students then use page 143 of the Student Book to describe the functions of each of xylem and phloem.

Supplement Explain to students that xylem vessels are made of empty columns of dead cells, whose cross walls have broken down, and outer walls have become thickened and strengthened with lignin. Ask students to think about the advantages of this structure as regards to their function. Comparing xylem vessels with drinking straws will help elucidate their transport function, and comparing them with scaffolding poles will help elucidate their support function.

3. Consolidation

Ask students to start a concept map on plant transport tissues, which they will add to during other activities in this section. This could be done as a group activity to encourage discussion of what should be added. They should begin with the words *Transport in plants* in the middle of the page and add any terms and knowledge they have gained from this learning episode. Save the concept maps for adding to later.

Technician's notes

Be sure to check the latest safety notes on these resources before proceeding.

The following resources are needed for task 2 on xylem and phloem:

prepared slides of transverse sections of plant roots, stems and leaves
The slides should show non-woody, herbaceous, dicotyledonous plants.

Answers

Page 144 Science in Context: Tree rings

In warmer years, when there was greater growth, the rings are wider than in years when it was cooler and there was less growth.

Page 145

1. In vascular bundles that form veins throughout the roots, stems, and leaves.
2. Phloem cells link together to form continuous phloem tissue in the vascular bundles. They carry dissolved food materials, such as sucrose and amino acids.
3. Supplement Xylem vessels are long continuous hollow tubes formed from columns dead cells with no cross walls. This allows water and dissolved substances to pass easily through the plant.

Learning episode B8.2 Water uptake

Learning objectives

- Identify in diagrams and images root hair cells and state their functions
- State that the large surface area of root hairs increases the uptake of water and mineral ions
- Outline the pathway taken by water through the root, stem and leaf as: root hair cells, root cortex cells, xylem, mesophyll cells
- Investigate, using a suitable stain, the pathway of water through the above-ground parts of a plant

Common misconceptions

Check that students learn the correct spelling *absorption*, as some spell it *absorbtion*.

This is a good opportunity to revise diffusion and absorption from earlier sections and check that students have not formed any misconceptions on that work.

Some students think that the ends of roots are open, like drinking straws. Ask students why water enters the roots – remind them of what they learned about osmosis in Section 3: *Movement into and out of cells.*

Resources

Student Book pages 145–146

Worksheet B8.2 Water movement through plants

B8.2 Technician's notes

Resources for demonstrations and class practicals (see Technician's notes)

Approach

1. Introduction

Remind students of their work on surface area and rate of diffusion from Worksheet B3.1c. If they did not carry out that practical work, present this here as a prepared demonstration (see Technician's notes).

If possible, show them a picture of a unicellular organism, such as *Paramecium* or *Amoeba*, from a slide or the internet. Ask them to apply what they learned from the practical to the diffusion of nutrients into and out of the unicellular organism. They should compare this with the problem of getting materials to and from diffusion surfaces and the external environment in larger organisms. Give them a minute or so to discuss in pairs how large/multicellular organisms, such as plants, might solve this problem. Take ideas from round the class. (If needed, remind students of their work in Section 6: *Plant nutrition,* and as then to think about which part of the plant absorbs mineral ions.) They should be able to suggest that the much-divided root system of the plant provides a large surface area for absorption.

2. Root hair cells

Students should look at a prepared slide of a plant root showing root hairs, or an image from the internet, to identify the root hair cells. In addition they should look at Fig. 8.5 on page 145 of the Student Book, which shows the root hair cells on the outside of a root. They should identify the root hair cells as the site of absorption of water and mineral ions. This can be reinforced by explaining that if a plant is lifted out of the soil, these delicate root hairs are usually damaged. So, even if the plant is immediately replanted, it will take a few days before it can absorb water and mineral ions effectively again.

Link this work to the introductory task to help students understand the role of root hair cells in providing a much-increased surface area for absorption of water and mineral ions.

Supplement Ask students to suggest another surface in organisms which is adapted for absorption by having features that greatly increase its surface area. They should be able to recall the villi and microvilli of the small intestine from their work in Section 7: *Human nutrition*.

3. Water movement through plants

Worksheet B8.2 provides a method for investigating the transport of water through the above-ground parts of a plant. This takes a few minutes to set up, but then needs to be left for about 24 hours before looking at the results. Alternatively, this could be set up by a technician the day before the lesson.

Celery stalks work well for this, as they have large vascular bundles that can easily be removed from the stalk tissue. If celery is not available, use a plant stem that is wide and fairly transparent, such as Busy Lizzie (*Impatiens*) or just wide, such as *Coleus*. If the plant stems have flowers, ideally choose white flowers as these will more clearly show the dye in the petals. If the plant stems have leaves, ideally choose those with pale or yellow leaves.

SAFETY INFORMATION: Remind students to take care with sharp knives.

Use the information on water uptake on pages 145–146 of the Student Book to help describe the whole pathway taken by water through the plant from the root hair cells to the leaf mesophyll cells.

4. Consolidation

Students could develop the concept maps they started in the last learning episode, to include what they have learned in this learning episode. Alternatively ask students to write a short paragraph for a web science encyclopaedia to explain how plants are adapted to take in water and mineral ions.

Technician's notes

Be sure to check the latest safety notes on these resources before proceeding.

The following resources are needed for the introduction:

picture / slide of a unicellular organism, such as *Paramecium* or *Amoeba* and the means to display it

The following resources are needed for task 2 on root hair cells, per student or group:

prepared slide of transverse section of a plant root, showing root hair cells, or a photograph of this

The following resources are needed for the class practical B8.2, per group:

stem or stalk of plant
1 cm^3 food colouring
teat pipette
water
250 cm^3 beaker
sharp knife and cutting board
forceps
hand lens

See note above about suitable plants to use.

Keep the ends of the stalks/stems sitting in water until needed, to prevent air entering the vascular tissue.

SAFETY INFORMATION: Remind students to take care when using sharp knives.

Answers

Page 146

1. It enters through the root hair cells, moves through the root cortex cells to the xylem in the centre of the root. It moves through the xylem, up the stem and into the leaves. In the leaves, it moves out of the xylem into the mesophyll cells.
2. Place a stem of a plant in water containing food colouring. The colour will travel through the xylem with the water, and show where the xylem is in the stem, leaves and flowers.
3. a) osmosis
   b) active transport

Worksheet B8.2 Water movement through plants

The movement of water through the parts of a plant that are above ground can be investigated using water containing a coloured dye, such as food colouring. As the water is taken up, it carries the dye with it. The dye stains tissues with which it comes into contact.

Apparatus

stem or stalk of plant

food colouring

pipette

water

beaker

sharp knife and cutting board

forceps

hand lens

SAFETY INFORMATION
Take care with sharp knives.

Method

1. Add about 1 cm³ dye to half a beaker of water and mix.
2. Use the knife to cut off about 1 cm from the end of the plant stem. Discard the end and place the stem immediately into the beaker of coloured water. Leave the stem in the beaker for 24 hours.
3. After 24 hours, use the hand lens to look carefully at the stem, leaves and petals of the flower (if present). Identify where the dye has been transported. Quickly sketch and annotate a diagram to show where the dye is visible.
4. Place the stem on the cutting board, and use the sharp knife to cut across the stem about half-way up.
5. Use the hand lens to look closely at the end of the stem. Is the dye evenly distributed across the stem or concentrated in some areas more than others? Sketch and annotate a diagram to show what you see.

Handling experimental observations and data

6. Using your knowledge of the structure of a plant, explain what you have seen of the distribution of dye.
7. If you had left the stem in the water for another two days, what would you expect to have happened to the distribution of the dye? Explain your answer.
8. Use your findings to explain the importance of a transport system to plants.

Learning episode B8.3 Transpiration

Learning objectives

- Describe transpiration as the loss of water vapour from leaves
- State that water evaporates from the surfaces of the mesophyll cells into the air spaces and then diffuses out of the leaves through the stomata as water vapour
- Investigate and describe the effects of variation of temperature and wind speed on transpiration rate
- Supplement Explain how water vapour loss is related to: the large internal surface area provided by the interconnecting air spaces between mesophyll cells and the size and number of stomata
- Supplement Explain the mechanism by which water moves upwards in the xylem in terms of a transpiration pull that draws up a column of water molecules, held together by forces of attraction between water molecules
- Supplement Explain the effects on the rate of transpiration of varying the following factors: temperature, wind speed and humidity
- Supplement Explain how and why wilting occurs

Common misconceptions

Sometimes, when students describe transpiration as the evaporation of water from leaves, they do not appreciate that the evaporation takes place inside the leaf and the actual loss of water from the leaves is by diffusion of water vapour through the stomata.

Resources

Student Book pages 146–150

Worksheet B8.3 The rate of transpiration

B8.3 Technician's notes

Resources for class practicals (see Technician's notes)

Approach

1. Introduction

Give students the following words and ask them to write sentences that use as many of the words as possible: *root hair cell, absorption, water, root xylem, stem xylem, osmosis, surface area, mineral ions*. Take examples from around the class and check that students have remembered all the key points from the previous learning episode.

2. Transpiration

On the day before you teach this section, place a potted plant in a plastic bag so that all the above-ground parts are enclosed, but the pot is not. Tie the bag loosely around the plant stem. Then place the plant in a bright place until the lesson. If the light is bright enough, and it is reasonably warm, condensation should be visible on the inside of the bag.

Show the plant in its bag to students and explain that the only water that plant has had was given to the roots. Ask them to explain as fully as they can why the condensation has formed inside the bag. They should be able to explain that the water in the pot has been absorbed by the roots and lost as water vapour from the leaves.

Introduce the term *transpiration* for the loss of water vapour from a leaf. Make sure that students identify *xylem* as the transport tissue through which water and dissolved mineral ions move through a plant.

3. Transpiration from leaves

Ask students to look at a transverse section of a plant leaf, either under a light microscope or displayed on a screen. Ask them to think about the movement of water, entering the leaf from the xylem that has carried the water from the roots and diffusing out of leaves as water vapour through stomata. They should identify the surfaces of the mesophyll cells from which water evaporates into the air spaces, and the connection between the air spaces and stomata through which the water molecules diffuse. Take the opportunity to check understanding of diffusion from their work in Section 3: *Movement into and out of cells.*

Ask students to sketch an outline of a leaf and its key structures (including veins), and to annotate it to show how water moves through the leaf.

Supplement Students should relate the large surface area of the cells surrounding the air spaces to its effect on the rate of evaporation of water from the mesophyll cells. The rate of loss of water vapour by diffusion from the leaf is also related to the size and number of stomata.

Supplement **4. Transpiration pull**

Explain that there are forces of attraction between water molecules, so that they have a tendency to stick together. This means that if there is a tension ('pull') on one end of a column of water, such as when you suck on a drinking straw, the whole column of water is pulled along rather than breaking into separate drops.

Ask them to apply this concept to the movement of water through a plant from where it enters through the roots and exits through the leaves. They could annotate a cartoon, or draw a flow chart, that explains how water moves through the roots, stems and leaves.

5. Factors that affect transpiration

Worksheet B8.3 is a planning sheet on the factors that affect the rate of transpiration. It suggests equipment for producing a simple potometer, but a commercial potometer is likely to give better results if set up correctly.

If you have the equipment, consider allowing students to test their plans once you have checked them for safety and suitability. You will need to set up the potometer for them. Details are given in the Technician's notes below.

This experiment could be extended to record the movement of the bubble when leaves are removed from the stem. Ask students to predict the effect of this when the apparatus is set up, and to explain any results from this extension.

If the practical set-up is not feasible, ask students to read page 148 of the Student Book about the factors that affect the rate of transpiration. They should write a test question and answer asking how transpiration would change under different conditions (such as on a warm, still day or a cool, windy day).

Next they should exchange their question and answer with another student who should first try to answer the question, then check their answer with the one given and identify ways of improving both the question and answer.

Supplement Students should be expected to explain the effects of temperature, humidity and wind speed on transpiration rate using their knowledge of the effect of temperature on the kinetic energy of particles, and the effect of concentration of particles on diffusion (humidity is a measure of the concentration of water in air, and wind will move humid air away from leaves).

SAFETY INFORMATION: Wash hands thoroughly at the end of the practical. Be careful with the possibility of water being spilled close to the electrical fan.

Supplement **6. Wilting**

Show students a picture of a wilted plant and remind them of the effect of water loss on plant cells from Section 3: *Movement into and out of cells.*

Ask students to apply what they now know about transpiration to explain why wilting occurs

7. Consolidation

Students could add to their concept maps from the last two activities, to include what they have learned in this learning episode. Alternatively, ask students to write a question about plant transport tissues and transpiration, that is worth 3 or 4 marks, and the mark scheme for it.

They should exchange their question and mark scheme with another student and look for weaknesses in the other student's answer. They should suggest any amendments to improve the question or answer, then return it to the other student to discuss and change.

Technician's notes

Be sure to check the latest safety notes on these resources before proceeding.

The following resources are needed for task 2, for demonstration:

potted plant
clear plastic bag
root section under a light microscope *or* an example displayed on a screen; transverse section of a plant leaf, either under a light microscope or displayed on a screen

On the day before the lesson, place a potted plant in a plastic bag so that all the above-ground parts are enclosed, but the pot is not. Tie the bag loosely around the plant stem. Then place the plant in a bright place until the lesson. If the light is bright enough, and it is reasonably warm, condensation should be visible on the inside of the bag.

The following resources are needed for the class practical B8.3, per group:

potometer (see note above)
bright light (as light bank or next to bright window)
dim and/or dark conditions
light meter or sensor
electric fan with several settings
thermometer
access to warm area
access to cooler area

If a potometer is to be set up for the demonstration or for use by students in an investigation, use the following instructions.

Use woody shoots from a bush or tree that does not have glossy leaves. Assemble the potometer underwater to prevent air bubbles entering the apparatus. (Refer to the manufacturer's instructions.) The shoot should also be cut and inserted into the potometer under water.

Allow the leaves to dry before the lesson and allow the plant to adjust to the new conditions. Just before measurement starts, insert an air bubble into the tube as described in the manufacturer's instructions, and adjust its position to sit within the scale.

If students are to carry out their plans from the worksheet, the following apparatus may be needed. However, students may suggest other apparatus. The plans should be checked to make sure that the apparatus suggested is available.

If using lamps, please note the following.

- The voltage of the bulb should not exceed the safety limits of the bench lamp. Low-energy bulbs should have some sort of protection so that they do not protrude beyond the protective metal of the bench lamp. Or they can be protected by a plastic or glass screen.
- Low-energy bulbs contain mercury and so can be hazardous if the bulb is broken.
- You should be vigilant during this practical because of water being close to electrical equipment.
- LED lights emit a very narrow range of wavelengths of light, which might not be absorbed by the plant. Halogen light sources are fine for photosynthesis, but they do get hot. A large beaker of tap water placed in front of the light source will absorb the heat.
- Potometers should be prepared in a different part of the room from the light sources.

SAFETY INFORMATION: Wash hands thoroughly after handling the plants and ventilate the room after the activity.

Answers

Pages 148–149 Developing Practical Skills

1. In each case you would need two potometers set up as identically as possible, or run the investigation twice with the same equipment.
   a) Take measurements at a low temperature, and also at a higher temperature (for example, with a heater nearby, but below 40 °C, when damage may start to occur to proteins/enzymes), keeping all other conditions identical.
   b) Take measurements in still air conditions, and also in windy conditions (such as using a fan), keeping all other conditions identical.
2. a) moving air/sunlight with hot/moving air/sunlight so that heat/temperature is the only factor that differs
   b) still air/sunlight compared with moving air/sunlight, or still air/dark cupboard compared with moving air/dark cupboard, because light intensity is the same in both and only the factor of wind speed differs
3. a) The rate of transpiration increases with higher temperature because the bubble moved 5 cm much more quickly (54 seconds for hot/moving/sunlight compared with 75 seconds with normal/moving/sunlight).
   b) The rate of transpiration is faster when wind speed is greater because the bubble moved 5 cm much more quickly (light: 75 seconds in moving air compared with 135 seconds in still air; dark: 122 seconds in moving air compared with 257 seconds in still air).
4. As water is transpired from the leaf, more water is drawn into the leaf from the stem, and more water is then drawn into the stem from the capillary tubing. The bubble moves with the water, so the movement of the bubble indicates how much water has been taken up by the shoot.
5. Apart from a fault in the connection between the shoot and the tubing, some water is used in photosynthesis in the leaves. (But this is usually minimal over the time of the experiment.)
6. The conclusions are based on comparing two results in each case. Carrying out repeats with the same shoot, and with shoots from the same plant, in each set of conditions would make it possible to identify any anomalous results and take averages of results, to produce more reliable conclusions.

Page 150

1. Evaporation from the surfaces of a plant, particularly from the stomata of a leaf into the air.
2. Diagram should include annotations like the following, at the appropriate point: water molecules evaporate from surfaces of spongy mesophyll cells into air spaces; water molecules from air spaces move into and out through stomata into the air – diffusion (net movement) usually from inside leaf to outside; osmosis causes water molecules to move from xylem into neighbouring cells until they reach a palisade cell or a spongy mesophyll cell; transpiration is the evaporation of water from a leaf.
3. Supplement Closing stomata reduces diffusion of water molecules out of the leaf. At night, carbon dioxide is not needed for photosynthesis, so keeping stomata open would lose water unnecessarily.
4. Supplement a) When temperature is higher, particles move faster, so water molecules will diffuse out of the leaf more quickly.
   Supplement b) When air humidity is lower, there is a lower concentration of water molecules in the air outside the leaf. This increases the concentration gradient between the inside of the leaf and the outside air. This means the rate of diffusion will be faster.
5. Supplement Forces of attraction between water molecules means they stick to each other. So as water moves out of the xylem in the leaves, down its potential gradient into the spongy mesophyll cells, more water molecules are drawn up the xylem tube through the plant because of the forces of attraction between water molecules. This causes a water potential gradient between the root cortex cells and the xylem in the root, causing more water to enter the xylem.

Worksheet B8.3 The rate of transpiration

Transpiration is the loss of water vapour from the leaves of a plant. The rate of transpiration from the leaves of a plant shoot can be measured using a potometer.

There are many designs of potometer, but the diagram shows a simple version. A shoot is inserted into the apparatus and sealed so that there are no air leaks.

A bubble is introduced to the side-arm of the apparatus. The movement of this bubble is used as a measure of the amount of water lost by transpiration.

At first the screw clip is left open to let the shoot adjust to the new conditions. When the experiment begins, the screw clip is closed and the bubble's rate of movement over a fixed distance is measured.

The shoot can be exposed to different conditions to test their effect on the rate of transpiration.

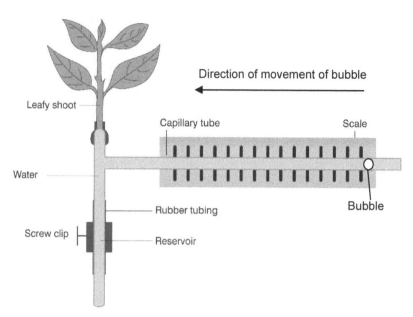

Apparatus

potometer	bright light (as light bank or next to bright window)
dim and/or dark conditions	light meter or sensor
electric fan with several settings	thermometer
access to warm area	access to cooler area

SAFETY INFORMATION
Wash hands thoroughly at the end of the practical.

Planning and evaluating investigations

1. You are going to plan an investigation, using this apparatus, to test the effect of different factors on the rate of transpiration.

You will need to consider:

- which factors you will test
- how you should set up the equipment to test each of the factors
- how you will adjust each factor and what you will measure in each test
- which other factors you will need to control and how to control them
- what risks there may be in carrying out your tests and how these can be managed
- how the limitations of the method suggested will affect the conclusions you could draw from any results.

2. Write out your plan for each factor, and make a prediction for each one.

3. Show your plan to your teacher. Your teacher will tell you whether you can now carry out your plan.

Supplement Learning episode B8.4 Translocation

Learning objectives

- Supplement Describe translocation as the movement of sucrose and amino acids in phloem from sources to sinks
- Supplement Describe:
  (a) sources as the parts of plants that release sucrose or amino acids
  (b) sinks as the parts of plants that use or store sucrose or amino acids
- Supplement Explain why some parts of a plant may act as a source and a sink at different times

Common misconceptions

Students may confuse the terms *transpiration* and *translocation*. Encourage them to find a way of remembering the difference, such as relating translocation to 'moving location/place' and transpiration to inspiration and expiration in human breathing. (Avoid linking it to *respiration*, as this could reinforce the misconception that respiration means breathing.)

Resources

Student Book pages 150–151

Approach

1. Introduction

Ask students what all plant cells need for growth. From earlier sections, they should be able to answer that they need the building blocks for making new substances such as proteins, large carbohydrates and fats and oils, as well as a source of energy. Then ask where these substances come from, and to suggest how they reach all cells. Students should identify the need for a transport system that is separate from xylem, and may suggest phloem as that system.

2. Translocation, sources and sinks

Introduce the term *translocation* as the transport of substances other than water and dissolved mineral ions around a plant. Ask students to suggest which substances might be translocated. Lead students to answers such as sucrose (produced from glucose made during photosynthesis) and amino acids (made by converting carbohydrates using nitrogen from nitrate ions absorbed from the soil).

Introduce the terms *source* and *sink* and give students a few minutes to consider where these substances may be made in a plant (their *sources*) and where they might be needed and for what purpose (their *sinks*). Provide hints such as 'growth', 'storage' and 'respiration', to help them do this. Students could use page 151 of the Student Book to help with this.

Students should consider how sources and sinks may change at different times of the year, and identify how translocation of substances may change in different seasons. Provide examples, such as the production of leaves in spring, and formation of storage organs during the autumn, for students to interpret.

3. Consolidation

Students complete their concept maps from the previous activities, to include what they have learned in this learning episode.

Answers

1. phloem
2. sucrose and amino acids
3. A source is a part of a plant where a substance is formed e.g. sucrose is made in leaf cells. A sink is a part of a plant where the substance leaves or is converted into something else, e.g. cells in the root or fruit may be sinks for sucrose.

Learning episode B8.5 Consolidation and summary

Learning aims

- Review the learning points of the topic summarised in the end of topic checklist
- Test understanding of the topic content by answering the end of topic questions

Resources

Student Book pages 152–154

Approach

Introduce the learning episode

Ask students to work with a partner to make a list of key words from this topic. They could then work together to produce a spider diagram showing how the different concepts are linked. They could compare their list with the list of key terms given on page 152 of the Student Book. Discuss the checklist on pages 152–153 and use questioning to see how much of the content they are comfortable with.

Students could make flashcards of the key content and then use the flashcards to quiz each other on the information.

Develop the learning episode

Ask students to work individually through the end of topic questions on page 154 of the Student Book without looking at the text. As they work, walk around the classroom observing their answers and asking questions as necessary to find out which questions are causing difficulties.

Finish the learning episode

After a set period, ask the students to stop working and discuss any areas of difficulty you observed as you walked round the class. Students should complete any unanswered questions for homework, but you should stress that they should try to answer the questions without looking at the text, so that they can see how much they have remembered.

Answers

End of topic questions mark scheme

The marks available for a question can indicate the level of detail you need to provide in your answer.

Question	Correct answer	Marks
1	C	1 mark
2	Clearly drawn, and fully labelled, plan diagrams for stem, root and leaf sections based on the photomicrographs in Fig. 8.3 on page 144 of the Student Book (1 mark for clear drawing, 1 mark for labelling, 1 mark for correctly labelling phloem or xylem per diagram.)	9 marks
3	They place the plant stem in water that contains soluble colouring.	1 mark
	Water is lost from the leaves of the plant due to transpiration.	1 mark
	This draws water out of the xylem in the leaves, which causes more water to be drawn up the xylem in the stem from the bottom of the stem.	1 mark
		1 mark

Question	Correct answer	Marks
	The soluble colouring is carried with the water up the xylem and into the petals.	
Supplement 4	The leaves of the plant on the windowsill receive more light than those on the shelf,	1 mark
	so the stomata in the leaves of the windowsill plant will be open wider so that photosynthesis can continue as quickly as possible.	1 mark
	Wide stomata allow water molecules to diffuse out of the air spaces in the leaves more quickly.	1 mark
	If the rate at which water is lost from the leaves is greater than the rate at which water is taken in from the soil, the cells in the plant will become flaccid, causing the plant to wilt.	1 mark
	(The temperature of the air around the windowsill plant and the temperature of its leaves are also likely to be higher than for the shaded plant, increasing the rate of transpiration for the light plant compared with the shaded plant.)	
Supplement 5	It is cooler at night	1 mark
	and transpiration rate is slower when the temperature is lower.	1 mark
	So only opening the stomata at night means the cactus loses much less water through transpiration than if the stomata opened during the hot day.	1 mark
	This increases the chance of the plant surviving in the very dry conditions.	1 mark
Supplement 6	In the spring, the shoots and roots are growing rapidly, and new leaves develop. These parts of the plant are the main sinks at this time,	1 mark
	needing sucrose to produce glucose in cells where needed for respiration	1 mark
	and to build the substances needed in the new tissue.	1 mark
	In spring and summer, photosynthesising leaves are the main sources.	1 mark
	In the summer, when the plant is flowering and reproducing, the flowers become a major sink for sucrose.	1 mark
	In the autumn, the leaves may be lost for winter and carbohydrates stored as starch in stems and special structures such as tubers, so these storage areas become the major sink for sucrose.	1 mark
	Next spring, the storage areas act as sources providing sucrose for growing roots, shoots and leaves.	
	Total:	28 marks

Exam-style questions and sample answers have been written by the authors. In examinations, the way marks are awarded may be different. References to assessment and/or assessment preparation are the publisher's interpretation of the syllabus requirements and may not fully reflect the approach of Cambridge Assessment International Education.

The marks available for a question can indicate the level of detail you need to provide in your answer.

Pages 155–171

Question	Correct answer	Marks
1	A	1 mark
2	C	1 mark
3	D	1 mark
Supplement 4	Answer given in Student Book	
5 a)	i) A (bird), C (fish)	2 marks
	ii) E (plant)	1 mark
	iii) C (fish)	1 mark
	iv) B (crustacean), D (insect)	2 marks
5 b)	i) hair/fur	1 mark
	ii) mammary glands/produces milk	1 mark
	iii) Most mammals give birth to live young.	1 mark
6 a)	**Nutrition** is the taking in of substances needed for energy and for **growth** and **development**. **Respiration** is a series of reactions that take place in living cells to release **energy** from nutrient molecules so that cells can use this to keep them alive. **Sensitivity** is the ability to **detect** and **respond** to changes in external and internal conditions. Movement causes a change in **location** or **position** of the organism. In animals, this involves their entire bodies. Plants often move parts of their body in response to external stimuli, such as gravity or **light**.	11 marks – 1 mark for each of the words in **bold**
6 b)	The removal of metabolic waste materials,	1 mark
	and substances in excess of requirements.	1 mark

6 c)	nutrition (requires fuel)	1 mark
	respiration (chemical reactions that release energy)	1 mark
	movement	1 mark
	excretion (release waste materials – exhaust fumes)	1 mark
	(accept sensitivity)	
	It's not considered living because it does not show all seven characteristics,	1 mark
	it does not grow/reproduce	1 mark
7 a)	Individuals are put in the same group if they share particular features.	1 mark
7 b)	It is an internationally agreed system,	1 mark
	so does not have the problem of organisms being called different names in different parts of the world/different areas of a country.	1 mark
7 c) i)	*Lumbricus terrestris* (the first letter of the genus is in upper case; the species in lower case, and it is written in italics).	1 mark
ii)	A group of organisms that can reproduce, to produce fertile offspring.	1 mark
iii)	The first part shows the genus name,	1 mark
	the second part shows the species name.	1 mark
8	Viruses are not cellular/ they consist of genetic material surrounded by a protein coat.	1 mark
	They cannot live (or at least not for very long) outside the cells/body of an organism/ they are parasitic.	1 mark
	They do not have all seven characteristics of living things (MRS GREN). (Plus one extra mark for mentioning any of the following:)	1 mark
	Some viruses **move** but many rely on other organisms to get them inside host cells or transport mechanisms in their host to transport them around. Viruses **reproduce** but they are dependent on being inside a host to do this and it is unlike reproduction in other organisms (it involves a copying of their DNA or RNA and assembling a protein coat around them) They are **sensitive** to their environment in that they can enter their host cells They do not **grow** in the normal sense (they are assembled); they do not **respire** or **excrete** and do not require **nutrition**/feed. (So most scientists do not regard them as living organisms.)	1 mark
Supplement 9 a)	To put them into groups,	1 mark
	to help to understand the evolutionary relationships between groups.	1 mark
9 b) i)	have a backbone/vertebral column	1 mark
ii)	The organisms have the same pattern of bones/types of bones (in their fore limbs).	1 mark

iii)	The relative proportions of the bones/ relative sizes of the different bones.	1 mark
	The bat's wing has skin stretched over the bones but the bird's wing has attached feathers.	1 mark
Supplement 9 c)	Scientists compare the organisms' DNA base sequences.	1 mark
	The more similar the sequences, the more closely related the organisms.	1 mark
10 a)	exoskeleton	1 mark
	limbs and bodies have joints to allow movement	1 mark
10 b)	Suitable dichotomous key, i.e. branching into two to identify each organism; one mark for the identification of each organism. It is best based on the following: Body parts – cephalothorax and abdomen (crayfish and spider); head, thorax and abdomen (locust); head and trunk (centipede and millipede). Number of antennae – none (spider); one pair (locust, centipede and millipede); two pairs (crayfish). Number of legs – three pairs (locust); four pairs (spider); five pairs (crayfish); one pair per segment (centipede); two pairs per segment (millipede).	5 marks
11 a)	cell membrane	1 mark
	cytoplasm	1 mark
	nucleus	1 mark
	chromosomes	1 mark
	cell wall	1 mark
	cellulose	1 mark
	vacuole	1 mark
	chloroplasts	1 mark
	chlorophyll	1 mark
	photosynthesis	1 mark
11 b)	Ticks in table as shown below. 1 mark for each one correct.	8 marks

Structure	Cell	Tissue	Organ
Blood		✓	
Brain			✓
Egg	✓		
Liver			✓
Muscle		✓	
Neurone	✓		
Skin			✓
Sperm	✓		

12 a)	X – cytoplasm,	1 mark
	where many different chemical processes happen	1 mark
	Y – cell membrane,	1 mark
	controls what substances enter and leave the liver cell	1 mark
	Z – nucleus,	1 mark
	contains genetic material in the chromosomes/ control how a cell grows and functions/ controls protein synthesis/ controls cell division.	1 mark
Supplement 12 b)	width of cell image is 40 mm = 40 000 μm	1 mark
	magnification = image size ÷ actual size	1 mark
	= 40 000 ÷ 60	1 mark
	= ×667	1 mark
Supplement 12 c)	1 cm = 10 000 μm	1 mark
	number of cells = 10 000 ÷ 60	1 mark
	= 167	1 mark
13 a)	diffusion	1 mark

13 b)					9 marks

Dimensions of cube / mm	i) Surface area of cube / mm²	ii) Volume of cube / mm³	iii) Surface area: volume ratio
1 x 1 x 1	6	1	6: 1
5 x 5 x 5	150	125	150: 125 = 1.2 : 1
10 x 10 x 10	600	1000	600: 1000 = 0.6 : 1

Supplement 13 c)	As the volume of the cube increases, so does the surface area, but not to the same extent.	1 mark
Supplement 13 d) i)	For the small and medium cubes, there is sufficient surface area (relative to the volume)/ the distance to the centre of cube is sufficiently short, for the acid to diffuse to the centre of the cube during the lesson	1 mark
	For the large cube, there is insufficient surface area (relative to the volume)/ the distance to the centre of cube is too great, for the acid to diffuse to the centre of the cube during the lesson	1 mark
ii)	As organisms increase in size, their volume will increase,	1 mark
	so useful substances will not diffuse into their bodies quickly enough,	1 mark
	so they need to develop organ systems to transport needed substances to all part of their bodies quickly enough.	1 mark

14 a)	+32.2	1 mark
	+21.8	1 mark
	−2.2	1 mark
	−13.5	1 mark
	−19.9	1 mark
	−20.4	1 mark
14 b) i)	axes drawn with concentration of sucrose on x-axis and percentage change in mass on the y-axis	1 mark
	axes labelled, with units	1 mark
	points plotted correctly	1 mark
	points joined appropriately	1 mark
ii)	student answer from graph:	1 mark
	accurate reading of intersection with x-axis	1 mark
	correct units [g per cm^3]	1 mark
iii)	In distilled water and low concentrations of sucrose, the potato cylinders increase in mass.	1 mark
	In higher concentrations of sucrose, the potato cylinders decrease in mass.	1 mark
	At a certain concentration/given concentration, there is no change in mass.	1 mark
iv)	Osmosis	1 mark
15 a)	1 mark for each row	5 marks

Large biological molecule	Smaller molecules that make up the molecule
Glycogen	glucose
Fat	fatty acids, glycerol
Protein	amino acids
Starch	glucose

15 b) i)	i) add Benedict's solution, to the sample to be tested (in a test tube)	1 mark
	heat to 95 °C.	1 mark
	colour change from blue	1 mark
	(to green to yellow to orange) to red-brown, indicates that glucose/a reducing sugar is present	1 mark
ii)	add iodine solution, to a food sample (on a spotting tile)	1 mark
	colour change from orange-brown	1 mark
	to blue-black, indicates that starch is present	1 mark

Supplement 16 a) i)	Enzyme activity increases as pH increases,	1 mark
	until it reaches a peak/optimum at pH 4.	
	Above pH 4 enzyme activity decreases as pH increases.	1 mark
	Enzyme activity depends on the substrate being able to fit into the	1 mark
	enzyme's active site/ being able to form an enzyme-substrate complex.	1 mark
	At pH levels above and below the optimum, the active site changes shape/	1 mark
	enzyme becomes denatured	1 mark
	so it can no long bind as well to the substrate.	
ii)	Enzyme activity increases as temperature increases,	1 mark
	until it reaches a peak/optimum at around 50 °C, and then starts to	1 mark
	decrease	1 mark
	As temperature increases, substrate and enzyme move more quickly/ have	
	more kinetic energy,	1 mark
	so there are more frequent effective collisions, increasing the rate of	1 mark
	reaction.	
	Above the optimum temperature, the active site changes shape/ enzyme	
	becomes denatured, so reducing the rate of reaction.	
Supplement 16 b) i)	pepsin	1 mark
	trypsin	1 mark
ii)	The student is correct for the pH range of 4–8,	1 mark
	but we do not have any evidence for papain's activity outside this range.	1 mark
17 a)	stomach pancreas small intestine	1 mark for each correct label
17 b i)	stomach protease: 1.6	1 mark
	small intestine protease 8.0	1 mark
17 b) ii)	suitable context, e.g. digestion of amylase by starch or digestion of gelatin on film by protease	1 mark
	description of suitable way of varying the temperature	1 mark
	description of suitable way of measuring reaction time/rate of reaction	1 mark
	description of how other variables are controlled	1 mark
	description of how to ensure repeatability	1 mark
Supplement 18	Answer given in Student Book	

Supplement		
19 a)	$6CO_2 + 6H_2O \rightarrow C_6H_{12}O_6 + 6O_2$	1 mark for reactants 1 mark for products 1 mark for balancing
19 b)	leaf has a large surface area for gas exchange	1 mark
	thin (usually less than 1 mm) so gases only need to diffuse over small distances	1 mark
	stomata to regulate the diffusion of gases in and out	1 mark
	mesophyll cells have large total surface area for the absorption and release of gases	1 mark
	large air spaces between the cells of the spongy mesophyll so that gases can move freely	1 mark
20 a)	A: cuticle	1 mark
	B: (upper) epidermis	1 mark
	C: palisade mesophyll	1 mark
	D: spongy mesophyll	1 mark
	E: guard cell (allow: stoma)	1 mark
20 b)	light energy carbon dioxide + water → glucose + oxygen chlorophyll	1 mark for reactants 1 mark for products 1 mark for light energy/ chlorophyll
20 c)	The loss of water vapour from the leaf,	1 mark
	by evaporation of water from the mesophyll cells,	1 mark
	followed by diffusion of water vapour through the stomata.	1 mark
20 d)	Place the cut end of a plant stem,	1 mark
	into the end of a potometer/length of capillary tubing,	1 mark
	under water,	1 mark
	ensuring an airtight seal.	1 mark
	Introduce an air bubble into the potometer/capillary tube.	1 mark
	Measure the rate at which the air bubble travels,	1 mark
	at different temperatures.	1 mark
	(Allow technique based on measurement of loss of mass)	
21 a) i)	A: root hair/root hair cell	1 mark
	B: xylem vessel	1 mark

Supplement 21 a) ii)	Soil water has a higher water potential than root hair cells,	1 mark
	so water moves by osmosis from the soil into the root hair cells.	1 mark
	Now these cells have a higher water potential than the cells further inside the root,	1 mark
	so water moves further into the root along a water potential gradient.	1 mark
	This continues until water reaches the xylem vessels which carry it up to the rest of the plant.	1 mark
		1 mark
	The water potential gradient is maintained as water molecules are removed from the top of the xylem vessels by transpiration by the leaves.	1 mark
		1 mark
Supplement 21 b)	Translocation is the movement of sucrose/amino acids through a plant,	1 mark
	through the phloem,	1 mark
	from sources to sinks.	1 mark
	Sources are parts of plants that release sucrose/amino acids,	1 mark
	sinks are parts of plants that use/store sucrose/amino acids.	1 mark
	Example(s) of sources (e.g. photosynthesising leaves) / sinks (e.g. substances being stored in storage organs)	1 mark

Introduction

In this section students will learn about the transport systems in animals.

Links to other topics

Topics	Essential background knowledge	Useful links
1 Characteristics and classification of living organisms	1.1 Characteristics of living organisms 1.3 Features of organisms	
2 Organisation of the organism	2.1 Cell structure and size of specimens	
3 Movement into and out of cells	3.1 Diffusion 3.2 Osmosis	
5 Enzymes	5.1 Enzymes	
7 Human nutrition	7.2 Digestive system 7.5 Absorption	
10 Diseases and immunity		10.1 Diseases and immunity
11 Gas exchange in humans		11.1 Gas exchange in humans
12 Respiration		12.1 Respiration 12.2 Aerobic respiration 12.3 Anaerobic respiration
13 Excretion in humans		13.1 Excretion in humans
14 Coordination and response		14.3 Hormones 14.4 Homeostasis
16 Reproduction		16.4 Sexual reproduction in humans 16.5 Sexual hormones in humans

Topic overview

B9.1	Circulatory systems and the heart
	This learning episode looks at the overall structure of circulatory systems and of the heart. It includes practical work on the effect of exercise on pulse rate, and a demonstration of the dissection of a heart. Students will also learn about the causes of coronary heart disease, some risk factors associated with it and how it may be prevented. Supplement There is the opportunity to study and compare the single circulatory system of fish and the double circulatory system of mammals. Students will also learn how structure is related to function in the heart, and why exercise affects pulse rate.
B9.2	Blood vessels and blood
	This learning episode looks at the structure and function of arteries, veins and capillaries. It also looks at the structure and function of the different components in human blood. Supplement Students will learn how vessel structure is related to function, as well as learning further detail of the functions of some blood components.

B9.3	**Consolidation and summary**
	This learning episode provides an opportunity for a quick recap on the ideas encountered in the section. Students can answer the end of topic questions in the Student Book.

Careers links

These are some scientific careers that focus on this area of biology but careers in many other fields use the knowledge and skills gained studying science. Many health professionals need to have an understanding of the human circulatory system. **Phlebotomists** take blood from people for clinical testing, research or blood donations. **Cardiologists** are doctors who specialise in treating heart conditions.

Starting points

The Student Book section opener puts the ideas in the section into context and sets the scene. It also allows students to acknowledge and value their prior learning, and provides a benchmark against which future learning can be compared.

- The questions provide a structure for introducing the section and can be used in a number of different ways:
- You could ask students to consider the questions as an introductory homework task.
- You could put students into groups to share their own ideas and understanding and then to report back to the whole class.
- Students could be given access to the Internet, preferably with a tight timescale, to find out the information required.

You could then use a spider chart or other form of wall chart to summarise everybody's ideas.

Recording these initial ideas allows you to retain them for reference as the individual topics are developed. In this way, your students' progress in learning can be readily acknowledged.

Learning episode B9.1 Circulatory systems and the heart

Learning objectives

- Describe the circulatory system as a system of blood vessels with a pump and valves to ensure one-way flow of blood
- Supplement Describe the single circulation of a fish
- Supplement Describe the double circulation of a mammal
- Supplement Explain the advantages of a double circulation
- Identify in diagrams and images the structures of the mammalian heart, limited to: muscular wall, septum, left and right ventricles, left and right atria, one-way valves and coronary arteries
- State that blood is pumped away from the heart in arteries and returns to the heart in veins
- State that the activity of the heart may be monitored by: ECG, pulse rate and listening to sounds of valves closing
- Investigate and describe the effect of physical activity on the heart rate
- Describe coronary heart disease in terms of the blockage of coronary arteries and state the possible risk factors including: diet, lack of exercise, stress, smoking, genetic predisposition, age and sex
- Discuss the roles of diet and exercise in reducing the risk of coronary heart disease
- Supplement Identify in diagrams and images the atrioventricular and semilunar valves in the mammalian heart
- Supplement Explain the relative thickness of:
  - (a) the muscle walls of the left and right ventricles
  - (b) the muscle walls of the atria compared to those of the ventricles
- Supplement Explain the importance of the septum in separating oxygenated and deoxygenated blood
- Supplement Describe the functioning of the heart in terms of the contraction of muscles of the atria and ventricles and the action of the valves
- Supplement Explain the effect of physical activity on the heart rate

Common misconceptions

Many students may be unaware that the mammalian circulatory system has a double circulation. It is important that students covering the Supplement content learn this.

Some students think that the heart does not need its own blood supply – the heart is full of blood, so why does it need its own blood supply? Coronary blood vessels are needed to supply the tissues in the heart. This can be understood best when the role of different blood vessels is introduced, because substances in the blood diffuse most effectively into and out of capillaries, whose walls are formed from a single layer of cells.

Resources

Student Book pages 175–183

B9.1 Technician notes

Supplement Worksheet B9.1a Comparing circulatory systems

Worksheet B9.1b Structure of a mammalian heart

Worksheet B9.1c Exercise and pulse rate

B9.1c Technician notes

Resources for class practicals (see Technician's notes)

Approach

1. Introduction

Remind students of their earlier work on levels of organisation in Section 2: *Organisation of the organism*. Using the circulatory system as an example, ask them to suggest an organ, a tissue and a cell in this system. List as many correct examples of each on the board and prompt for others that may not be immediately obvious to students. Examples are: organ – heart, artery, vein; tissue – blood, muscle in heart and blood vessel walls; cell – red blood cell, white blood cell, muscle cell. You could return to these lists at the end of this learning episode to see whether students can add more examples of each.

Note that the content of the first learning objective in the syllabus, 'describing the term *circulatory system* as containing a pump (the heart) and valves that ensure one-way flow of blood', will be covered in the study of the heart and blood vessels.

`Supplement` Students covering this content will be introduced to these details in task 2 below.

`Supplement` 2. Comparing circulatory systems

Worksheet B9.1a is a classwork sheet that provides some detail on the circulatory system of a fish, with questions that will help students make comparisons to the double circulatory system of mammals. If you have access to a fish, you could demonstrate a dissection of the body cavity to display the heart (the dark red organ just behind the gills), although it is unlikely that students will see much more than the basic structure (see Technician's notes).

3. Mammalian heart dissection demonstration

If you can get a mammalian heart from a butcher, and you are willing to do so, demonstrate the dissection of a heart (see Technician's notes). If possible, get a heart that still has the end of the blood vessels attached. You may want to wear disposable gloves during the dissection, and afterwards all surfaces and tools that have been in contact with the blood will need disinfecting, using a disinfectant such as 1% Virkon, for at least ten minutes. Wash your hands thoroughly after finishing the dissection and clearing up.

Start by pointing out where the blood vessels are attached (to the 'top' of the heart). Students will not be able to identify which vessel is which at this point. Use a sharp scalpel to cut down through each of the blood vessels into the heart, and then between the atria and ventricles, to open the heart out completely.

Help students to identify and distinguish the two atria and the two ventricles, by looking at differences in the thickness of the muscle walls. Also point out the valves and 'heart strings' in the ventricles that prevent the valves opening the wrong way when the heart contracts. Discuss how this ensures that blood only flows one way through the heart.

Students should make notes of the structures and their relationships during the dissection. Worksheet B9.1b could be used to support this.

`Supplement` Students also need to be able to: identify the atrioventricular and semilunar valves; explain the differences in the thickness of the walls of the right and left ventricles, and differences in the thickness of the atria and ventricles. They also need to explain the importance of the septum. These points should be covered during the dissection if possible.

4. Exercise and pulse rate

Worksheet B9.1c gives students an opportunity to investigate the effect of exercise on pulse rate. This work could be combined with the investigation of the effect of exercise on breathing in Worksheet B11.2, Section 11: *Gas exchange in humans*. The data collected on breathing could then be saved for analysis in the appropriate lesson. If you intend to do this, please check the technician's notes for that investigation and explain to students how to carry out the additional measurements safely.

Ideally students should take repeat measurements at each level of exercise and calculate an average from their results. Also, if the same method is used by all students, the data could be collated and compared, to identify that different people have different pulse rates, but that they all respond to an increase in level of exercise with an increased pulse rate.

Check that students have drawn up a suitable table for their results before they carry out the tests.

Supplement Students should be expected to explain these results in terms of delivering oxygen and glucose (food molecules) for respiration more rapidly to the actively respiring muscle cells, which need to use energy for contraction.

SAFETY INFORMATION: Be considerate of students' concerns about exercise. If some students are unable or unwilling to carry out the exercise, pair them with other students so that one is the test subject and the other is the recorder of results.

Over-exertion may be a hazard, especially for those with some medical conditions. Competitive situations can lead to careless behaviour and accidents.

Students who are exercising should be appropriately dressed, such as in gym kit (for example, shorts or tracksuit bottoms, T-shirt and training shoes), and the exercise should take place under supervision. If any student shows signs of difficulty, they should stop exercising immediately and sit quietly until they recover. For further advice on this, consult your PE department.

A suitable form of exercise is doing step-ups on stable equipment. This is much better than students running upstairs.

If taking the pulse rate in the neck, be careful that blood flow to the brain is not restricted.

Students should also research and report on other ways of monitoring heart activity using ECG or by listening to the sounds of valves closing. They could use the internet or information from pages 179–180 of the Student Book to collect their information.

5. Coronary heart disease

Discuss the importance of a separate, dedicated blood supply to the heart tissues through the coronary blood vessels. Although the heart is continually permeated with blood, the exchange of substances between the blood and tissues is most effective in through capillaries, which have very thin walls.

Ask students to consider the impact of a blockage of a coronary artery on the tissues that are beyond the blockage. From what students already know, they should be able to suggest that the tissues will be deprived of substances that they need, such as oxygen and glucose. They may also suggest that waste substances such carbon dioxide may build up in the tissues. The combined effects of this will quickly lead to heart damage or even death.

Ask students to read pages 181–182 of the Student Book to identify the direct cause of coronary heart disease (for example, deposition of cholesterol) and associated risk factors including diet, lack of exercise, stress, smoking, genetic predisposition, age and sex. They could extend this by completing research into the relationships between these risk factors and the causes of coronary heart disease, and then produce a poster or health leaflet giving guidance on how to reduce the risks.

6. Consolidation

Give students two pieces of paper. Ask them to write the word *true* on one piece and *false* on the other. Then give a series of true and false statements on what students have learned from this section, and ask students to hold up the correct piece of paper each time. Choose the statements to match the ability of the students. For example:

- The heart has four chambers. [true] (simple form)
- The mammalian heart has four chambers, two ventricles at the top and two atria at the bottom. [false] (more challenging)

The responses will help you identify any areas that students are still not sure about.

Technician's notes

Be sure to check the latest safety notes on these resources before proceeding.

The following resources are needed for the heart dissection demonstration:

heart
sharp knife or scalpel and dissecting board

disposable gloves
disinfectant such as 1% Virkon for disinfecting all tools and work surfaces
diagram of the heart

You will need one heart for this demonstration. Ideally the heart should still have the end of the blood vessels attached – they are best sourced from a butcher with a request to prepare them like this.

SAFETY INFORMATION: You may want to wear disposable gloves during the dissection, and afterwards all surfaces and tools that have been in contact with the blood will need disinfecting, using a disinfectant such as 1% Virkon, for at least 10 minutes. Wash your hands thoroughly after finishing the dissection and clearing up.

The following resources are needed for the class practical B9.1c, per group:

stopwatch

For this activity, students will need to exercise at different levels. Make sure that students are dressed appropriately and behave in a safe manner.

Answers

Page 177

1. To pump blood around the body.
2. Valves in the heart and veins.
3. Supplement Blood passes twice through the heart for every full circulation around the body.
4. Supplement The blood being pumped twice for each complete circulation allows higher blood pressure to be maintained around the body. Also, pressure in the two circulations can be different, so the higher blood pressure needed to get blood through most of the body doesn't damage the delicate capillaries in lung tissue.

Page 179

1. Left atrium, right atrium, left ventricle, right ventricle
2. Arteries carry blood away from the heart, whereas veins carry blood towards the heart.
3. Supplement Vena cava, right atrium, right ventricle, pulmonary artery, pulmonary vein, left atrium, left ventricle, aorta

Pages 180–181 Developing Practical Skills

1. a) Allowing exercise to continue for 2 minutes before taking the pulse rate gives time for the heart rate to adjust to conditions inside the body as a result of the altered level of exercise.
   b) Allowing sufficient time to rest between exercise levels gives the heart time to recover from the previous exercise and conditions inside the body to return to a resting state. So, all tests start from the same base point.
2. The data shows that heart rate increases from the resting rate as the level of exercise increases.
3. As the level of exercise increases, the heart rate also increases.
4. Supplement As the level of exercise increases, the muscles are working harder so need more oxygen and glucose for respiration to release the energy needed for contraction. The muscle cells also produce waste products, such as carbon dioxide, faster. As the heart pumps faster, materials are circulated more rapidly in the blood to supply what the muscle cells need and to remove waste products fast enough.
5. The data only comes from one subject during one set of tests. The investigation should be repeated several times with the same subject, to check that the measurements are consistent, and to average the results because heart rate varies a lot even in the same person. The investigation should also be repeated with several other test subjects, to make sure the results apply generally and do not just show the effect on one individual.

Page 181

1. Taking a pulse count, listening to the heart, or taking an ECG.
2. Resting heart rate varies widely due to many factors, including age, health and fitness, so a single value for the average is too limited.
3. As level of physical activity increases, so heart rate increases.
4. Supplement Heart rate increases with exercise so the blood can circulate faster round the body, delivering oxygen and glucose to muscle cells for the increased rate of respiration to generate the energy needed for contraction. It also removes waste carbon dioxide from muscle tissue more rapidly to prevent it building up and affecting cells.

Page 183

1. To supply the oxygen and glucose needed for the heart muscle cells to respire and to remove waste carbon dioxide.
2. Any four from: smoking, diet containing a lot of saturated fat, stress, genetic factors, lack of exercise.
3. Eat a diet that is relatively low in saturated fat, don't smoke, try to reduce stress and make sure they get enough exercise.

Supplement Worksheet B9.1a

1. Table similar to the following:

	Fish	Mammal
Circulation	closed	closed
Heart structure	one atrium, one ventricle	two atria, two ventricles
Heart function	to pump blood around body	to pump blood around body
Direction of blood flow	one way	one way
Order of organs	heart, gills, other body organs, heart	heart, lungs, heart, other body organs, heart

2. Blood flows twice through the heart for every complete circulation of the body in a double circulation, but only once in a single circulation.
3. a) gills
   b) The blood pressure through the mammalian lungs is much lower than the pressure of blood passing to the rest of the body.
   c) A lower pressure of blood to the lungs reduces the risk of damaging the delicate tissues in those organs while allowing a higher blood pressure to push blood to other organs in the body.

Supplement Worksheet B9.1a Comparing circulatory systems

The diagram below shows part of the circulatory system of a bony fish. The diagram is simplified because the fish has more blood vessels and gills than this. However, the structure of the heart and plan of the circulation are correct.

Note that the gills in a fish are where oxygen is absorbed from the water into the blood, and carbon dioxide released from the blood into the surrounding water.

Use the diagram and your Student Book to help you answer the questions below.

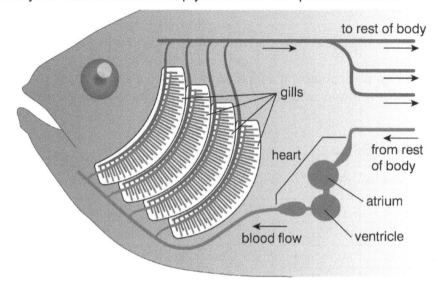

1. Draw up a table to show the similarities and differences between fish circulation and mammal circulation. Your table should include the following:

    o if it is an open or closed circulation (in an open circulation, blood can move anywhere in the body; in a closed system, blood is always contained within vessels or heart)
    o the structure of the heart (the chambers from which it is formed)
    o the function of the heart
    o the direction of blood flow through the system (one way or multiple directions)
    o the order of organs through which blood flows, starting with the heart.

2. Fish have a single blood circulation, whereas mammals have a double circulation. Use your table to help you to explain what this means.

3. When the heart contracts, the pressure of the blood passing out of it increases. As the blood flows through the rest of the circulatory system, its pressure decreases.
   a) Apart from the heart, in which organ is blood pressure highest in the fish circulation?
   b) Compare your answer to part a) with the blood pressure in the two parts of the human circulation described on pages 177–179 of your Student Book.
   c) Use your answer to part b) to explain the advantages of a double circulation.

Worksheet B9.1b Structure of a mammalian heart

The diagram shows a section through a mammalian heart.

Label the diagram to show the key features, and add notes to describe their structure and their function.

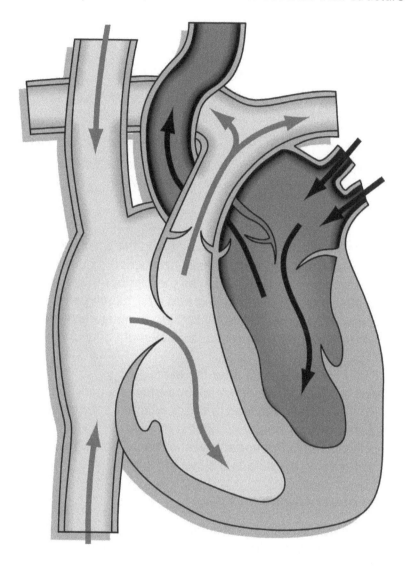

Worksheet B9.1c Exercise and pulse rate

This investigation looks at the effect of different levels of exercise on pulse rate. Work in pairs for this investigation, with one of you exercising and the other recording the results.

Apparatus

stopwatch

SAFETY INFORMATION

If you are exercising, wear appropriate clothing (such as gym kit, for example shorts or tracksuit bottoms, T-shirt and training shoes.)

Do not run around the laboratory.

If at any time you feel uncomfortable during exercise, stop immediately, inform the teacher and sit quietly until you have recovered.

Method

1. Check that you can measure pulse rate successfully using the wrist pulse point. It is often better to count pulses for 15 or 30 seconds and then multiply up to get the number of pulses per minute.
2. Decide what levels of exercise you will test, such as the number of steps made in a fixed time, or sitting or standing still, after walking, after jogging and after running as fast as possible.
3. Decide how long to carry out each level of exercise before measurements are taken. You could try a couple of test exercises to see whether one minute is enough, or two minutes produces higher results.
4. Decide when you will take measurements after the exercise.
5. Decide how long to wait between levels of exercise to allow the body to recover fully before the next test. You could try a couple of test exercises to check whether three minutes is sufficient for pulse rate to return to the resting rate, or whether longer is needed.
6. Decide whether or not you need to do repeat measurements at each level of exercise.
7. Carry out your tests and record your data in a suitable table.

Handling experimental observations and data

8. If you have taken repeat measurements, check that there are no anomalous values. If there are, try to explain them. Then, ignoring anomalous values, calculate average values for each level of exercise.
9. Use your results to draw a suitable graph or chart.
10. Describe the shape of your graph, and try to explain any pattern in your results.
11. If other students have used the same method as you have, compare your graph with theirs. Describe any similarities and differences between the graphs and try to explain them.
12. Draw a conclusion from your results about the effect of exercise on pulse rate.

Planning and evaluating investigations

13. Explain fully why the student who was exercising needed to rest between tests.
14. Describe any problems that you had with your investigation and suggest how the method could be adjusted to reduce the effect of these problems.
15. This investigation produced semi-quantitative data (the exercise was described in levels, and the pulse rates were measured on continuous scales). Suggest how you could adapt the method to make the exercise levels quantitative, so that you could produce a more reliable conclusion.

Learning episode B9.2 Blood vessels and blood

Learning objectives

- Describe the structure of arteries, veins and capillaries, limited to: relative thickness of wall, diameter of the lumen and the presence of valves in veins
- State the functions of capillaries
- Identify in diagrams and images the main blood vessels to and from the:
    - (a) heart, limited to: vena cava, aorta, pulmonary artery and pulmonary vein
    - (b) lungs, limited to: pulmonary artery and pulmonary vein
    - (c) kidney, limited to: renal artery and renal vein
- Supplement Explain how the structure of arteries and veins is related to the pressure of the blood that they transport
- Supplement Explain how the structure of capillaries is related to their functions
- Supplement Identify, in diagrams and images, the main blood vessels to and from the liver as: hepatic artery, hepatic veins and hepatic portal vein
- List the components of blood as: red blood cells, white blood cells, platelets and plasma
- Identify red and white blood cells in photomicrographs and diagrams
- State the functions of the following components of blood:
    - (a) red blood cells in transporting oxygen, including the role of haemoglobin
    - (b) white blood cells in phagocytosis and antibody production
    - (c) platelets in clotting (details are not required)
    - (d) plasma in the transport of blood cells, ions, nutrients, urea, hormones and carbon dioxide
- State the roles of blood clotting as preventing blood loss and the entry of pathogens
- Supplement Identify lymphocytes and phagocytes in photomicrographs and diagrams
- Supplement State the functions of:
    - (a) lymphocytes – antibody production
    - (b) phagocytes – engulfing pathogens by phagocytosis
- Supplement Describe the process of clotting as the conversion of fibrinogen to fibrin to form a mesh

Common misconceptions

Some students think that blood is a red liquid. Blood is more complex than this: it is composed of a straw-coloured liquid, called plasma, and many cells, including the red blood cells that give it its overall colour.

Resources

Student Book pages 184–191

B9.2 Technician notes

Resources for class practicals (see Technician's notes)

Approach

1. Introduction

Give students one minute to write down three things they remember about blood and blood vessels. Take examples from around the class, and compile a list of remembered facts. Encourage discussion of points that have been misremembered, so as not to reinforce these, but do not worry if there is no resolution now. Instead, make sure these points are resolved correctly by the end of the lesson.

2. Blood vessels

Using the diagram of the human circulatory system on page 184 of the Student Book, point out the distinction between arteries, which carry blood away from the heart, and veins, which carry blood back towards the heart (except the hepatic portal vein). (Note that, for Core, the syllabus only requires naming of blood vessels related to the heart, lungs and kidneys. For Supplement Students also need to name those related to the liver.)

Discuss the names of different pairs of arteries and veins and how they are named according to the organ they supply: *pulmonary* for lung and *renal* for kidneys. (Supplement Also, *hepatic* for liver.) Give students five minutes to learn as many names as they can, then ask them to close the book and answer question requiring them to name some blood vessels. For example:

- Which vessel carries blood away from the heart to the rest of the body (apart from the lungs)? [aorta]
- Which blood vessel carries blood away from the kidneys back towards the heart? [renal vein].

If appropriate, include some more difficult questions, such as:

- Which artery carries deoxygenated blood? [pulmonary artery].

3. Structure of blood vessels

If available, students could look at, and draw, prepared slides of a vein, an artery and a capillary. Alternatively, display slides of these using a digital microscope, or use images from the internet. Students should annotate their drawing to show the key features of each type of blood vessel.

Remind students that all these vessels are linked by asking them to describe the route blood might take from the heart through the kidneys and back to the heart. They should mention the capillaries within the kidneys between the renal artery and renal veins.

If possible, show students a video clip of blood flowing through capillaries.

Supplement Students should explain how each of the key features of each type of blood vessel relates to its function.

4. Looking at blood

Use pages 187–189 of the Student Book to introduce the components of blood. If possible, give students the opportunity to study prepared slides of human blood. Alternatively, display such a slide using a digital microscope, or use a suitable image from the internet.

Ask students to draw examples of each of the different kinds of cell that they can see. This should include red blood cells, and white blood cells if visible. It is unlikely that they will identify platelets, but there may be more than one kind of white blood cell. Ask them to compare the size, structure and number of the different kinds of cell.

Students should also make notes to describe the functions of all the components of blood: red blood cells, white blood cells, platelets and plasma.

Supplement When looking at white blood cells, introduce the distinction between lymphocytes and phagocytes. Students should include notes on their drawings, where appropriate, to describe the differences between the cells and their functions.

Supplement When discussing platelets, introduce the process by which platelets cause fibrin to form from fibrinogen. Explain that this process is highly complex, and that the description given on page 190 of the Student Book is highly simplified. Ask students to suggest why the process is so highly controlled. They should be able to suggest that it makes sure blood clots can be formed anywhere they are needed in the blood system, but not where they are not needed (this is dangerous and can cause the death of tissues).

5. Red blood cells

Using the photograph of an electron micrograph of red blood cells on page 189 of the Student Book, ask students to sketch one red blood cell and label it to explain how its structure is adapted to its function.

Students should use their knowledge of factors that affect the rate of diffusion from Section 3: *Movement into and out of cells*, to help explain the adaptations of red blood cells.

6. Consolidation: crossword clues

Ask students to write clues for a crossword on the structure and function of all of the components in blood, and the different kinds of blood vessel. They should write at least ten clues. Then ask them to test their clues on another student, and to amend the clues if they prove difficult to answer.

Technician's notes

Be sure to check the latest safety notes on these resources before proceeding.

The following resources are needed for the demonstration on blood vessels:

prepared slides of a vein, an artery and a capillary *or* slides of these using a digital microscope, *or* use images from the internet
video clip of blood flowing through capillaries

The following resources are needed for a demonstration on blood:

prepared slides of human blood; display such a slide using a digital microscope *or* a suitable image from the internet

Answers

Page 187

1. a) renal arteries   b) aorta   c) pulmonary veins
2. Arteries are large vessels with thick, elastic muscular walls; capillaries are tiny blood vessels with very thin walls that are often only one cell thick; veins are large vessels with a large lumen and valves to prevent backflow of blood.
3. Supplement The walls stretch as blood enters them, and slowly recoil as the blood flows through, evening out the pressure so that the change in pressure is reduced.

Page 191 Science in context: Blood tests

White blood cells defend the body against pathogens so the patient may be more vulnerable to infection.

Page 191

1.

Blood component	Function
plasma	carries dissolved substances, such as carbon dioxide, glucose, urea and hormones; also transfers heat energy from warmer to cooler parts of the body
red blood cells	carry oxygen
white blood cells	protect against infection
platelets	cause blood clots to form when a blood vessel is damaged

2. Supplement Phagocytes engulf pathogens inside the body and destroy them. Lymphocytes produce antibodies that attack pathogens.
3. Supplement Any cut that damages a blood vessel can create an easy route of infection into the body. So forming a blood clot where there is damage, as quickly as possible, helps to reduce the risk of infection. The platelets do this by converting soluble fibrinogen to insoluble fibrin which forms a mesh to trap blood cells to produce the clot.

Learning episode B9.3 Consolidation and summary

Learning aims

- Review the learning points of the topic summarised in the end of topic checklist
- Test understanding of the topic content by answering the end of topic questions

Resources

Student Book pages 192–195

Approach

Introduce the learning episode

Ask students to work with a partner to make a list of key words from this topic. They could then work together to produce a spider diagram showing how the different concepts are linked. They could compare their list with the list of key terms given on page 192 of the Student Book. Discuss the checklist on pages 192–193 and use questioning to see how much of the content they are comfortable with.

Students could make flashcards of the key content and then use the flashcards to quiz each other on the information.

Develop the learning episode

Ask students to work individually through the end of topic questions on pages 194–195 of the Student Book without looking at the text. As they work, walk around the classroom observing their answers and asking questions as necessary to find out which questions are causing difficulties.

Finish the learning episode

After a set period, ask the students to stop working and discuss any areas of difficulty you observed as you walked round the class. Students should complete any unanswered questions for homework, but you should stress that they should try to answer the questions without looking at the text, so that they can see how much they have remembered.

Answers

End of topic questions mark scheme

The marks available for a question can indicate the level of detail you need to provide in your answer.

Question	Correct answer	Marks
1	B	1 mark
2 a)	Electrical activity of the heart as it contracts.	1 mark
2 b)	The heart of patient B has been damaged in the heart attack and is not working properly.	1 mark
2 c)	The pulses would be closer together,	1 mark
	because pulse rate increases with activity.	1 mark
2 d)	Narrowing of a coronary artery,	1 mark
	would reduce blood flow to some of the heart muscle,	1 mark
	and those cells would not work properly.	1 mark
3	Red blood cells contain haemoglobin, which carries oxygen around the body.	1 mark
	If the red blood cell count is lower than usual, the amount of oxygen that can be carried by the blood will be reduced.	1 mark
	So the rate at which oxygen can be supplied to cells is reduced, so respiration in the cells will be slower.	1 mark
	Respiration releases energy for all cell processes, including activity, so the cells of an anaemic person will not be able to supply energy as quickly as normal and they will feel tired more rapidly.	1 mark
4 a)	They help in the formation of blood clots,	1 mark
	which block damaged blood vessels	1 mark
4 b)	It stops large amounts of blood escaping from the blood vessel,	1 mark
	and prevents pathogens getting into the body, where they could cause illness.	1 mark
4 c)	It will stop blood reaching cells beyond the clot,	1 mark
	which will prevent the cells getting the oxygen and sugars needed for respiration, so the cells may die.	1 mark
		1 mark
4 d)	Aspirin prevents platelets from functioning and producing clots that could cause thrombosis.	1 mark
5 a)	Smoking, diet, stress, genetic tendency (others are possible)	1 mark per factor to a max of 4 marks
5 b)	A person can give up smoking,	1 mark
	change their diet to reduce the amount of saturated fat,	1 mark
	and try to reduce stress,	1 mark
	but they cannot change their genes.	1 mark

6 a)	Factor that correlates with the occurrence of the disease/as the factor increases so the risk of having the disease increases	1 mark
6 b)	Many studies suggest that the greater the proportion of saturated fat in the diet, the greater the risk of the coronary arteries becoming narrowed or blocked.	1 mark
6 c)	The countries are arranged according to saturated fat consumption, from Germany with about 13.5% on the left to France with about 15.5% on the right. The death rate curve shows no obvious pattern.	1 mark 1 mark 1 mark
6 d)	It is possible that the graph indicates no correlation because the two curves do not have a similar shape. However, there may be other factors involved that have not been measured that are also affecting death rate from CHD, because the people in one country may have different diets and/or lifestyles to the people in another country. So this graph is not proof whether saturated fat consumption and death from CHD are correlated or not.	1 mark 1 mark 1 mark 1 mark
Supplement 7	The muscle in the left ventricle wall is much thicker than in the right ventricle wall, so it can contract more strongly and create a greater pressure. Blood from the right ventricle goes only to the lungs, where a high pressure would damage the capillaries that flow past the alveoli. Blood from the left ventricle will eventually travel around the body, through capillaries in organs, before returning to the heart, so a much larger initial pressure is needed to help it move this distance.	1 mark 1 mark 1 mark 1 mark
8 a)	Fish have a single circulation,	1 mark
	blood goes through the heart once on each complete circuit.	1 mark
	Mammals have a double circulation,	1 mark
	blood goes through the heart twice on each complete circuit.	1 mark
8 b)	A double circulation maintains a higher pressure around the body,	1 mark
	so blood can be transported more efficiently/quickly.	1 mark
	Total:	47 marks

Introduction

In this section, students will learn about transmissible diseases, hygiene and immunity.

A visit from a health professional could be a useful source of information for some of the tasks in Learning episode 10.1 on transmissible diseases and hygiene, and in Learning episode 10.2 on immunity and vaccination. Before the visit, work with students to produce suitable questions to give greatest benefit from this opportunity.

Links to other topics

Topics	Essential background knowledge	Useful links
1 Characteristics and classification of living organisms	1.1 Characteristics of living organisms 1.3 Features of organisms	
2 Organisation of the organism	2.1 Cell structure and size of specimens	
3 Movement into and out of cells	3.2 Osmosis	
4 Biological molecules	4.1 Biological molecules	
5 Enzymes	5.1 Enzymes	
7 Human nutrition	7.2 Digestive system	
9 Transport in animals	9.4 Blood	
11 Gas exchange in humans		11.1 Gas exchange in humans
15 Drugs		15.1 Drugs
16 Reproduction		16.1 Asexual reproduction 16.6 Sexually transmitted infections
17 Inheritance		17.1 Chromosomes, genes and proteins
19 Organisms and their environment		19.2 Food chains and food webs 19.4 Populations
20 Human influences on ecosystems		20.3 Pollution

Topic overview

B10.1	Disease, defence and hygiene
	This learning episode covers transmissible diseases, their causes and methods of transmission. It also looks at the natural defences of the body and ways in which the risk of transmission can be reduced. Note that vaccination is only briefly mentioned, but it is covered in more detail in the next learning episode.
Supplement B10.2	The immune system
	This learning episode looks in detail at how the immune system provides immunity to transmissible disease, through active and passive immunity, including the way in which vaccination can enhance immunity. The cause and effects of cholera are also covered.
B10.3	Consolidation and summary
	This learning episode provides an opportunity for a quick recap on the ideas encountered in the section. Students can answer the end of questions in the Student Book.

Careers links

These are some scientific careers that focus on this area of biology but careers in many other fields use the knowledge and skills gained studying science. **Epidemiologists** study patterns of health and disease in large populations to identify causes and suggest treatment methods. **Biotechnologists** and **pharmaceutical technicians** are involved in developing new vaccines.

Starting points

The Student Book section opener puts the ideas in the section into context and sets the scene. It also allows students to acknowledge and value their prior learning, and provides a benchmark against which future learning can be compared.

- The questions provide a structure for introducing the section and can be used in a number of different ways:
- You could ask students to consider the questions as an introductory homework task.
- You could put students into groups to share their own ideas and understanding and then to report back to the whole class.
- Students could be given access to the Internet, preferably with a tight timescale, to find out the information required.

You could then use a spider chart or other form of wall chart to summarise everybody's ideas.

Recording these initial ideas allows you to retain them for reference as the individual topics are developed. In this way, your students' progress in learning can be readily acknowledged.

Learning episode B10.1 Disease, defence and hygiene

Learning objectives

- Describe a pathogen as a disease-causing organism
- Describe a transmissible disease as a disease in which the pathogen can be passed from one host to another
- State that a pathogen is transmitted:
  (a) by direct contact, including through blood and other body fluids
  (b) indirectly, including from contaminated surfaces, food, animals and air
- Describe the body defences, limited to: skin, hairs in the nose, mucus, stomach acid and white blood cells
- Explain the importance of the following in controlling the spread of disease:
  (a) a clean water supply
  (b) hygienic food preparation
  (c) good personal hygiene
  (d) waste disposal
  (e) sewage treatment (details of the stages of sewage treatment are **not** required)

Common misconceptions

The term *germ* is commonly used to mean something microscopic that can cause disease. In a scientific sense, *germ* means something that gives rise to life and growth, such as the germ cells within a seed. So it is important that students learn to use the correct term when talking about disease-causing microorganisms: *pathogen* (literally translated as 'something that gives rise to illness').

Many people confuse the relationship between microorganisms, bacteria and disease. Students should become aware that:

- microorganisms include many groups other than bacteria, such as viruses, protoctists and some fungi (e.g. yeasts) (see Section 1: *Characteristics and classification of living organisms.*)
- not all bacteria inside the body are pathogens – the majority have no effect on health and some are essential for health (particularly those in the digestive system, see Section 7: *Human nutrition.*)
- many kinds of bacteria and other microorganisms live freely in the environment and play an important role in the recycling of nutrients (see Section 19: *Organisms and their environment*).

Resources

Student Book pages 199–203

Worksheet B10.1 Methods of transmission

Approach

1. Introduction

Ask students to define the term *disease*. Give them a minute or two to discuss their definitions in pairs or small groups and then take examples from around the class. Try to elicit a class definition of the word. Then write the word hyphenated like this: *dis-ease*. Explain that the word, quite literally, means the lack of ease or comfort. This meaning can be developed into 'something that causes the body to function abnormally, often causing symptoms (signs) of illness'.

Take examples of diseases from around the class and, without explaining why, write the names on the board in groups according to their types, such as transmissible or infectious diseases, disorders caused by malfunction of the body (e.g. cancer, diabetes, coronary heart disease), deficiency diseases (e.g. scurvy, rickets – Section 7: *Human nutrition*), genetic or inheritable diseases (such as sickle cell anaemia – Section 17: *Inheritance*). Challenge students to identify why the names are grouped as they are. Explain that this section will focus on diseases that can be passed from person to person, and introduce the term *transmissible* (or infectious) *disease* as the name for this group of diseases. (If students seem unsure about the term, make it clear that the other kinds of disease are not transmitted from person to person. Although genetic diseases are inherited from a parent, they are not considered transmissible diseases where infection can pass from an infected person to any uninfected person.) Also introduce the term *pathogen* (see common misconceptions above).

2. Transmissible diseases – methods of transmission

Worksheet B10.1 provides an outline diagram of Fig. 10.2 on page 200 of the Student Book. Students could annotate the diagram to give examples of diseases that are transmitted by each of the methods shown. Pages 200–201 of the Student Book could help with this. Alternatively students could use other sources, such as books or the internet, to provide examples. (Suggest that students leave room for additional notes, which they will add in Task 4.)

Challenge students to explain the difference between direct and indirect contact methods of transmission.

3. Natural barriers to infection

Introduce this task by stating that we are surrounded by microorganisms, in the air and on every surface. (Take this opportunity, if needed, to reinforce the fact that only some of these are pathogens.) Ask why we are not ill all the time. Students should identify that the body is adapted to protect against infection much of the time.

Give students the headings: *Mechanical barriers*, *Chemical barriers*, *Immune system*. Then ask them to make notes of different examples under each of the headings. Pages 203–205 of the Student Book can help with this, and supplemented by research in other books or the internet if there is time. Note that detail of the immune system and vaccination is beyond the core syllabus. This detail will be covered in learning episode 10.2 for those students studying supplement content.

4. Controlling the spread of disease

Ask students to return to their annotated diagrams on transmissible diseases and to add notes that show how the risk of transmission by each method could be reduced. Page 202–205 of the Student Book can help with this.

Explain that methods that protect against the transmission of disease are known as good *hygiene*. Ask students to write a list of hygiene rules for students at the school, in order to reduce the risk of spreading disease.

5. Health poster

Give students the opportunity to bring together all that they have learned in this learning episode by focusing on one particular transmissible disease. Ask them to produce a health poster about the disease, which includes the following:

- identifies what causes the disease
- describes how the disease is transmitted
- describes how the body normally defends against the disease
- explains how the risks of transmission can be reduced.

Allow time for students to swap posters and mark them. They should identify two points that are well covered, and two points that could be improved. Posters should be returned so that students can identify how they could improve the weaker points.

6. Consolidation

Ask students to compile a list of ten questions that could be used as crossword clues on what they have learned from this learning episode. They could then test out their clues on another student and use the responses to identify any weaknesses in their learning.

Answers

Page 200

1. A disease caused by a pathogen that can be passed from one host to another.
2. The host is a human, the pathogen is a virus.
3. Any three from: bacteria, viruses, fungi, protoctists.

Page 202

1. Direct: any one from blood (HIV, hepatitis B, hepatitis C) or semen (HIV, syphilis, gonorrhoea, other STI). Other examples may be correct.
   Indirect: water droplets in air (colds, flu); drinking water (cholera, typhoid, dysentery); contaminated surfaces (athlete's foot, food poisoning pathogens); insect bite (malaria, dengue fever).
2. Any transmissible disease with a suitable description and explanation of how to control its spread.

Page 203

1. A barrier that physically prevents entry into the body, e.g. nose hairs that filter air, or skin.
2. A chemical that destroys the pathogen, e.g. lysozymes in mucus or acid in stomach.
3. White blood cells attack pathogens, either by engulfing and destroying them, or by producing antibodies that pathogens.

Page 205 Science in context: Mutating pathogens

1. The antigens on the viruses are all (slightly) different.
2. EITHER :
   No, because you already have antibodies against the particular antigens on that virus

   OR

   Possibly, if the virus has changed/the genes for the antigen proteins have mutated.

Worksheet B10.1 Methods of transmission

The diagram below shows the ways in which pathogens that cause transmissible (infectious) diseases may enter a person's body.

- Add notes to the diagram to explain clearly how pathogens enter the body in each of these ways.
- Then add examples of pathogens (or the diseases they cause) that enter in each way. Pages 202–203 of the Student Book may help you.

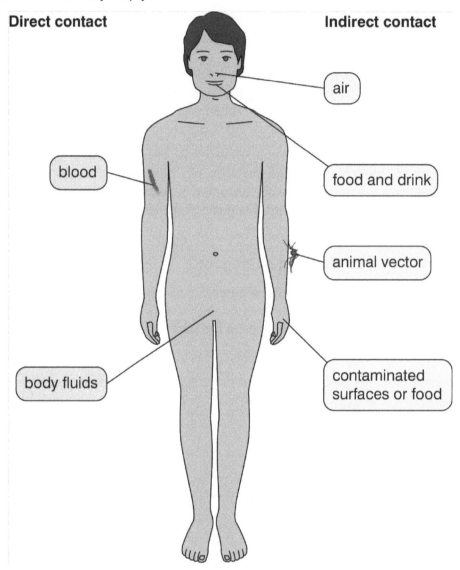

Supplement Learning episode B10.2 The immune system

Learning objectives

- Supplement Describe active immunity as defence against a pathogen by antibody production in the body
- Supplement State that each pathogen has its own antigens, which have specific shapes
- Supplement Describe antibodies as proteins that bind to antigens leading to direct destruction of pathogens or marking of pathogens for destruction by phagocytes
- Supplement State that specific antibodies have complementary shapes which fit specific antigens
- Supplement Explain that active immunity is gained after an infection by a pathogen or by vaccination
- Supplement Outline the process of vaccination:
  (a) weakened pathogens or their antigens are put into the body
  (b) the antigens stimulate an immune response by lymphocytes which produce antibodies
  (c) memory cells are produced that give long-term immunity
- Supplement Explain the role of vaccination in controlling the spread of diseases
- Supplement Explain that passive immunity is a short-term defence against a pathogen by antibodies acquired from another individual, including across the placenta and in breast milk
- Supplement Explain the importance of breast-feeding for the development of passive immunity in infants
- Supplement State that memory cells are not produced in passive immunity
- Supplement Describe cholera as a disease caused by a bacterium which is transmitted in contaminated water
- Supplement Explain that the cholera bacterium produces a toxin that causes secretion of chloride ions into the small intestine, causing osmotic movement of water into the gut, causing diarrhoea, dehydration and loss of ions from the blood

Common misconceptions

It is important that students realise from this learning episode that immunity to one pathogen does not usually confer immunity to a different pathogen. For example, immunity to polio does not confer immunity to measles, or vice versa. The development of immunisation by Edward Jenner against smallpox, using a vaccine made from cowpox virus, was unusual and possible because the two viruses are very similar in structure. So the immune response to one virus also protected against infection by the other.

There are many misconceptions about vaccination, many of which are beyond the requirements of the syllabus but may come up in discussion. One is that because a vaccine often contains the pathogen, then it must be dangerous. Students must realise that the form of the pathogen in a vaccine is harmless, or produces such a weak form of the illness that the person vaccinated may show few or no symptoms. In very rare circumstances, a person may react badly to vaccination, either to the weakened pathogen or to something else in the vaccine. The risk of a bad reaction to vaccination is far lower than the risk of harm from infection by the pathogen, which is why health professionals recommend vaccination.

Resources

Student Book pages 204–209

Supplement Worksheet B10.2 How antibodies defend the body

Approach

1. Introduction

Give students the saying, 'prevention is better than cure' and ask them to suggest as many reasons as they can think of how this could apply to transmissible diseases. They could do this individually, or in small groups to encourage discussion.

Take examples from around the class and try to encourage examples that relate not just to the individual, but also to their families, the health system and the country (which usually relates to income at a personal and national level, and the cost to healthcare).

2. The immune response – how antibodies defend the body

Worksheet 10.2 provides a task and some images that will help students clarify their knowledge and understanding of how the immune system responds to infection by pathogens. Make sure the students identify the role of antigens and antibodies, and the two possible outcomes for a pathogen tagged by antibodies.

3. Active and passive immunity

Ask students to draw up a table that compares active and passive immunity. They could use information from pages 204–209 of the Student Book to help them complete their table. They should identify not only how the two types of immunity arise, but also the differences in prolonged immunity.

Although this is beyond the requirements of the syllabus, it is important that students understand the role of lymphocytes and memory cells in the development of active immunity, as preparation for their learning about vaccination.

4. Vaccination

Ask students to find out, either by questioning a health professional, looking at a health leaflet on vaccination, or by researching on the internet, what the current recommendations for vaccination are where they live. They could extend their research to look at how recommendations vary because of the presence of different pathogens in different places, and how travellers abroad may need to consider vaccinations against additional diseases. They should realise that recommendations also vary at different times, as new vaccines become available or pathogens change.

Ask students to compare the effect of vaccination with the development of active immunity after infection, and to state the advantages of vaccination.

5. Cholera

Students could use the information on cholera on pages 208–209 in the Student Book, or from other research, to prepare an information poster about this disease, its cause, its transmission and its effects. Remind students that the most effective health information posters present their information in a clear and attractive way. At the end, students could vote on which poster they think is the most effective.

Students should include a simple and clear explanation of how the cholera bacterium causes diarrhoea, leading to loss of ions and water from the body, and include references to osmosis.

6. Consolidation

Ask students to write bullet notes as planning for a health leaflet for expectant mothers on protecting their baby against infection. The notes should cover the importance of breast-feeding for conferring passive immunity, the need for vaccination for conferring active immunity to common diseases, and the need for good hygiene to help prevent transmission of disease.

Answers

Page 207 Science in context: Smallpox eradication

With herd immunity, if most people had been vaccinated, and few people had smallpox, the chances of an unvaccinated person catching the disease might be low. However, unless the disease had been completely eradicated, there was still a chance that an unvaccinated person might catch smallpox. As smallpox was such a serious disease it was worthwhile vaccinating everyone.

Page 208

1. a) Chemical/protein on the surface of a pathogen cell to which antibodies attach.
   b) Protection from infection as a result of antibodies in the body.
2. Part of the response of lymphocytes to an infection is to release memory cells into the blood. On a second infection, the memory cells respond by causing large quantities of antibodies to be released quickly. This destroys the pathogens before they cause disease.
3. Active immunity is caused by making the immune response produce antibodies and memory cells – it lasts a long time. Passive immunity is caused by giving a person antibodies – no memory cells are produced, so protection is only short-term.

Page 209

1. The production of large quantities of watery faeces.
2. In drinking water contaminated with untreated sewage.
3. The toxin produced by the cholera bacterium causes chloride ions to be secreted into the small intestine, which causes water to move into the gut by osmosis. Water and ions are lost from the body through the diarrhoea.

Supplement Worksheet B10.2 How antibodies defend the body

Your task is to draw a series of images that could be used as a storyboard to brief an animation showing how the immune system attacks pathogens. The diagrams below may help you.

Cut out the diagrams and arrange them in a way that shows:

- where antibodies are found in the body
- how only antibodies of the right shape can attach to the antigens on a particular pathogen
- how antibodies may cause the pathogen to break open or die
- how antibodies attract phagocytes that engulf the labelled pathogens and destroy them.

You may wish to add other images to your storyboard, to make the animation clearer.

Remember to add a script to the storyboard, and any labelling, to make sure that the viewer can understand what is happening as clearly as possible.

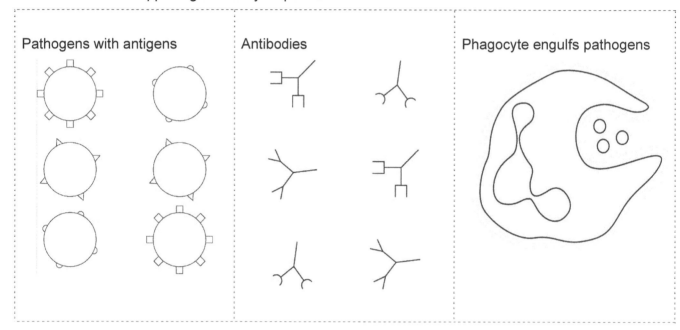

Pathogens with antigens   Antibodies   Phagocyte engulfs pathogens

Learning episode B10.3 Consolidation and summary

Learning aims

- Review the learning points of the topic summarised in the end of topic checklist
- Test understanding of the topic content by answering the end of topic questions

Resources

Student Book pages 210–213

Approach

Introduce the learning episode

Ask students to work with a partner to make a list of key words from this topic. They could then work together to produce a spider diagram showing how the different concepts are linked. They could compare their list with the list of key terms given on page 210 of the Student Book. Discuss the checklist on pages 210–211 and use questioning to see how much of the content they are comfortable with.

Students could make flashcards of the key content and then use the flashcards to quiz each other on the information.

Develop the learning episode

Ask students to work individually through the end of topic questions on pages 212–213 of the Student Book without looking at the text. As they work, walk around the classroom observing their answers and asking questions as necessary to find out which questions are causing difficulties.

Finish the learning episode

After a set period, ask the students to stop working and discuss any areas of difficulty you observed as you walked round the class. Students should complete any unanswered questions for homework, but you should stress that they should try to answer the questions without looking at the text, so that they can see how much they have remembered.

Answers

End of topic questions mark scheme

The marks available for a question can indicate the level of detail you need to provide in your answer.

Question	Correct answer	Marks
1	D	1 mark
2 a)	Keeping things clean of pathogens.	1 mark
2 b)	To reduce the risk of transmission of pathogens,	1 mark
	to people who eat the food they prepare.	1 mark
2 c)	Cleans most of the pathogens off the hands before touching food.	1 mark
	Makes sure pathogens are not transmitted from dirty equipment to food during preparation.	1 mark
	Prevents transmission of pathogens from uncooked food to cooked food,	1 mark
	which will not be heated enough to kill pathogens before people eat it.	1 mark
	Prevents transmission of pathogens in blood (if there are any) to the food.	1 mark
3 a)	Pathogen – disease-causing organism.	1 mark
	Microscopic – too small to be seen with the naked eye/needs a microscope to see it.	1 mark
3 b)	Before we had microscopes it was difficult to know how some diseases were transmitted because we could not see the pathogens,	1 mark

	so sometimes the methods used to try to cure them did not work.	1 mark
Supplement 4 a)	The white blood cells that produce antibodies to attack a pathogen also produce memory cells.	1 mark
	These rapidly produce more antibodies on a subsequent infection of the same pathogen.	1 mark
Supplement 4 b)	Antibodies respond only to antigens on the specific pathogen that they were produced to deal with,	1 mark
	so the measles antibodies will not attack the chickenpox pathogen.	1 mark
Supplement 5 a)	A harmless version of the polio virus is given as the vaccine.	1 mark
	This causes the body to respond as if infected with the harmful virus by producing lymphocytes that attack the virus by producing antibodies against it.	1 mark
	The lymphocytes also produce memory cells that remain in the body after infection.	1 mark
	If the body is infected with polio virus again, the memory cells cause a rapid release of antibodies against the virus, which kill the pathogens before they can cause disease.	1 mark
Supplement 5 b)	Active immunity,	1 mark
	because the immune system actively responds by producing antibodies.	1 mark
Supplement 5 c)	The baby receives antibodies from the mother through the placenta before birth and through breast milk after birth.	1 mark
	These give only temporary protection / no memory cells produced,	1 mark
	so the baby will need vaccination when old enough so that it can develop life-long immunity.	1 mark
Supplement 5 d)	Any suitable suggestion that indicates isolation of cases to prevent further transmission,	1 mark
	and vaccination of people in the surrounding area to prevent them catching and transmitting it.	1 mark
Supplement 6 a)	There is a tiny risk that the child might react to the vaccine and become ill	1 mark
Supplement 6 b)	The risk of getting the disease is much greater than the risk of a reaction to a vaccination,	1 mark
	and the child may be more badly affected by the disease than the vaccination.	1 mark
Supplement 6 c)	Compulsory vaccination increases the possibility of eradicating a disease in an area,	1 mark
	because there will be no individuals that can act as host.	1 mark
	However, some people may not be able to have the vaccination because of other health reasons,	1 mark
	so there should be some discretion over who does or does not get immunised. (Other arguments may also be suitable if supported by an appropriate justification.)	1 mark
	Total:	35 marks

Introduction

This section covers gas exchange surfaces in relation to the lungs in humans, and looks at the effect of exercise on breathing.

Links to other topics

Topics	Essential background knowledge	Useful links
1 Characteristics and classification of living organisms	1.1 Characteristics of living organisms 1.3 Features of organisms	
2 Organisation of the organism	2.1 Cell structure and size of specimens	
3 Movement into and out of cells	3.1 Diffusion	
5 Enzymes	5.1 Enzymes	
9 Transport in animals	9.1 Circulatory systems 9.2 Heart 9.3 Blood vessels 9.4 Blood	
10 Diseases and immunity	10.1 Diseases and immunity	
12 Respiration		12.1 Respiration 12.2 Aerobic respiration 12.3 Anaerobic respiration
13 Excretion in humans		13.1 Excretion in humans
14 Coordination and response		14.4 Homeostasis

Topic overview

B11.1	**Gas exchange system** This learning episode gives students an understanding of the structure of the human gas exchange system. This learning episode provides good opportunities to revise and reinforce earlier learning on adaptations for effective diffusion (see Section 3: *Movement into and out of cells*). Supplement Students will also learn how ventilation of the lungs occurs.
B11.2	**The effects of gas exchange** This learning episode gives students an understanding of how gas exchange changes the composition of air. Students will also investigate the effect of exercise on breathing. Supplement Students will learn how to explain the differences between inspired and expired air, and how to explain the changes in breathing associated with physical activity.
B11.3	**Consolidation and summary** This learning episode provides an opportunity for a quick recap on the ideas encountered in the section. Students can answer the end of topic questions in the Student Book.

Careers links

These are some scientific careers that focus on this area of biology but careers in many other fields use the knowledge and skills gained studying science. **Sports physiologists** work with athletes to improve their performance and need to understand how exercise affects gas exchange. **Anaesthetists** have to closely monitor patients' breathing, oxygen and carbon dioxide levels during operations.

Starting points

The Student Book section opener puts the ideas in the section into context and sets the scene. It also allows students to acknowledge and value their prior learning, and provides a benchmark against which future learning can be compared.

- The questions provide a structure for introducing the section and can be used in a number of different ways:
- You could ask students to consider the questions as an introductory homework task.
- You could put students into groups to share their own ideas and understanding and then to report back to the whole class.
- Students could be given access to the Internet, preferably with a tight timescale, to find out the information required.

You could then use a spider chart or other form of wall chart to summarise everybody's ideas.

Recording these initial ideas allows you to retain them for reference as the individual topics are developed. In this way, your students' progress in learning can be readily acknowledged.

Learning episode B11.1 Gas exchange system

Learning objectives

- Describe the features of gas exchange surfaces in humans, limited to: large surface area, thin surface, good blood supply and good ventilation with air
- Identify in diagrams and images the following parts of the breathing system: lungs, diaphragm, ribs, intercostal muscles, larynx, trachea, bronchi, bronchioles, alveoli and associated capillaries
- Supplement Identify in diagrams and images the internal and external intercostal muscles
- Supplement State the function of cartilage in the trachea
- Supplement Explain the role of the ribs, the internal and external intercostal muscles and the diaphragm in producing volume and pressure changes in the thorax leading to the ventilation of the lungs
- Supplement Explain the role of goblet cells, mucus and ciliated cells in protecting the breathing system from pathogens and particles

Common misconceptions

It is important to reinforce the difference between the terms *breathing (ventilation)*, which is the movement of gases into and out of the gas exchange system, *gas exchange*, which is the exchange of gases across the gas exchange surface between the air and the body, and *respiration* (as in cellular respiration). Respiration will be covered in Section 12.

Resources

Student Book pages 217–222

Worksheet B11.1 Structures of the human gas exchange system

B11.1 Technician's notes

Resources for demonstrations (see Technician's notes)

Approach

1. Introduction

Remind students of their work on diffusion in Section 3: *Movement into and out of cells*, by asking them to remember what features an effective exchange surface has. If needed, use the example of the leaf that they learned about in Section 6: *Plant nutrition*. Give them a short while to discuss these features in pairs or small groups, and then take examples from around the class. They should be able to remember a large surface area, thin surface across which diffusion occurs, and a good supply of the gases that are being exchanged. (In the case of the leaf, that means carbon dioxide from the air or respiration, oxygen from the air or photosynthesis, and water vapour from water supplied to the leaf by the xylem.)

Then get students to apply this knowledge to the human gas exchange system. They should be aware from earlier learning that gas exchange takes place between the air and the blood inside the lungs. Ask them what features they would expect to see in the lungs in order that the exchange of gases between the blood and air in the lungs by diffusion is as effective as possible. They should realise that the internal surface of the lung should be thin and have as large a surface area as possible.

2. Structures of the human gas exchange system

Use pages 217–219 of the Student Book to introduce the structures of the gas exchange system in humans. This involves a lot of terms, some familiar, but also others that are new. An unlabelled version of Fig.11.3 is given on Worksheet 11.1. Students could use this to identify and label the key features of the human gas exchange system.

To help with learning, ask students to create their own glossary definitions for each of the bold words in the text. They could test out their definitions on other students to see whether they can guess the term correctly.

If available from a butcher, and if you are willing to do so, dissect lungs from a sheep to help students understand their three-dimensional structure.

SAFETY INFORMATION: You will need to wear disposable gloves while carrying out the demonstration, and wash down all surfaces, and soak all tools used in disinfectant (e.g. 1% Virkon) for at least 10 minutes after the demonstration. Wash your hands thoroughly after removing the gloves.

If present, students should also identify the larynx, trachea, diaphragm, ribs and intercostal muscles.

Supplement Students should identify the internal and external intercostal muscles and cartilage in the trachea, and be able to describe their functions, as well as those of the ribs and diaphragm.

Supplement 3. Modelling breathing demonstration

Ask students to place their hands on the lower part of their ribcage and to breathe in and out gently. They should note which parts of the lower thorax and abdomen change as they inhale, and again when they exhale. Then ask them to repeat this, taking deep breaths. They should compare how different parts of the lower thorax and abdomen move during gentle and deep breathing.

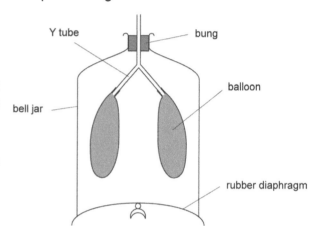

If you have it available, use the apparatus shown on the right to demonstrate the effect of the diaphragm on breathing, by pulling down on the rubber diaphragm (rubber sheeting) and then pushing up on it (see Technician's notes). Ask students first to describe what they see, then try to explain it. Talk first in simple terms of pulling and pushing.

This should be translated into consideration of changing volume and pressure within the jar and within the 'lungs'.

Ask students to compare the model with the real system, and to identify the strengths and weaknesses of the model in explaining how we breathe. They should explain the differences in terms of the roles of intercostal muscles and the ribcage.

4. Alveoli

If available, allow students to study prepared microscope slides of alveoli, and to draw and label what they see. Alternatively, display such a slide using a digital microscope, or a suitable image from the internet, and discuss what is visible with the class.

Focus on the key adaptations that make alveoli well adapted for their role in gas exchange: the thin layer of tissue between the air in the lungs and the capillaries; the large number of capillaries supplying lots of blood; the large number of alveoli providing a large surface area for exchange.

Supplement 5. Cilia and mucus

Ask students to look at prepared slides of lung tissue showing cilia and mucus in the bronchioles, or display such a slide using a digital microscope or from the internet.

Explain that the mucus is produced by goblet cells lining the tubes of the lungs and that the cilia are extensions of ciliated cells and continually move the mucus up out of the lungs to the back of the throat, where it can be swallowed and broken down in the digestive system.

Ask students to draw a plan diagram of the cells they see, and to label the diagram to show how these actions protect the lungs from pathogens and particles, and why this is important.

6. Consolidation

Ask students to prepare a quick quiz of ten simple questions and answers on what they have learned from this learning episode. They should test out their questions on other students and identify those that they do not fully understand or could be improved. Allow a few minutes for students to check anything they are unsure of in the Student Book and discuss any areas that are showing general weakness.

Technician's notes

Be sure to check the latest safety notes on these resources before proceeding.

The following resources are needed for the demonstration on the structures of the human gas exchange system:

sheep lungs
sharp knife or scalpel
dissecting board
disposable gloves
disinfectant such as 1% Virkon for disinfecting all tools and work surfaces

If available from a butcher, and if you are willing to do so, dissect lungs from a sheep to help students understand their three-dimensional structure. You should wear disposable gloves while carrying out the demonstration. You will need to wash down all surfaces, and soak all tools used in disinfectant (e.g. 1% Virkon) for at least ten minutes after the demonstration. Wash your hands thoroughly after removing the gloves.

The following resources are needed for the demonstration on breathing:

If you have the equipment, use this apparatus to demonstrate the effect of the diaphragm on breathing.

rubber sheet for diaphragm (e.g. from a large rubber balloon)
Y tube
bung
2 small balloons
bell jar

Answers

Page 220

1. The lungs are adapted by having: many alveoli to provide a large surface area; thin alveoli walls to reduce the length of the diffusion pathway; good ventilation with air and a good blood supply to maintain concentration gradients for oxygen and carbon dioxide.
2. The trachea carries air from the mouth down to the bronchi; the bronchi carry air into the lungs where they branch into many bronchioles; the bronchioles carry the air to the alveoli; the alveoli are where gas exchange with the blood takes place.
3. Sketch similar to Fig. 11.4, with annotations showing: thin lining of alveolar wall and wall of capillary allows rapid diffusion; high concentration gradients for gases between blood and air in alveolus due to continuous blood flow through capillary and ventilation of alveolus (lungs); large surface area of alveoli, maximising area of contact between capillary and alveoli, maximising area over which diffusion can occur.
4. The mucus traps particles and microorganisms that are in the air breathed in, and the cilia move the mucus and anything trapped in it up out of the lungs to the throat, where it can be swallowed. This protects the lungs from damage and infection.

Page 222 Science in context: Ventilation by machine

The external air pressure may be so low that it may be difficult by normal breathing to reduce the air pressure inside the thorax enough to cause enough air to enter from outside.

Supplement Page 222

1. a) The muscle surrounding the diaphragm contracts so that the diaphragm flattens, pulling downwards on the thorax; the intercostal muscles contract, lifting the ribs out and up.
   b) These movements of the diaphragm and intercostal muscles increase the volume of the thorax, causing the volume of the lungs to increase and so decreasing the air pressure inside the lungs causing air from outside to enter the lungs.
2. a) The diaphragm muscle relaxes, and is pushed upwards by the organs below it; the external intercostal muscles relax, so the ribs fall back and down. If you are breathing out hard, the internal intercostal muscles will also contract, helping the ribs move down quickly.
   b) These movements of the diaphragm and intercostal muscles reduce the volume of the thorax, so reducing the volume of the lungs, so increasing the air pressure inside the lungs which forces air out of the lungs.

Worksheet B11.1 Structures of the human gas exchange system

The diagram below shows the main structures of the human gas exchange system. Label the diagram to identify the key features.

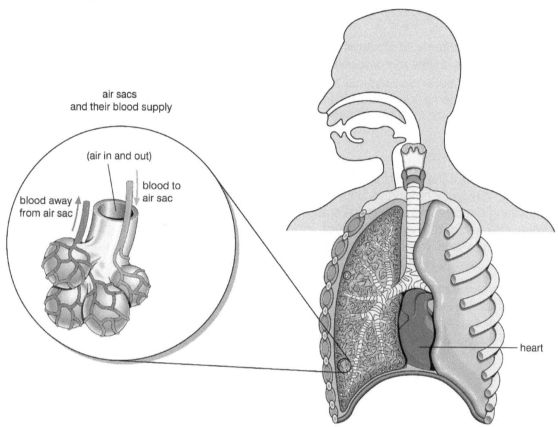

Learning episode B11.2 The effects of gas exchange

Learning objectives

- Investigate the differences in composition between inspired and expired air using limewater as a test for carbon dioxide
- Describe the differences in composition between inspired and expired air, limited to: oxygen, carbon dioxide and water vapour
- Investigate and describe the effects of physical activity on the rate and depth of breathing
- Supplement Explain the differences in composition between inspired and expired air
- Supplement Explain the link between physical activity and the rate and depth of breathing in terms of: an increased carbon dioxide concentration in the blood, which is detected by the brain, leading to an increased rate and greater depth of breathing

Common misconceptions

Students may get confused about the meaning of the terms *inspiration* and *expiration*, because they have different meanings in everyday language from those in scientific language. It is potentially even more problematic as linking them to *respiration* risks reinforcing the confusion between respiration and ventilation, which should be avoided. It may help if you include the definition of each word when you use it (e.g. 'inspiration, breathing in' and 'expiration, breathing out', or 'inspiration, inhaling' and 'expiration, exhaling') until students get used to making these associations themselves.

Students sometimes interpret diagrams of gas exchange to say that *all* the oxygen from air in the lungs moves into the blood, and *all* the carbon dioxide in the blood moves into the air in the lungs. If students are clear that the process of exchange is diffusion, then they should realise from their work in Section 3: *Movement into and out of cells*, that this is not possible. Work in this learning episode should help to clarify this.

Resources

Student Book pages 222–225

Worksheet B11.2 The effect of exercise on breathing; B11.2 Tech notes

Resources for demonstration and class practical (see Technician's notes, following)

Approach

1. Introduction

Display a copy of Fig. 11.3 from page 218 of the Student Book, or the diagram on Worksheet B11.1, or ask them to look at the diagram in their book. If using the diagram from the worksheet, briefly revise the main structures, including the alveoli, capillaries, bronchioles, bronchi and trachea. Ask students to identify the main role of each of these structures. In particular, make sure they identify the alveolar surface as the gas exchange surface.

2. Gas exchange in the alveoli

Ask students to sketch a simple outline diagram of the alveoli in an air sac and its related capillaries. They should then label the diagram to show what happens to the gases in air and blood at the alveolar surface.

Remind students of their work on diffusion in Section 3: *Movement into and out of cells*, and discuss why oxygen and carbon dioxide are exchanged between the blood and alveolar air in terms of diffusion. If students have difficulty remembering this, try reinforcing the idea with a simple model using beads or counters of two colours (one for oxygen and one for carbon dioxide) and a ruler or piece of string to model the alveolar surface. Arrange the beads so that there are some of each colour on each side, but that there are more of one colour on one side (i.e. oxygen on the alveolar side) and more of the other colour on the other side (i.e. carbon dioxide in the blood). Students should then move the beads to describe diffusion of the gases. They should remember that diffusion is the net movement of particles from a region of their higher concentration to their lower concentration, and if the system is static the two sides will end up with equal numbers of beads on each side. They could then consider the importance of ventilation and blood flow in maintaining the concentration differences that drive diffusion.

Ask students to write their own explanation on their diagram of how exchange of gases takes place in the lungs by diffusion.

3. Inspired and expired air

From the previous task, students should be able to suggest differences in the concentrations of oxygen and carbon dioxide in inspired air and expired air.

Students should be familiar with the limewater test for carbon dioxide, and so be able to suggest how the difference in carbon dioxide concentration could be tested. If the apparatus used to investigate the composition of inspired air and expired air, shown in Fig. 11.7 on page 223 of the Student Book, is available, either demonstrate the test or ask a student to do so (see Technician's notes). (Note that it is essential that the apparatus is set up correctly so that there is no risk of breathing in limewater.) Alternatively, show students the diagram instead.

Ask students to sketch the apparatus and annotate it to explain how the apparatus works and why there is a change in the limewater in one of the tubes but not the other.

Point out that there are other gases in inspired air and expired air, but that most do not change in concentration. Ask students to explain why (they are not used or produced in the body). The only one that does change is water vapour. Although students do not need to know a test for this, a demonstration of breathing on to a cold surface (such as a mirror or shiny metal) to produce a film of condensation will help to reinforce the idea.

Supplement Students should be expected to explain the differences in concentration of oxygen, carbon dioxide and water in inspired and expired air.

4. The effect of exercise on breathing

Worksheet B11.2 provides a method for investigating the effect of exercise on breathing. (This may be combined with gathering data for the investigation in Worksheet B9.1c on the effect of exercise on pulse rate.) Students may have produced their own methods when answering questions in the developing practical skills box on page 224 of the Student Book. If their plans are suitable, consider allowing them to carry those out instead of using the method on the worksheet. They should, however, complete the questions on the sheet using their own data.

Check that students have drawn up a suitable table for their results before they carry out the tests.

Students may need help with analysing their results, including identifying anomalous data. If students are unsure what is meant by 'quantitative', explain this before they answer the final question on the worksheet.

If a spirometer is not available, students can carry out the investigation measuring the rate of breathing only.

The method lets students decide how and when to take measurements, but guides them towards particular ways of carrying out the investigation. If class results are to be collated, you may wish to control the method so that all students work in the same way.

Students should find that the rate and depth of breathing increase as the level of exercise increases. However, there will probably be significant variation between students because of different levels of fitness and the way that students carry out their investigations. A quantitative method for measuring level of exercise would be using exercise equipment, such as a treadmill, which can be adjusted (for example, for speed).

SAFETY INFORMATION: Be considerate of students' concerns about exercise. If some students are unable or unwilling to carry out the exercise, pair them with other students so that one is the test subject and the other the recorder of results.

Over-exertion may be a hazard, especially for those with some medical conditions. Competitive situations can lead to careless behaviour and accidents.

Students who are exercising should be appropriately dressed, such as in gym kit, and the exercise should take place under supervision. If any student shows signs of difficulty, they should stop exercising immediately and sit quietly until they recover. For further advice on this, consult your PE department.

A suitable form of exercise is doing step-ups on stable equipment. This is much better than students running upstairs.

Mouthpieces should be sterilised before use and should not be shared.

Supplement Students should be able to explain the changes in breathing as a response to physical activity. They could use page 225 of the Student Book to help them describe the role of the brain in monitoring and responding to an increased carbon dioxide concentration in the blood. You could briefly introduce the notion of homeostasis but explain that this will be covered in more detail in Section 14: *Coordination and response.*

5. Consolidation: making notes

Ask students to write five bullet point notes for a friend who has missed this work on the effects of gas exchange. Give them two minutes to do this and then ask them to compare their notes with a neighbour and select the best five from the two lists. They could then work with another pair to produce the best five points from these. Then take examples from groups and ask students to select the best to create a class list of five key points.

Technician's notes

Be sure to check the latest safety notes on these resources before proceeding.

The following resources are needed for the demonstration using limewater:

apparatus set up as shown in Fig. 11.7 on page 223 of the Student Book

Note that is essential that the apparatus is set up correctly so that there is no risk of breathing in limewater.)

Sterilise the mouthpiece using Milton® solution (sodium dichloroisocyanurate) made up according to the manufacturer's instructions. It should be soaked in the solution for at least 30 minutes before use.

The following resources are needed for the class practical B11.2, per group:

spirometer (see note above)
nose clip
stopwatch

If you do not have a commercial spirometer, the following simple apparatus can be constructed. This will not produce as reliable results, as it depends on the student breathing out normally once (rather than breathing in and out using the spirometer), and the tendency will be to breathe out more heavily than normal on each test. However, it may still produce useful results.

Calibrate a 2 litre plastic bottle by adding 500 cm^3 of water at a time, and marking the volume on the side of the bottle with a waterproof marker. When the bottle is full of water, invert it into a water trough without allowing any air into the bottle.

Insert a flexible plastic tube into the neck of the bottle and secure the bottle and tube in position. Clean the other end of the tubing with antiseptic solution. Alternatively, add a mouthpiece to the end of the tubing that can easily be removed and sterilised after each test. You will need a sterilised one of these for every student who is measured.

Sterilise mouthpieces using Milton® solution (sodium dichloroisocyanurate) made up according to the manufacturer's instructions. They should be soaked in the solution for at least 30 minutes before use.

Answers

Page 224 Developing Practical Skills

1. Plan should include:
   - several people (because of variation between individuals)
   - some form of exercise that can be controlled, so that each individual is exercising as much as the others
   - stopwatch to measure number of breaths over a particular time
   - spirometer to measure depth of breathing – the people being tested will have to be instructed to breathe normally because forced breaths easily increase volume
   - breathing rate and volume need to be measured immediately after the 2 minutes.

2. Test subjects need to be reasonably fit, so that exercise is not a risk to health, and should be wearing suitable clothing for exercise. Exercise ideally should be carried out in open space or in a gym, where the risks of tripping over obstacles are minimised.
3. First check for outliers (errors in measurements), such as the volume of breath for C at rest, which seems much higher than the others. Then, ignoring outliers, calculate the average for each factor in the two conditions.

Page 225 Science in context: How breathing rate is controlled

An increased in carbon dioxide concentration lowers pH which can affect enzyme activity by altering the shape of their active sites, in some cases denaturing them.

Page 225

1. The percentage of oxygen is less in expired air than inspired air. The percentage of carbon dioxide is greater in expired air than inspired air. The percentage of water vapour is higher in expired air than inspired air.
2. As level of exercise increases, rate and depth of breathing increase.
3. Supplement There is less oxygen in expired air than inspired air because oxygen in the body is used for respiration. There is more carbon dioxide in expired air than inspired air because the body produces carbon dioxide in respiration. There is more water vapour in expired air than inspired air because water molecules evaporate from the surface of the alveoli due to warmth of the body.
4. Supplement More exercise means more carbon dioxide is produced from an increased rate of respiration. Carbon dioxide is a soluble acidic gas so could cause the body tissues and blood to become more acidic. A change in pH can affect many enzymes and so affect the rate at which life processes are carried out in the body. Slowing down the rate of life processes may harm the body.

Worksheet B11.2 The effect of exercise on breathing

You will probably have noticed that when you run for a while, or rush up a long flight of stairs, your breathing becomes more rapid and possibly deeper. But how strongly is breathing affected by exercise? This investigation looks at the effect of different levels of exercise on rate of breathing, and (if a spirometer is available) on depth of breathing. Work in pairs for this investigation, with one of you exercising and the other recording the results.

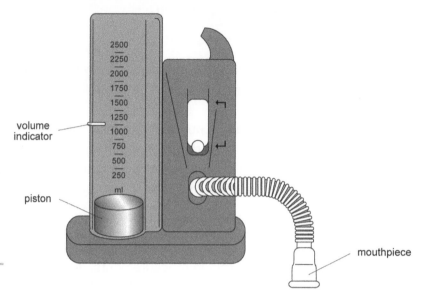

Apparatus

spirometer and nose clip (if available)

stopwatch

Method

1. Decide what measurements you will take, such as rate or depth of breathing and how you will take them.

   - Rate of breathing is usually measured as number of breaths per minute, but you can measure just for half a minute and multiply by two.

   - Depth of breathing will be measured using the spirometer. This may measure only one breath; in which case you must try to make it representative of your breathing at the time.

2. Decide what levels of exercise you will test, such as the number of steps made in a fixed time, or sitting or standing still, after walking, after jogging, after running as fast as possible.
3. Decide how long each level of exercise will be carried out before measurements are taken. For example, you could try a couple of test exercises to see whether one minute is enough, or if two minutes produces higher results.
4. Decide when you will take measurements after the exercise.
5. Decide how long to wait between levels of exercise to allow the body to recover fully before the next test.
6. Decide whether or not you need to do repeat measurements at each level of exercise.
7. Carry out your tests and record your data in a suitable table.

Page 1 of 2

Handling experimental observations and data

8. If you have taken repeat measurements, check first that there are no anomalous values (values that do not fit the pattern of results). If there are, try to explain them. Then, ignoring anomalous values, calculate average values for each level of exercise.
9. Use your results to draw a suitable graph or chart.
10. Describe the shape of your graph, and try to explain any pattern in your results.
11. If other students have used the same method as you have, compare your graph with theirs. Describe any similarities and differences between the graphs and try to explain them.
12. Draw a conclusion from your results about the effect of exercise on breathing.

Planning and evaluating investigations

13. Explain fully why the student who was exercising needed to rest between tests.
14. Describe any problems you had with this investigation. How do you think they affected your results?
15. Suggest how the method could be adjusted to avoid these problems.
16. Some investigations produce semi-quantitative data (e.g. if the exercise was described in levels, and the breathing measurements were measured on continuous scales). If appropriate, suggest how you could adapt your method to make the exercise levels quantitative, so that you could produce a more reliable conclusion.

Page 2 of 2

Learning episode B11.3 Consolidation and summary

Learning aims

- Review the learning points of the topic summarised in the end of topic checklist
- Test understanding of the topic content by answering the end of topic questions

Resources

Student Book pages 226–227

Approach

Introduce the learning episode

Ask students to work with a partner to make a list of key words from this topic. They could then work together to produce a spider diagram showing how the different concepts are linked. They could compare their list with the list of key terms given on page 226 of the Student Book. Discuss the checklist on page 226 and use questioning to see how much of the content they are comfortable with.

Students could make flashcards of the key content and then use the flashcards to quiz each other on the information.

Develop the learning episode

Ask students to work individually through the end of topic questions on page 227 of the Student Book without looking at the text. As they work, walk around the classroom observing their answers and asking questions as necessary to find out which questions are causing difficulties.

Finish the learning episode

After a set period, ask the students to stop working and discuss any areas of difficulty you observed as you walked round the class. Students should complete any unanswered questions for homework, but you should stress that they should try to answer the questions without looking at the text, so that they can see how much they have remembered.

Answers

End of topic questions mark scheme

The marks available for a question can indicate the level of detail you need to provide in your answer.

Question	Correct answer	Marks
1	A	1 mark
2 a)	Diffusion is the net movement of particles from a region of their higher concentration to a region of their lower concentration.	1 mark
	Gas exchange is the movement of gases across a surface, such as into or out of an organism from the air or from water.	1 mark
2 b)	Diffusion causes more oxygen to be taken into the blood because there is a lower concentration of oxygen in the blood than in the alveoli.	1 mark
	Diffusion causes more carbon dioxide to leave the blood in the lungs because there is a higher concentration of carbon dioxide in the blood than in the alveoli.	1 mark
2 c)	The alveoli create as large a surface area as possible for diffusion.	1 mark
	The large supply blood around the alveoli and the ventilation movements, maintain high concentration gradients between the blood and air in the lungs.	1 mark
		1 mark
		1 mark

Question	Correct answer	Marks
	The thin cell walls of the capillaries and alveoli minimise the diffusion distance.	
Supplement 3 a) i)	The balloons will expand.	1 mark
	This is because pulling down on the rubber diaphragm increases the volume inside the bell jar,	1 mark
	so decreasing the pressure inside the bell jar,	1 mark
	so air will move into the balloons through the pipe.	1 mark
ii)	The balloons will deflate.	1 mark
	This is because the volume inside the bell jar has decreased,	1 mark
	so increasing the pressure inside the bell jar,	1 mark
	so air will be pushed out of the balloons into the air.	1 mark
Supplement 3 b)	The intercostal muscles are not modelled.	1 mark
	During deeper or forced breathing the external muscles contract, pulling the ribs out and upwards,	1 mark
	increasing the volume of the thorax even further, and increasing the volume of air entering the lungs.	1 mark
	During exhalation, the external muscles relax (and the internal muscles may contract), letting the ribs move down and in,	1 mark
	reducing the volume of the thorax and resulting in a larger breath out.	1 mark
Supplement 4 a)	To support the trachea, keeping it open,	1 mark
	even if the air pressure inside decreases.	1 mark
Supplement 4 b)	A is a ciliated cell.	1 mark
	It has small hairs/cilia on its surface that waft to move mucus in one direction across its surface.	1 mark
Supplement 4 c)	B is a goblet cell.	1 mark
	It secretes mucus on to the surface of the cell/on to the lining of the airway.	1 mark
Supplement 4 d)	The mucus traps particles and pathogens carried in the air that enters the trachea from the air outside.	1 mark
	The cilia waft the mucus and trapped particles and pathogens up to the throat, where they pass into the oesophagus and to the stomach, where pathogens are destroyed.	1 mark
	This helps to prevent the infection of the lungs by pathogens.	1 mark
	Total:	31 marks

Introduction

This section covers aerobic respiration and anaerobic respiration and the similarities and differences between them.

Links to other topics

Topics	Essential background knowledge	Useful links
1 Characteristics and classification of living organisms	1.1 Characteristics of living organisms	
2 Organisation of the organism	2.1 Cell structure and size of specimens	
3 Movement into and out of cells	3.1 Diffusion	
4 Biological molecules	4.1 Biological molecules	
5 Enzymes	5.1 Enzymes	
7 Human nutrition	7.5 Absorption	
9 Transport in animals	9.1 Circulatory systems 9.2 Heart 9.4 Blood	
11 Gas exchange in humans	11.1 Gas exchange in humans	
13 Excretion in humans		13.1 Excretion in humans
14 Coordination and response		14.4 Homeostasis
19 Organisms and their environment		19.1 Energy flow 19.2 Food chains and food webs 19.3 Nutrient cycles
20 Human influences on ecosystems		20.3 Pollution
21 Biotechnology and genetic modification		21.2 Biotechnology

Topic overview

B12.1	**Aerobic respiration**
	This learning episode gives students the opportunity to explore the importance of respiration, and find out in detail about aerobic respiration.
	Supplement Students will also encounter the balanced chemical equation for aerobic respiration.
B12.2	**Anaerobic respiration**
	In this learning episode, students compare anaerobic respiration in muscle cells with that in yeast cells. They also investigate the effect of temperature on respiration in yeast.
	Supplement Students will encounter the balanced chemical equation for anaerobic respiration in yeast, and learn about changes in lactic acid concentration and oxygen debt resulting from anaerobic respiration in muscles.

B12.3	**Consolidation and summary**
	This learning episode provides an opportunity for a quick recap on the ideas encountered in the section. Students can answer the end of topic questions in the Student Book.

Careers links

These are some scientific careers that focus on this area of biology but careers in many other fields use the knowledge and skills gained studying science. **Bakers** use yeast and the product of respiration (carbon dioxide) during baking. **Fermentation scientists** research how yeast could be used in industrial processes such as the production of biofuels.

Starting points

The Student Book section opener puts the ideas in the section into context and sets the scene. It also allows students to acknowledge and value their prior learning, and provides a benchmark against which future learning can be compared.

- The questions provide a structure for introducing the section and can be used in a number of different ways:
- You could ask students to consider the questions as an introductory homework task.
- You could put students into groups to share their own ideas and understanding and then to report back to the whole class.
- Students could be given access to the Internet, preferably with a tight timescale, to find out the information required.

You could then use a spider chart or other form of wall chart to summarise everybody's ideas.

Recording these initial ideas allows you to retain them for reference as the individual topics are developed. In this way, your students' progress in learning can be readily acknowledged.

Learning episode B12.1 Aerobic respiration

Learning objectives

- State the uses of energy in living organisms, including: muscle contraction, protein synthesis, cell division, active transport, growth, the passage of nerve impulses and the maintenance of a constant body temperature
- Describe aerobic respiration as the chemical reactions in cells that use oxygen to break down nutrient molecules to release energy
- State the word equation for aerobic respiration as:
    glucose + oxygen → carbon dioxide + water
- **Supplement** State the balanced chemical equation for aerobic respiration as:
    $C_6H_{12}O_6 + 6O_2 \rightarrow 6CO_2 + 6H_2O$

Common misconceptions

In common language, the term *respiration* is used to mean breathing or ventilation. It is important to establish its correct scientific meaning: the release of energy from food molecules within cells. This should be developed through the definitions and comparison of aerobic respiration and anaerobic respiration.

There are many misconceptions around the understanding of the concept of energy. If students are also studying physics or chemistry, it can be helpful to develop a standard language and phraseology with colleagues. This can then be used when discussing aspects of energy and energy transfers to avoid the risk of developing misunderstandings about the conservation of energy in systems.

Note that the equations used to describe aerobic and anaerobic respiration are summaries of complex series of reactions. At this level, students do not need to know any of the intermediate stages although it can help prepare students going on to higher-level courses to realise that there is much more going on than shown here.

Resources

Student Book pages 231–233

Approach

1. Introduction

Following on from their work on human nutrition, ask students to explain why we need to eat. They should link food to the need for energy for life processes (if not, remind them of Section 1.1: *Characteristics of living organisms*). Ask them to write down each of the life processes, and then an example to show how energy is used in each life process. Take examples from around the class for discussion. Try to get as wide a range of examples from students as you can.

2. Uses of energy in the human body

Give students a few minutes to list as many uses of energy in the human body as they can think of. Take examples from around the class and try to elicit all the examples in the syllabus (muscle contraction, protein synthesis, cell division, active transport, growth, passage of nerve impulses, maintenance of stable core body temperature – although the last two of these are covered in later sections). This may require prompting to think at the cellular level.

Ask students then to write a sentence for each example to show how the energy is used, for example:

- In muscles, energy is used when the muscle contracts.
- In protein synthesis, energy is used to form bonds between amino acids and so build a protein.

For examples that students have not yet come across in this course, provide sentences (or allow students to research on their own using later sections of the Student Book or on the internet) such as:

- In nerve cells, energy is used when electrical impulses pass along them.
- As energy is transferred from a warm object to its cooler surroundings, our body has to continually release energy from respiration to maintain a constant internal temperature.

Then ask how energy is released from food. From earlier work, students should be able to answer *respiration*.

3. Aerobic respiration

Give students the terms *nutrient molecule*, *energy*, and *respiration*, and ask them to use the words to develop a definition for respiration. Students may well include *oxygen* in their definitions. In which case, point out that the general definition does not require this, and that they will be studying two forms of respiration, which will explain why.

Use pages 231–232 of the Student Book to introduce *aerobic respiration* and the word equation that summarises the reaction. Make sure students understand *aerobic* and how it is related to the equations.

Ask students to use their knowledge of body systems to explain how each of the reactants is delivered to each cell, and how waste products are removed. Also ask about why every human cell can carry out aerobic respiration.

Supplement Students should also learn the balanced chemical equation for aerobic respiration.

4. Consolidation

Give students a few minutes to write down three key points to answer a web query, 'What is aerobic respiration and what do we need it for?' They should compare their points with another student, selecting the three best, and repeat this in fours. Then take examples from around the class, and encourage discussion to select the best three for compiling a response to the query.

Answers

Page 233

1. a) b) and c):
   glucose *(from digested food from alimentary canal)* + oxygen *(from air via lungs)* →
   carbon dioxide *(excreted through lungs)* + water *(used in cells or excreted through kidneys)*
   (+ energy *(transferred to other chemicals in cell processes)*)
   d) Glucose replaced by fats from hump, and very little water excreted through kidneys.
2. Inside cells.
3. Any three from: muscles cells for contraction; synthesis of new molecules, such as proteins, for growth; active transport across cell membranes; passage of nerve impulses; maintenance of core body temperature.
4. $C_6H_{12}O_6 + 6O_2 \rightarrow 6CO_2 + 6H_2O$

Page 233 Science in context: Water from respiration

They use water released from the respiration of their food.

Page 234 Science in context: Respiration in athletics

The build-up of lactic acid may cause so much fatigue and pain that they cannot run any more. This may not matter if the race is finished, but they would not want to have to stop during the race.

Learning episode B12.2 Anaerobic respiration

Learning objectives

- Describe anaerobic respiration as the chemical reactions in cells that break down nutrient molecules to release energy without using oxygen
- State that anaerobic respiration releases much less energy per glucose molecule than aerobic respiration
- State the word equation for anaerobic respiration in yeast as:
  glucose → alcohol + carbon dioxide
- Investigate and describe the effect of temperature on respiration in yeast
- State the word equation for anaerobic respiration in muscles during vigorous exercise as:
  glucose → lactic acid
- Supplement State the balanced chemical equation for anaerobic respiration in yeast as:
  $C_6H_{12}O_6 \rightarrow 2C_2H_5OH + 2CO_2$
- Supplement State that lactic acid builds up in muscles and blood during vigorous exercise causing an oxygen debt
- Supplement Outline how the oxygen debt is removed after exercise, limited to:
  (a) continuation of fast heart rate to transport lactic acid in the blood from the muscles to the liver
  (b) continuation of deeper and faster breathing to supply oxygen for aerobic respiration of lactic acid
  (c) aerobic respiration of lactic acid in the liver

Common misconceptions

Anaerobic respiration in animals is often discussed as a 'bad thing' because it produces lactic acid (more precisely, lactate ions are produced as everything is in solution). This is based on the idea that acid is harmful. However, this misinterprets the role of lactate. When oxygen is limited, energy can still be released from glucose to produce an intermediate that builds up in cells. Hydrogen ions also increase in cells, because they cannot be removed by later stages of aerobic respiration. This causes 'acidosis' which affects enzymes. The combination of the intermediate and hydrogen ions produces lactic acid.

Without the production of lactate ions, muscle fatigue (as in the failure of muscles to produce their full effort) would happen even more rapidly than it does. This indicates that anaerobic respiration has an essential part to play in survival (such as when trying to escape from a predator). Students do not need to understand this detail, though it could be provided if you feel there is a risk of students developing misconceptions through their research.

The syllabus uses the term 'oxygen debt' to refer to the oxygen needed after anaerobic exercise to oxidise lactate. Sports scientists now use the acronym EPOC (excess post-exercise oxygen consumption) to mean the oxygen that is needed after any vigorous exercise, including that which is fully aerobic. The oxygen is used to return many body systems to their resting state. Students may come across this in research.

Resources

Student Book pages 233–237

Worksheet B12.2 Effect of temperature on respiration in yeast

B12.2 Technician's notes

Resources for class practical (see Technician's notes)

Approach

1. Introduction

Explain that *anaerobic* means 'without oxygen', and ask students to apply what they have learned about aerobic respiration to define the term *anaerobic respiration*. Give them a minute or two to discuss this in pairs or small groups and take answers. If necessary, lead the discussion towards the definition given in the Student Book on page 234.

2. Anaerobic respiration in muscle cells

Ask students for situations in which anaerobic respiration might occur in animals. Students may suggest times when oxygen is not available, such as in diving mammals. Encourage them to think of what happens in muscles during vigorous exercise and to suggest the benefit of anaerobic respiration in muscle cells.

Ask students to read page 234 of the Student Book to find the word equation for anaerobic respiration in animal cells, and to identify two disadvantages compared with aerobic respiration (less energy released per glucose molecule, production of lactic acid that needs breaking down after exercise ends).

Supplement Students should describe lactic acid building up in muscles and blood during vigorous exercise, and explain what is meant by the 'oxygen debt' and how it is removed.

3. Anaerobic respiration in yeast cells

Explain that yeast cells carry out a different form of anaerobic respiration compared with animals. Ask students to use pages 233–237 of the Student Book to identify the similarities and differences between the two forms of anaerobic respiration.

Then ask them to research how yeast is used to make bread. They should use their research to explain how any products of anaerobic respiration contribute to the process and the quality of the bread produced.

Supplement Students should also find out about the balanced chemical equation for anaerobic respiration in yeast.

4. Investigating the effect of temperature on respiration in yeast

Worksheet B12.2 provides a method for investigating the effect of temperature on (anaerobic) respiration in yeast. If there is sufficient equipment, students could carry this out in groups. Alternatively, set up the equipment as a demonstration, and ask students to use the results to answer the questions on the worksheet (see Technician's notes).

SAFETY INFORMATION: Be careful when working with the hottest water bath. Do not allow balloons to become over-inflated.

5. Consolidation

Write the word equations (or Supplement balanced chemical equations) for aerobic respiration, and anaerobic respiration in muscle cells and yeast cells, on the board. Then give students a few minutes to write true or false statements about the different kinds of respiration. Allow time for students to test their statements out on others, and identify any areas of weakness. Use your own statements, if needed, to help students clarify any misunderstandings.

Technician's notes

Be sure to check the latest safety notes on these resources before proceeding.

The following resources are needed for the class practical B12.2, per group:

yeast solution (5–10 g/litre of fresh yeast, or 2–5 g/litre of dried yeast)
sugar solution (50 g/litre)
4 water baths at different temperatures, e.g. 20 °C, 30 °C, 40 °C, 50 °C
4 test tubes and racks
4 balloons
10 cm^3 measuring pipettes or cylinders
vegetable oil
marker pen
long strip of paper, long enough to wrap around an inflated balloon and overlap
ruler

Ideally use fresh yeast made up into solution with a small amount of sugar and left for a few hours to activate. Alternatively, use a fresh packet of dried yeast, made up according to the manufacturer's instructions. Old yeast will not work as well.

Answers

Page 237

1. During vigorous exercise, they may not be able to get enough oxygen from the blood for all the energy they need for contracting. So the additional energy comes from anaerobic respiration.
2. Similarities: use glucose as substrate, produce energy, don't need oxygen.
   Differences: animals produce lactic acid; plants produce ethanol and carbon dioxide.
3. The amount of energy released from a mole of glucose molecules is much greater during aerobic respiration (c. 2900 kJ) than in anaerobic respiration (c. 150 kJ) in muscle cells.
4. Carbon dioxide is released by anaerobic respiration of glucose by yeast. Bubbles of the gas trapped in the dough makes it rise.

Worksheet B12.2 Effect of temperature on respiration in yeast

Yeast is a single-celled fungus that respires anaerobically, releasing carbon dioxide from the breakdown of sugars. The reactions of respiration are affected by temperature because they are controlled by enzymes. To measure the effect of temperature on the rate of respiration, we can measure the volume of carbon dioxide produced over time by yeast kept at different temperatures.

Apparatus

yeast solution

10 cm^3 measuring pipettes or cylinders

sugar solution vegetable oil

4 water baths at different temperatures

marker pen

4 test tubes and rackslong strip of paper

4 balloons

ruler

SAFETY INFORMATION

Be careful when working with the hottest water bath.

Do not allow balloons to become over-inflated.

Wash hands thoroughly after the practical work.

Method

5. Mark each of the tubes with your initials.
6. Measure 10 cm3 of the yeast solution into each test tube. Then work quickly through steps 3 and 4.
7. With a clean pipette or cylinder, measure 10 cm3 sugar solution into one of the tubes and mix it with the yeast using a long spoon or stick.
8. Carefully pour in about 1–2 cm3 oil so that it sits on top of the solution. Immediately attach a balloon to the top of the tube, so that it will catch any gas produced and inflate. Place the tube in a tube rack.
9. Repeat step 2 as quickly as possible for each of the other tubes, placing each one in a different rack.
10. Place each rack in a different water bath, again being careful when interacting with the hottest bath.
11. Leave the tubes for 20 minutes, but being careful not to let the balloons over-inflate.
12. Measure the circumference of each balloon as follows:
    - Wrap a strip of paper around the widest point and mark the point where the strips overlap.
    - Take the strip of paper off the balloon.
    - Measure the length of the paper strip up to that point that you have marked.
13. Draw up a suitable table to record your results

Handling experimental observations and data

14. Use your table to draw a suitable chart or graph.
15. Describe any pattern shown in the graph.
16. Explain any pattern using your knowledge of the effect of temperature on living organisms.

Planning and evaluating investigations

17. This apparatus only gives an approximate measure of the carbon dioxide produced. Explain how you could adjust the apparatus to get a more accurate result. Do you think this would change your results? Explain your answer.

Learning episode B12.3 Consolidation and summary

Learning aims

- Review the learning points of the topic summarised in the end of topic checklist
- Test understanding of the topic content by answering the end of topic questions

Resources

Student Book pages 238–239

Approach

Introduce the learning episode

Ask students to work with a partner to make a list of key words from this topic. They could then work together to produce a spider diagram showing how the different concepts are linked. They could compare their list with the list of key terms given on page 238 of the Student Book. Discuss the checklist on page 238 and use questioning to see how much of the content they are comfortable with.

Students could make flashcards of the key content and then use the flashcards to quiz each other on the information.

Develop the learning episode

Ask students to work individually through the end of topic questions on page 239 of the Student Book without looking at the text. As they work, walk around the classroom observing their answers and asking questions as necessary to find out which questions are causing difficulties.

Finish the learning episode

After a set period, ask the students to stop working and discuss any areas of difficulty you observed as you walked round the class. Students should complete any unanswered questions for homework, but you should stress that they should try to answer the questions without looking at the text, so that they can see how much they have remembered.

Answers

End of topic questions mark scheme

The marks available for a question can indicate the level of detail you need to provide in your answer.

Question	Correct answer	Marks
1	D	1 mark
2 a)	breathing/respiratory system (lungs) – oxygen	1 mark
	digestive system (small intestine) – glucose	1 mark
	circulatory system (blood) – oxygen and glucose	1 mark
2 b)	circulatory system (blood) – carbon dioxide and water	1 mark
	breathing/respiratory system (lungs) – carbon dioxide and water	1 mark
	excretory system (kidneys) – water	1 mark
3	any suitable table that shows these facts clearly:	8 marks

	Aerobic respiration	Anaerobic respiration
Similarities	use glucose as reactant	
	release energy for cell processes	
Differences	always uses oxygen produces carbon dioxide and water releases a lot of energy from each glucose molecule	oxygen is not needed produces either lactic acid (animals) or alcohol and carbon dioxide (yeast) releases a little energy from each glucose molecule

Question	Correct answer	Marks
4	At the start the muscles will use the oxygen available for aerobic respiration.	1 mark
	Oxygen levels will drop over time, because the whale cannot breathe air again until it returns to the surface.	1 mark
	During this time, the muscle cells will respire anaerobically,	1 mark
	to provide the energy needed for diving and swimming.	1 mark
5 a)	The sugar is broken down by the yeast in respiration,	1 mark
	and the energy released is used for making new cells.	1 mark
5 b)	i) The rate of expansion of the dough would be less,	1 mark
	because the yeast would be respiring more slowly.	1 mark
	This is because the enzymes that control the respiration reactions would be working more slowly.	1 mark
	ii) If the temperature were much higher than this, the enzymes would be denatured,	1 mark
	so that respiration would slow down or stop,	1 mark
	and the dough would not rise anymore because no carbon dioxide would be produced.	1 mark
	Total:	27 marks

Introduction

This section introduces students to a fuller description of excretion, and describes the role of the kidneys in the excretion of urine.

Links to other topics

Topics	Essential background knowledge	Useful links
1 Characteristics and classification of living organisms	1.1 Characteristics of living organisms	
2 Organisation of the organism	2.1 Cell structure and size of specimens	
3 Movement into and out of cells	3.1 Diffusion 3.2 Osmosis	
4 Biological molecules	4.1 Biological molecules	
7 Human nutrition	7.1 Diet	
9 Transport in animals	9.1 Circulatory systems 9.3 Blood vessels 9.4 Blood	
12 Respiration	12.1 Respiration 12.2 Aerobic respiration 12.3 Anaerobic respiration	
14 Coordination and response		14.4 Homeostasis
19 Organisms and their environment		19.3 Nutrient cycles

Topic overview

B13.1	**Excretion** This learning episode introduces excretion, including the roles of the kidneys and lungs in excretion in humans. Supplement This follows with the role of the liver in the formation of urea.
Supplement B13.2	**Kidney function** This learning episode looks at the detailed structure of the human kidney and kidney nephron, and the processes that take place in the kidney which result in the production of urine.
B13.3	**Consolidation and summary** This learning episode provides an opportunity for a quick recap on the ideas encountered in the section as well as time for the students to answer the end of topic questions in the Student Book.

Careers links

These are some scientific careers that focus on this area of biology but careers in many other fields use the knowledge and skills gained studying science. **Nephrologists** are doctors who treat patients who have diseases which affect the kidney. **Transplant nurses** specialise in supporting patients and families that can offer or need organ donations.

Starting points

The Student Book section opener puts the ideas in the section into context and sets the scene. It also allows students to acknowledge and value their prior learning, and provides a benchmark against which future learning can be compared.

- The questions provide a structure for introducing the section and can be used in a number of different ways:
- You could ask students to consider the questions as an introductory homework task.
- You could put students into groups to share their own ideas and understanding and then to report back to the whole class.
- Students could be given access to the Internet, preferably with a tight timescale, to find out the information required.

You could then use a spider chart or other form of wall chart to summarise everybody's ideas.

Recording these initial ideas allows you to retain them for reference as the individual topics are developed. In this way, your students' progress in learning can be readily acknowledged.

Learning episode B13.1 Excretion

Learning objectives

- State that carbon dioxide is excreted through the lungs
- State that the kidneys excrete urea and excess water and ions
- Identify in diagrams and images the kidneys, ureters, bladder and urethra
- Supplement Describe the role of the liver in the assimilation of amino acids by converting them to proteins
- Supplement State that urea is formed in the liver from excess amino acids
- Supplement Describe deamination as the removal of the nitrogen-containing part of amino acids to form urea
- Supplement Explain the importance of excretion, limited to toxicity of urea

Common misconceptions

The terms *egestion* and *excretion* are commonly confused. Reinforce the fact that substances that are excreted must have come from inside the body, such as from metabolic processes or absorbed in excess from the small intestine.

Some students confuse *urea*, which is made in the liver, with *urine*, which is produced in the kidneys.

Resources

Student Book pages 243–244

Approach

1. Introduction

Ask students what humans excrete and which organs are involved in their excretion. If necessary, reinforce the distinction between *excretion* and *egestion* at this point.

Give students two minutes to work in pairs on this, then take answers. They should be able to answer with carbon dioxide (lungs) and urine (kidneys). Remind students that urine is a mixture of substances and ask them to state some of these substances. They may mention urea, water and ions (salts), but it does not matter at this stage if they do not – note any omissions and make sure they are well covered in the learning episode.

2. The human urinary system

Give students one minute to write down all the structures of the urinary system that they can remember, their position in the body and what role each structure plays in the system. Then ask them to compare their list with the text and diagram of the urinary system in Fig. 13.2 on page 244 of the Student Book. They should identify any gaps and fill them in using the details in the book. If there were significant gaps in knowledge, give students a few minutes in their pairs, taking it in turns to ask questions to reinforce what they have just learned.

Supplement 3. Role of the liver

Students must grasp that the kidneys' role is as a filtration/reabsorption system – the waste materials they excrete are (mostly) produced elsewhere. Introduce this by referring to the liver as a 'processing area' – where substances that are potentially toxic to the body are broken down to less toxic substances.

Ask students to make notes from page 243 of the Student Book on the role of the liver in the formation of urea. They should include the terms *assimilation* and *deamination* in their notes, and an explanation of why urea must be excreted from the body.

4. Consolidation: definition

Ask students to write a definition for a web answers site or encyclopaedia for the term *excretion*, including examples that clarify the answer. They should compare their definition with another student and work together to improve their definition.

Answers

Page 243

1. The removal from the body of waste substances from metabolic reactions in cells and substances which are in excess of requirements.
2. They excrete carbon dioxide, which is the waste product of respiration.
3. They excrete urea, and excess water and excess ions.
4. **Supplement** Urea is toxic in the body and so will cause harm if it is not excreted.

Page 244

1.

Structure	Function
kidneys	produce urine by filtering waste substances from the blood
ureters	carry urine from kidneys to bladder
bladder	stores urine until it is released to the environment
urethra	short tube linking bladder to environment

2. water, urea, ions

Supplement Learning episode B13.2 Kidney function

Learning objectives

- Supplement Identify in diagrams and images the structure of the kidney, limited to the cortex and medulla
- Supplement Outline the structure and function of a nephron and its associated blood vessels, limited to:
    - (a) the role of the glomerulus in the filtration from the blood of water, glucose, urea and ions
    - (b) the role of the nephron in the reabsorption of all of the glucose, some of the ions and most of the water back into the blood
    - (c) the formation of urine containing urea, excess water and excess ions
    (details of these processes are **not** required)

Common misconceptions

In order to understand fully how urine is produced in the nephron, students need a good understanding of the processes of diffusion, osmosis and active transport (see Section 3: *Movement into and out of cells*).

Students may come across the term *kidney tubule*, which is another name for a nephron.

Resources

Student Book pages 245–247

B13.2 Technician's notes

Supplement Worksheet B13.2a Kidney dissection

Supplement Worksheet B13.2b Structure of a nephron

Resources for demonstration and class practical (see Technician's notes)

Approach

1. Introduction – Excretion demonstration

Ask students how much water/liquid they drink each day. Then ask them to consider the colour of their urine in the morning and during the day.

Show students two flasks of 'urine', one darker yellow than the other. (These should not be real urine, but prepared using water and different concentrations of yellow food colouring.) There should be about twice the volume of pale 'urine' as dark 'urine'. Explain that the darker 'urine' was collected when the body was hot and had not drunk much water, and the pale 'urine' was collected after drinking a lot of water (see Technician's notes).

Ask students to describe the differences between the liquid in the flasks, and try to explain any differences. They should link the idea of a well-hydrated body to the production of a large amount of dilute urine. If students are not certain about the term, introduce the word *excess* and ask them to use the examples of the two flasks to explain what it means. Lead students to the idea that excess water in the body is removed from the body in the urine, and this is why the volume of urine varies with the state of hydration.

2. Functions of the kidney

Following on from the introduction, ask students which substances might be found in urine. If they are uncertain, ask them to read pages 245–247 in the Student Book, and to explain the excretion of urea and excess ions in urine. They should link the presence of ions and urea in the blood to the absorption of mineral ions and amino acids from food by the digestive system. Explain that usually glucose is not found in urine – that its presence in urine indicates a disorder such as diabetes. If available, show students urine test sticks (or an image of these from the internet) and explain that the sticks test for a range of substances. They are simple and important diagnostic tools for doctors, as changes in urine composition can indicate many types of disorder.

3. Structure of the kidney

Worksheet B13.2a gives students the opportunity to dissect and identify the main structures in a kidney. Alternatively, this could be done as a class demonstration (see Technician's notes). Use the diagram of the structure of a kidney in Fig. 13.3 on page 245 of the Student Book to help them identify all the key structures.

SAFETY INFORMATION: Clean any splashes of blood with disinfectant immediately.
Soak all surfaces and equipment that have come into contact with blood for at least ten minutes with disinfectant, such as 1% Virkon, at the end of the practical. Warn students about handling knives or scalpels with care.

Wear disposable gloves during the dissection and wash your hands thoroughly after the practical.

4. Producing urine in a nephron

There are many video clips on the internet of the structure of a nephron and how it relates to the production of urine. However, most of these are aimed at a higher level than covered by Cambridge IGCSE. If you have access to displaying video in class, find one that is suitable for your students because it will help their understanding.

Before showing the video clip, ask students to write down five questions that they want to answer from watching it. Discuss those questions and help them to select the five most suitable for the video and for what they need to know. This will help them to focus on the key points that they need to learn. Make sure these include the filtration of substances from the glomerulus, and reabsorption of glucose, water and ions in the nephron.

Give students the large diagram of a nephron on Worksheet 13.2b and ask them to use what they learn from the video to label it with the names of the parts and descriptions of what happens in the key parts.

Be aware that the video may use additional or alternative terms that students do not need to know. Provide a list of alternatives for words that they do need to know, but tell them to ignore the rest.

If you do not have access to a suitable video, students can complete the worksheet using information from the Student Book or from their own research.

5. Consolidation: true or false

Ask students to write down three to five statements about the human kidney, some true and some false. Take examples from around the class, asking other students to put up their hands if they think a statement is true.

Help students to clarify their understanding by choosing a student who has the right answer to explain why they made that choice.

Technician's notes

Be sure to check the latest safety notes on these resources before proceeding.

The following resources for the demonstration in the introduction:

two flasks
water
yellow food colouring

Colour water with yellow food colouring to match the colours of pale urine and dark urine. There should be about twice as much pale 'urine' as dark 'urine'.

The following resources for the class practical 13.2a, per group:

kidney
sharp knife or scalpel
dissecting board
disinfectant such as 1% Virkon for disinfecting all tools and work surfaces

You will need at least one kidney for each two groups/pair of students. Ideally the kidneys should still have their outer capsule and ureter attached – they are best sourced from a butcher with a request to prepare them like this. If you are concerned about students using sharp knives or scalpels, then present the kidneys already cut in half vertically and ask students to start at step 4 of the instructions on the worksheet.

Answers

Page 247 Science in context: Surviving without water

They still need to get rid of some substances, such as urea.

Page 247

1. Supplement Filtration and reabsorption.
2. Supplement Reabsorbed – any from: water, glucose, mineral salts. Not reabsorbed: urea.

Supplement Worksheet B13.2a Kidney dissection

In this dissection, you will get the opportunity to identify the key structures in a kidney.

If you have been given a whole kidney, then initially work in pairs until step 4, at which point you should work individually.

Apparatus

kidney

sharp knife or scalpel

dissecting board

disinfectant

SAFETY INFORMATION

Clean any splashes of blood with disinfectant immediately.

Soak all surfaces and equipment that have come into contact with blood for at least ten minutes with disinfectant at the end of the practical.

Wash your hands thoroughly after the practical.

Method

1. Look at the external shape and structure of the kidney, and note the point where the ureter attaches to the kidney. This is also the point where the renal artery and vein attach to the kidney, but it is unlikely that these will still be in place.
2. If the kidney has an outer capsule, note the colour and toughness of the capsule. This helps to protect the kidney from damage from the outside. Use the knife or scalpel to make a small cut in the capsule. Then carefully cut and remove the capsule without damaging the kidney inside.
3. Carefully cut the kidney into two halves vertically.
4. Note where the ureter is attached to the pelvis in the middle of the kidney. The pelvis is the area into which all the collecting ducts drain, delivering urine that will flow down the ureter to the bladder.
5. Note that there are two distinct areas of different shades of red in the kidney. The outer area is the cortex. This is the part of the kidney where the renal capsules containing the glomerulus of each nephron lie. You will not be able to see individual nephrons because they are too small, but you will be able to see that this area is the brighter red of the two areas, because the parts of the nephron found here are well supported by the capillary network and so contain a lot of blood.
6. Look at the medulla (think of 'middle' to help you remember which area is which), and note that it is paler than the cortex. This area consists of the loops in the middle of the nephrons, and the collecting ducts from all the nephrons. The collecting ducts end at the inner edge of the medulla and drain into the pelvis. There are blood capillaries in the medulla (which is why it is red) but they are not as dense as in the cortex.
7. Carefully draw the structure of the kidney that you can see, and label it to explain which parts of a nephron occur in which areas.
8. Dispose of the kidney as instructed by your teacher, and thoroughly clean everything that has come into contact with the kidney.

Supplement Worksheet B13.2b Structure of a nephron

1. Label the diagram to identify the different parts of a nephron.
2. Then indicate in which parts of the nephron the following processes take place:
   a) filtration
   b) reabsorption of water and ions
   c) reabsorption of glucose.
3. Add notes about how these processes happen. Remember to include scientific terms where appropriate in your explanations.

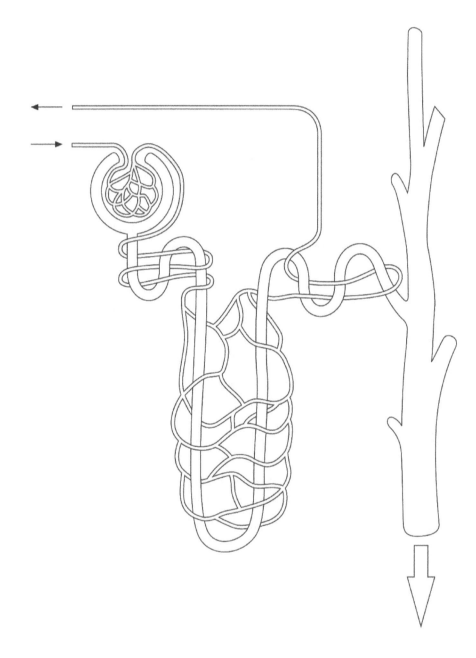

Learning episode B13.3 Consolidation and summary

Learning aims

- Review the learning points of the topic summarised in the end of topic checklist
- Test understanding of the topic content by answering the end of topic questions

Resources

Student Book pages 248–249

Approach

Introduce the learning episode

Ask students to work with a partner to make a list of key words from this topic. They could then work together to produce a spider diagram showing how the different concepts are linked. They could compare their list with the list of key terms given on page 248 of the Student Book. Discuss the checklist on page 248 and use questioning to see how much of the content they are comfortable with.

Students could make flashcards of the key content and then use the flashcards to quiz each other on the information.

Develop the learning episode

Ask students to work individually through the end of topic questions on page 249 of the Student Book without looking at the text. As they work, walk around the classroom observing their answers and asking questions as necessary to find out which questions are causing difficulties.

Finish the learning episode

After a set period, ask the students to stop working and discuss any areas of difficulty you observed as you walked round the class. Students should complete any unanswered questions for homework, but you should stress that they should try to answer the questions without looking at the text, so that they can see how much they have remembered.

Answers

End of topic questions mark scheme

The marks available for a question can indicate the level of detail you need to provide in your answer.

Question	Correct answer	Marks
1	D	1 mark
2 a)	lungs	1 mark
	kidneys	1 mark
2 b)	carbon dioxide from lungs	1 mark
	urea and excess water/ions from kidney	1 mark
Supplement 3	The glucose has been filtered out of the blood.	1 mark
	Normally it is all reabsorbed back into the blood from the nephron, so this suggests that something is wrong with the body.	1 mark
Supplement 4	Some amino acids are converted to proteins,	1 mark
	this is called assimilation.	1 mark
	Excess amino acids are broken down to form urea,	1 mark
	from the nitrogen-containing part of the amino acids / by deamination.	1 mark
Supplement 5 a)	Filtration removes many small molecules from the blood, including water, ions, urea and glucose.	1 mark
		1 mark

	These pass into the nephron from the glomerulus and form the filtrate, most of which will eventually form urine.	
Supplement 5 b)	Reabsorption removes useful molecules, such as glucose, some water and some ions, from the filtrate back into the blood. What is left in the nephron forms urine.	1 mark 1 mark
Supplement 6	Glucose is reabsorbed because it is essential in the body for cellular respiration.	1 mark
	Some water and ions are reabsorbed because they are necessary for cell processes.	1 mark
	Urea is excreted because it is toxic to the body.	1 mark
	Excess water and ions are excreted because they are not needed and may cause problems if they remain in the body.	1 mark
	Total:	19 marks

B14 Coordination and response

Introduction

This section covers nervous and hormonal control in humans, homeostasis in humans, and tropic responses to gravity and light in plants.

Links to other topics

Topics	Essential background knowledge	Useful links
1 Characteristics and classification of living organisms	1.1 Characteristics of living organisms	
2 Organisation of the organism	2.1 Cell structure and size of specimens	
3 Movement into and out of cells	3.1 Diffusion	
5 Enzymes	5.1 Enzymes	
9 Transport in animals	9.1 Circulatory systems 9.2 Heart 9.3 Blood vessels 9.4 Blood	
11 Gas exchange in humans	11.1 Gas exchange in humans	
12 Respiration	12.3 Anaerobic respiration	
13 Excretion in humans	13.1 Excretion in humans	
16 Reproduction		16.4 Sexual reproduction in humans 16.5 Sexual hormones in humans

Topic overview

B14.1	The human nervous system
	In this learning episode students learn about the structure of the human nervous system, the roles of the different kinds of neurones, and what happens during reflex actions. There is also practical work to complete on the speed of reaction. Supplement Students learn about how synapses work.
B14.2	The human eye
	In this learning episode, students learn about the eye as an example of a sense organ, and its responses to light. Supplement Students learn about the mechanism of the pupil reflex, how accommodation happens and the role of rods and cones in vision.
B14.3	Human hormones
	This learning episode looks at how some responses are controlled by the hormones, particularly adrenaline. Supplement Students learn about glucagon and further details about adrenaline.

B14.4	**Homeostasis**
	This learning episode looks at homeostasis in humans, for example, the control of blood sugar levels through insulin.
	Supplement Students investigate how homeostatic mechanisms maintain a constant internal body temperature. There is a practical opportunity to model the effect of sweat on cooling the body. Students will learn more about the control of blood glucose concentration and the concept of negative feedback in homeostasis. They will also learn about the treatment of Type 1 diabetes.
	If possible, arrange a visit from a health professional to talk to students about Type 1 diabetes.
B14.5	**Tropic responses**
	This learning episode gives students the opportunity to investigate the growth responses of plants to light and gravity.
	Supplement Students will learn about the role of auxins in controlling growth.
B14.6	**Consolidation and summary**
	This learning episode provides an opportunity for a quick recap on the ideas encountered in the section, as well as time for the students to answer the end of topic questions in the Student Book.

Careers links

These are some scientific careers that focus on this area of biology but careers in many other fields use the knowledge and skills gained studying science. **Ophthalmologists** test people's eyesight and can diagnose diseases by looking at their retinas during examinations. **Medical product designers** have developed devices which can people with diabetes can use to test for blood sugar levels at home.

Starting points

The Student Book section opener puts the ideas in the section into context and sets the scene. It also allows students to acknowledge and value their prior learning, and provides a benchmark against which future learning can be compared.

The questions provide a structure for introducing the section and can be used in a number of different ways:

- You could ask students to consider the questions as an introductory homework task.
- You could put students into groups to share their own ideas and understanding and then to report back to the whole class.
- Students could be given access to the Internet, preferably with a tight timescale, to find out the information required.

You could then use a spider chart or other form of wall chart to summarise everybody's ideas.

Recording these initial ideas allows you to retain them for reference as the individual topics are developed. In this way, your students' progress in learning can be readily acknowledged.

Learning episode B14.1 The human nervous system

Learning objectives

- Describe sense organs as groups of receptor cells responding to specific stimuli: light, sound, touch, temperature and chemicals
- State that electrical impulses travel along neurones
- Describe the mammalian nervous system in terms of:
  (a) the central nervous system (CNS) consisting of the brain and the spinal cord
  (b) the peripheral nervous system (PNS) consisting of the nerves outside of the brain and spinal cord
- Describe the role of the nervous system as coordination and regulation of body functions
- Identify in diagrams and images sensory, relay and motor neurones
- Describe a simple reflex arc in terms of: receptor, sensory neurone, relay neurone, motor neurone and effector
- Describe a reflex action as a means of automatically and rapidly integrating and coordinating stimuli with the responses of effectors (muscles and glands)
- Describe a synapse as a junction between two neurones
- Supplement Describe the structure of a synapse, including the presence of vesicles containing neurotransmitter molecules, the synaptic gap and receptor proteins
- Supplement Describe the events at a synapse as:
  (a) an impulse stimulates the release of neurotransmitter molecules from vesicles into the synaptic gap
  (b) the neurotransmitter molecules diffuse across the gap
  (c) neurotransmitter molecules bind with receptor proteins on the next neurone
  (d) an impulse is then stimulated in the next neurone
- Supplement State that synapses ensure that impulses travel in one direction only

Common misconceptions

Some students think that a nerve contains a single connection between one part of the body and another. Encourage them to think more of a multi-core cable, such as one that carries phone or digital connections, to help them realise that many connections can take place between cells at opposite ends of a nerve at the same time.

Resources

Student Book pages 254–259

Worksheet B14.1a Speed of reaction

B14.1 Technician's notes

Supplement Worksheet B14.1b Action at synapses

Resources for demonstration and class practical (see Technician's notes)

Approach

1. Introduction: human senses

Introduce students to the concepts of stimulus and response in general terms: a change in the surroundings or in the body and the way the body changes as a result. Ask them for a few examples of stimulus and response in humans. Extend the discussion to the concepts of *receptors* (sense organs that detect stimuli) and *effectors* (organs that carry out responses). Use the suggested examples and ask students to identify the receptor and effector in each case.

Give students three minutes to work individually, or in pairs, to find examples of responses to as many different kinds of stimuli that they can think of; identifying the sensor and receptor in each case. Take examples from around the class until you have covered all the sense organs mentioned in the table on page 259 of the Student Book.

2. Structure of the nervous system

Ask students what connects sensors and effectors in the human body – they should be able to give the answer *nerves*. Use pages 254–256 of the Student Book to introduce the nervous system of the human body. Students should be able to identify the role of sense organs, peripheral nerves, the central nervous system and effectors in responding to changes in the environment.

Explain that a nerve is made of many nerve cells, called neurones, and that there are different neurones in different parts of the nervous system. Ask them to use pages 255–256 in the Student Book to identify the three types of neurone and to compare their structures and functions. They should combine this information to describe a reflex arc.

3. Reflex responses

Use pages 257–259 of the Student Book to introduce the idea of reflex actions. Students could also research examples using books or the internet. They should be able to explain the response in terms of the reflex arc, so ask them to draw a reflex arc for one of their examples using the diagram of the response to the touch of a hot object on page 258 of the Student Book to help them.

The response of the pupil to light can be demonstrated easily with a small torch in a dimly lit room. Allow a minute for eyes to adjust to the light conditions and then shine the torch towards the face (not directly into the eyes). If you have sufficient small mirrors, students could investigate their own response (see Technician's notes). (Note, the pupil response is also discussed in Learning episode 14.2.)

SAFETY INFORMATION: A bright light shone close to the eye may produce an after-image that lingers for several minutes and may be alarming for some students. Students with albinism should not have bright lights shone near their eyes.

Discuss the difference between a reflex response and a response that involves a thought, such as the response to catching the ruler on Worksheet B14.1a, or trying to catch a ball. They should identify not only a difference in speed (reflex responses generally being faster), but also a difference in purpose (reflex responses are generally ones that protect us from harm, such as the heat/touch reflex and the light/pupil reflex).

4. Speed of reaction

Worksheet B14.1a gives students the opportunity to test the speed of their reaction to a stimulus, in this case catching a falling ruler. The method is very simple and quick, so students can easily collect a large amount of data for analysis. This makes it particularly useful for looking at the effect of the number of repeats on the reliability of a result, as described on the worksheet.

In addition, if it is possible, students can collect the class data on a spreadsheet, together with data on other variables (such as gender, athletic achievement, types of computer games played, time of day tested). They can then analyse whether these variables have any effect on speed of reaction. Allow time before the investigation for students to discuss what might affect speed of reaction, and for them to suggest variables on which to collect data for testing later.

The graph of averages should show that around five or six repeats of an investigation produce an acceptable approximation of an average from a larger number of repeats. This should help students when planning investigations where repeat measurements will be needed.

The comparison of speed of reaction with other factors – such as gender, sporting ability, computer game preferences – may or may not show differences. For extra challenge, introduce the idea of *reliability of difference*, such as how big a difference is needed for it to be considered a real difference. With students who are good at maths, this is an opportunity to discuss statistical significance and introduce the use of standard error although it should be made clear to students that these aspects are beyond the requirements of the syllabus.

SAFETY INFORMATION: Remind students to take care with the falling ruler. Students should not wear open-toed shoes.

Supplement **5. Synapses**

Worksheet B14.1b provides a simple image that students can develop into a sequence of images that describe how a synapse works. The information on synapses on pages 256–257 of the Student Book can help with this.

6. Consolidation: key points

Ask students to make a list of five key points from their work on the human nervous system, for a friend who has missed this work. They should then compare their points with another student and decide on the best five points to cover everything in the work. Take examples from around the class and encourage discussion to produce a class list of the key points on this section.

Technician's notes

Be sure to check the latest safety notes on these resources before proceeding.

The following resources are needed for the demonstration of pupil response to light:

torch

The following resources are needed for the class practical B14.1a, per group:

ruler, at least 50 cm long, ideally of shatterproof plastic or wood

Answers

Page 256

1. The ability to detect and respond to changes in the external environment and internal conditions of the body.
2. A receptor detects a change in the environment, which is the stimulus. This causes a response from an effector in the body.
3. Muscles and glands
4. Sensory neurones have long dendrons and axons that link the sense organ with the central nervous system.
   Relay neurones are short neurones with many dendrites, found in the central nervous system, that link sensory neurones to motor neurones or other relay neurones.
   Motor neurones have many dendrites to link with relay neurones and end on the effector, such as a muscle.

Page 257

1. A synapse is the junction between two neurones.
2. **Supplement** Neurotransmitters are molecules released at a synapse that diffuse across the synaptic gap and trigger an electrical impulse in the next neurone.

Page 259

1. A reflex action is an automatic, often rapid, response to a stimulus.
2. Reflex actions are usually very fast, which makes it possible to respond to a stimulus very quickly. Reflex actions are usually important in survival, e.g. to protect you from touching something very hot, or blinking to protect the eye if something comes toward it.

Worksheet B14.1a Speed of reaction

In this investigation you will test the speed of your reaction to a stimulus: a falling ruler. You will use the repeat results to estimate the fewest number of repeats that will give you a reliable answer.

Before you collect any data, consider which factors may affect the results. For example, you may think you respond faster in the morning than in the afternoon. Also consider factors that vary between different people that might affect the speed of reaction. As a group or class, decide which factors you wish to test and collect data on these to add to your class or group results.

Apparatus

long ruler (at least 50 cm)

SAFETY INFORMATION
Take care with the falling ruler.
Do not wear open-toed shoes for this practical.

Method

1. Work with a partner. One of you is the test subject and the other is the researcher. Change places halfway through the investigation so that each of you will have your reactions tested.
2. The test subject should stand with their hand in front of them with an open grip in anticipation of catching the ruler. The researcher should hold the ruler just above the open hand, with the 0 cm point on the ruler between their fingers and thumb.
3. The researcher should then drop the ruler without warning. The test subject should close their hand as quickly as possible to catch the ruler. The researcher should wait different times before the ruler is dropped, to make it more difficult for the test subject to anticipate when it will fall.
4. Record the point on the ruler where it was caught, to the nearest millimetre. If the ruler is dropped, or caught in a very short distance, disregard that result and repeat the trial.
5. Carry out the test 15 times, and record all valid results.
6. Then change places and repeat steps 2–5.

Handling experimental observations and data

7. Using your own results, calculate averages as follows:

   - take the first two results and calculate and record the average

   - take the first three results and calculate and record the average.

   Repeat this until you have calculated and recorded the average for all 15 results.

8. Plot your averages on a graph, with number of tries on the *x*-axis and average values on the *y*-axis. For a try of one, plot the value of your first result.
9. Describe the shape of your graph and explain what it means.
10. Use your graph to determine the smallest number of repeats that gives a reasonable estimate of the average of all results. Explain how you can apply this finding to other investigations.

Page 1 of 2

11. If possible, collect the final average from all students on a class spreadsheet. Collect data on all the factors discussed and agreed by the class before the investigation.
12. Use the spreadsheet to investigate whether there is any link between any of the factors and speed of reaction. For example, if you wish to test the effect of sporting ability, you could calculate an average for each group of students who specialise in a particular sport, and compare them with an average for those who do not play a sport regularly.

Planning and evaluating investigations

13. Describe any problems you had with this investigation. How do you think they affected your results?
14. Suggest how the method could be adjusted to avoid these problems and get a more reliable way of measuring speed of reaction.

Supplement Worksheet B14.1b Action at synapses

Use the grid below to draw a sequence of images that could be used to brief an animation of how a nerve impulse is transmitted across a synapse. A simple image of a synapse before a nerve impulse arrives is given in the top left box.

In the box below each image, describe what should happen at each stage of the animation.

Add notes for the voice-over to go with each image.

Remember to identify every key part of the diagram, and to explain how the action at synapses ensures that impulses can only pass in one direction through the nervous system.

Notes	**Notes**
Notes	**Notes**

Learning episode B14.2 The human eye

Learning objectives

- Identify in diagrams and images the structures of the eye, limited to: cornea, iris, pupil, lens, retina, optic nerve and blind spot
- Describe the function of each part of the eye, limited to:
    (a) cornea – refracts light
    (b) iris – controls how much light enters the pupil
    (c) lens – focuses light on to the retina
    (d) retina – contains light receptors, some sensitive to light of different colours
    (e) optic nerve – carries impulses to the brain
- Explain the pupil reflex, limited to changes in light intensity and pupil diameter
- Supplement Explain the pupil reflex in terms of the antagonistic action of circular and radial muscles in the iris
- Supplement Explain accommodation to view near and distant objects in terms of the contraction and relaxation of the ciliary muscles, tension in the suspensory ligaments, shape of the lens and refraction of light
- Supplement Describe the distribution of rods and cones in the retina of a human
- Supplement Outline the function of rods and cones, limited to:
    (a) greater sensitivity of rods for night vision
    (b) three different kinds of cones, absorbing light of different colours, for colour vision
- Supplement Identify in diagrams and images the position of the fovea and state its function

Common misconceptions

Many people think that we see with our eyes and hear with our ears. This is incorrect. Our eyes and ears collect information from the environment (because they contain receptor cells), responding to changes in light or sound by sending electrical impulses to the brain through nerve cells. It is the responses in the brain that produce what we 'see' or 'hear'. For example, a person with perfectly functioning eyes may not 'see' if their optic nerves have been damaged as a result of an accident.

Resources

Student Book pages 259–263

B14.2 Technician's notes

Worksheet B14.2a Structure of the eye

Supplement Worksheet B14.2b Accommodation

Resources for demonstration (or class practical) (see Technician's notes)

Approach

1. Introduction

Give students a few minutes to sketch a diagram of how we see something. Ask them to include in their diagrams the labels *light source, eye* and any other labels that they think are important. They should compare their diagram with that of a neighbour to help identify any weaknesses. After a chance to improve their diagrams, take examples from around the class to build a diagram on the board. Make sure that students understand that light from a source is reflected off an object into the eye, and that light causes a response in the light-sensitive cells at the back of the eye. If needed, reinforce the convention of drawing light rays as arrowed straight lines using a ruler.

2. Structure of the eye

Worksheet B14.2a can be used for labelling the structures of the eye. Students should also add notes to explain the role of each key structure in the way the eye works. Page 260 in the Student Book can help with this. Remind students of the pupil reflex they looked at in Learning episode 14.1 that involves the iris and pupil.

To reinforce learning, ask students to compose five questions based on their diagram about the structures in the eye and their functions. They should exchange their questions with another student to try them out. Use the responses to identify any weaknesses, and address these using a question and answer session.

Supplement Students should also distinguish between the function and distribution of rods and cones, including their relation to the fovea. Students should be able to explain the pupil reflex in terms of the circular and radial muscles of the iris and their antagonistic action – they can use page 261 in the Student Book to help with this.

3. Dissection of an eye

This practical should be done by students under careful supervision because of the use of a sharp scalpel and scissors. Alternatively, do this as a demonstration, allowing time for students to observe the key parts at each stage (see Technician's notes).

You may be able to get a whole eye from a local butcher or abattoir. Try to get an eye that still has muscles attached, and ideally the day before the lesson, as dissection is easier when the eye is fresh.

Show students the outside of the eye and ask them what they can identify. This should include the cornea, the tough outer layer (sclera), and the muscles that attach the eye to the bony socket and allow the eye to move. Students should also note how firm the eye is, as a result of the jelly-like humour.

Using a scalpel, carefully cut into the cornea and release the more watery aqueous humour. Then insert the scalpel into the sclera, to produce a small incision. Insert the scissors into the incision and cut all around the eye to remove the back of the eye. Students should note the more jelly-like vitreous humour, the darkness of the retina, the position and attachment of the lens to the ciliary muscles by the suspensory ligaments, and the relation between the lens and pupil.

Although you may wish to use the terms *sclera*, *aqueous humour*, and *vitreous humour* during the dissection, students should be aware that they do not need to learn these terms as they are beyond the requirements of the syllabus and that only **Supplement** students need to know about *ciliary muscles* and *suspensory ligaments*.

SAFETY INFORMATION: Take care when using sharp scalpels. Wash down working surfaces and equipment with disinfectant, such as 1% Virkon, after completing the dissection.

Supplement 4. Accommodation in the eye

Worksheet B14.2b provides two diagrams of the eye. Students can complete these to show accommodation by drawing in the light rays from a near or far object, and describing the roles of the ciliary muscles and supporting ligaments in changing the shape of the lens.

5. Consolidation

Give students a few minutes to jot down definitions for the following terms: cornea, iris, pupil, lens, retina, optic nerve, blind spot, **Supplement** radial and circular muscles in the iris, antagonistic action, ciliary muscles, suspensory ligaments, rod, cone, fovea, accommodation.

Select a student to read out one of their definitions for other students to guess. Then ask another student to read out a different definition for others to guess. Repeat as needed to cover all the words.

Technician's notes

Be sure to check the latest safety notes on these resources before proceeding.

The following resources are needed for the dissection of the eye if done as a demonstration:

eye, e.g. from sheep
sharp knife or scalpel
sharp scissors
dissecting board
disinfectant such as 1% Virkon for disinfecting all tools and work surfaces

You will need one eye for this demonstration, ideally with muscles still attached. If students are to carry out the dissection, then allow one eye per group. Eyes are best sourced from a butcher or abattoir, with a request to prepare them like this. If possible, collect the eyes the day before the lesson, as they are best dissected fresh.

SAFETY INFORMATION: After the dissection all surfaces and tools that were used will need disinfecting, such as with 1% Virkon, for at least 10 minutes. Wash your hands thoroughly after finishing the dissection and clearing up.

Answers

Page 261

1. a) The cornea is transparent, so light passes through it easily into the eye.
    b) The pupil is a hole surrounded by the iris that lets light pass through to the back of the eye.
    c) The retina contains the light-sensitive cells that respond to light. The retina is also very dark, to absorb as much light as possible.
2. **Supplement** As light intensity increases, the pupil gets smaller, reducing the amount of light that can enter the eye – this protects the eye from any damage caused by too much light entering. As light intensity decreases, the pupil gets larger, increasing the amount of light that can enter the eye – this ensures enough light enters the eye to be able to see clearly.
    Pupil constriction in bright light happens because the radial muscles in the iris relax and the circular muscles contract. Pupil dilation in dim light happens because the radial muscles in the iris contract and the circular muscles relax.
3. **Supplement** Rod cells are found more around the periphery of the retina, and respond to light intensity, so we use them most when light levels are low. Cone cells are found more in the centre of the retina and respond to different colours of light, so we use them to distinguish colour when light levels are high.

Page 263

1. Light entering the eye from a near object needs to be refracted more than light from a distant object in order to focus it on the retina. The ciliary muscles contract, which reduces the tension on the suspensory ligaments that are attached to the lens. This allows the lens to become thicker and more rounded, which increases its focusing power.
2. For light from a distant object to be focused on the retina, the lens needs to be thin. So the ciliary muscles relax and the suspensory ligaments pull harder on the lens, causing it to flatten.

Worksheet B14.2a Structure of the eye

Label the main structures of the eye on the diagram below.

Then add notes to each label to explain the role of each structure in vision.

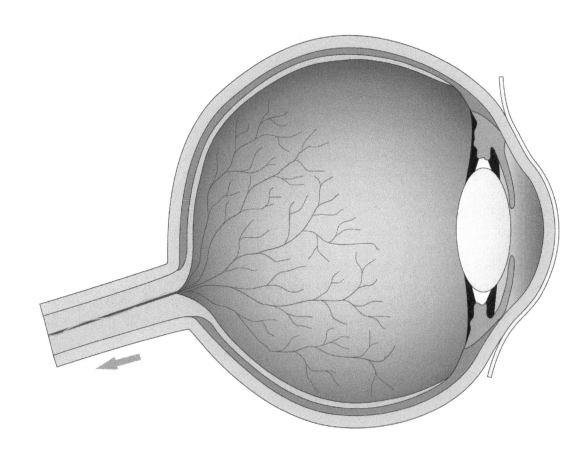

Supplement Worksheet B14.2b Accommodation

The diagrams below show how the lens of the eye changes shape depending on whether it focuses light from a near object or a far object onto the retina.

Using a ruler and sharp pencil, draw light rays to complete the diagrams.

Add notes to each diagram to explain how the lens shape is changed to focus light onto the retina.

Focusing light from a distant object

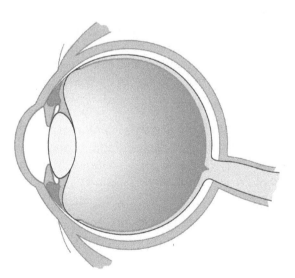

Focusing light from a near object

Learning episode B14.3 Human hormones

Learning objectives

- Describe a hormone as a chemical substance, produced by a gland and carried by the blood, which alters the activity of one or more specific target organs
- Identify in diagrams and images specific endocrine glands and state the hormones they secrete, limited to:
  (a) adrenal glands and adrenaline
  (b) pancreas and insulin
  (c) testes and testosterone
  (d) ovaries and oestrogen
- Describe adrenaline as the hormone secreted in 'fight or flight' situations and its effects, limited to:
  (a) increased breathing rate
  (b) increased heart rate
  (c) increased pupil diameter
- Compare nervous and hormonal control, limited to speed of action and duration of effect
- Supplement State that glucagon is secreted by the pancreas
- Supplement Describe the role of adrenaline in the control of metabolic activity, limited to:
  (a) increasing the blood glucose concentration
  (b) increasing heart rate

Common misconceptions

Many students do not grasp that all of the body is exposed to a particular hormone because it is transported in the blood, but that it only exerts its effects on the target cells in the effector.

Resources

Student Book pages 263–265

Approach

1. Introduction

Hormones have been briefly mentioned in earlier sections (see Section 9: *Transport in animals*). Ask students whether they can suggest any examples of hormones and what they do in the body. If they cannot, offer them some names of hormones with which they may be familiar (for example, adrenaline, oestrogen, testosterone, insulin) and ask what they have in common. Introduce the idea that hormones are chemicals that control how the body works.

2. A range of hormones

Present students with a set of cards naming endocrine glands and their secretions. Each card should contain the name of one endocrine gland or one hormone as follows:

adrenal glands	testes	ovaries	pancreas
testosterone	adrenaline	insulin	oestrogen

Ask students to pair up the gland with the hormone that it secretes. They should then identify where in the body the endocrine gland is found, using Fig. 14.10 on page 264 of the Student Book. (Make sure students are aware of where the kidneys are, and that the adrenal glands sit just above these.)

Supplement Also include cards naming glucagon and pancreas.

3. Adrenaline

Ask students to read pages 263–264 of the Student Book on adrenaline. They should make notes on when the hormone is produced, where it affects the body and what each effect is.

Supplement Ask students to use what they have learned to answer the question, 'Why is adrenaline important in survival in dangerous situations?'

4. Comparing systems

Students should use the adrenaline and other hormones to compare and contrast the hormonal system and nervous system in coordinating responses to stimuli.

The final list should include at least the first two of the following:

nervous system fast; hormonal system usually slower
nervous system responses do not usually last long; hormonal system responses generally last longer
nervous system uses electrical impulses; hormonal system uses chemical signals/hormones
(Supplement note, however, that students are also expected to know about the chemical transmission of nervous impulses across synapses)
nervous system usually responds with a small group of effectors (such as muscles), hormonal system may respond to changes with several different effectors, such as changes to heart rate and breathing, dilation of pupils, Supplement release of glucose by liver and muscle cells.

5. Consolidation

Give students a few minutes to jot down up to five key points on hormones for a short encyclopaedia article. Take examples from around the class and encourage discussion to select the best points.

Answers

Page 265

1. a) A chemical messenger in the body, produced by a gland and carried by the blood, that alters the activity of one or more target organs.
    b) A gland that secretes hormones.
    c) An organ that contains cells that are affected by hormones.
2. When faced with attack, or when suddenly frightened.
3. It prepares the body for action, for example, by increasing the amount of oxygen and glucose delivered to muscle cells for rapid respiration.
4. The nervous system produces fast, short-term responses as a result of electrical impulses that pass along neurones between the receptor, central nervous system and effector.
    The hormonal system produces longer-term responses that are slower, by the secretion of chemical hormones from endocrine glands into the blood to travel to effector organs, where they cause a change in activity.

Learning episode B14.4 Homeostasis

Learning objectives

- Describe homeostasis as the maintenance of a constant internal environment
- State that insulin decreases blood glucose concentration
- Supplement Explain the concept of homeostatic control by negative feedback with reference to a set point
- Supplement Describe the control of blood glucose concentration by the liver and the roles of insulin and glucagon
- Supplement Outline the treatment of Type 1 diabetes
- Supplement Identify in diagrams and images of the skin: hairs, hair erector muscles, sweat glands, receptors, sensory neurones, blood vessels and fatty tissue
- Supplement Describe the maintenance of a constant internal body temperature in mammals in terms of: insulation, sweating, shivering and the role of the brain
- Supplement Describe the maintenance of a constant internal body temperature in mammals in terms of vasodilation and vasoconstriction of arterioles supplying skin surface capillaries

Common misconceptions

When discussing homeostasis, it is important to talk in terms of *core body temperature* rather than just *body temperature*, because temperature varies greatly between different parts of the body. Too much variation in core body temperature (particularly of the brain, heart and liver) is a risk to health.

Resources

Student Book pages 263–269

B14.4 Technician's notes

Supplement Worksheet B14.4 Investigating a model for sweating

Resources for demonstration and class practical (see Technician's notes)

Approach

1. Introduction

Remind students of the role of the kidney in controlling body water content (and ion content), which they covered in Section 13: *Excretion in humans*. Explain that this is an example of homeostasis and ask them to use what they know to define the word in terms of body water content.

Discuss the definitions and agree a general definition that could apply generally to conditions inside the body. Remind students why homeostasis is needed by asking what would happen if there were no homeostatic control in the body.

Ask students to suggest other variables in the body that might need to be controlled. They should be able to suggest pH and temperature from their work on enzymes (Section 5), breathing (Section 11) and excretion earlier.

Supplement 2. Body temperature control demonstration

Choose a student who will not mind the attention. If available, attach a temperature sensor to a fingertip on one hand and then the other hand. If you do not have sensors, use a thermometer (e.g. a temperature strip thermometer). Measure the temperature of both fingertips – they should register similar values.

Show students a large bowl of cold water (around 10 °C) and tell them that the test subject will place one hand in the bowl. Ask them to predict how the skin temperature of the other hand will change over the following five minutes. It is likely that they will suggest no change since that hand is not in cold water, but do not give any response to predictions (see Technician's notes).

Once you have a prediction from the class, ask the test subject to place the hand without the sensor into the water and leave it for five minutes. Record the change in skin temperature of the other hand. It should go down.

Then ask students to read the section on temperature control in the Student Book (pages 265–268) and to apply what it says about temperature falling too far to what they have seen in the demonstration. They should be able to explain that cold blood returning to the brain from the cold hand triggers a response resulting in general vasoconstriction. This takes warm blood away from the surface of both hands, so the surface of both hands will be cooler.

SAFETY INFORMATION: Allow students to remove their hands from the cold water when they feel uncomfortable. Some students have reduced sensitivity, so do not allow the student to submerge their hands for longer than is sensible (you need to test the water temperature first).

Supplement 3. Investigating a model for sweating

Worksheet B14.4 provides a method that could be used to produce similar results to those in the Developing practical skills box on pages 267–268 of the Student Book. It uses wet and dry paper towels surrounding tubes of warm water to model the effect of sweating on heat loss.

Students should find that the tube wrapped in a wet paper towel loses heat more rapidly than the one in the dry paper towel. They should be able to explain this in terms of heat transfer – the water molecules in the paper towel gain energy from the warm tube and evaporate, removing the energy more rapidly than the transfer from the 'dry' tube by conduction and convection. This shows that sweating increases the rate of heat transfer from a hot body. The water should about 50 °C but not much hotter than this.

Check that students have drawn up a suitable table for their results before they carry out the tests.

SAFETY INFORMATION: Remind students to take care handling warm water.

Supplement 4. Homeostasis of body temperature

Ask students to draw two flow diagrams for the control of core body temperature, one starting with 'body too hot' on the left-hand side of a blank piece of paper and the other starting with 'body too cold' on the right-hand side. In their diagrams they should include not only the receptor, but also the effectors and how their response returns the core body temperature to a set point of around 37 °C.

Students should realise that their two diagrams can be combined, as on page 268 of the Student Book. Encourage them to look at temperature control as a continuous process that keeps core body temperature within tight limits. Explain that this is an example of negative feedback. Ask them to use their diagrams to help them define the meaning of the term. Make sure they are able to explain why keeping core body temperature within tight limits is so important to the body.

5. Control of blood glucose content

Explain that the control of blood glucose is another example of homeostasis. In this case the control partly comes from the secretion of insulin which is released when blood glucose content is too high.

Supplement Explain that control also comes from the secretion of glucagon when blood glucose content is too low. Give students a few minutes to consider how the hormones might act – if necessary, remind them to think what a *negative feedback* mechanism does. Then ask them to read page 269 of the Student Book and to draw a flowchart diagram for blood glucose control like the one they did for body temperature.

Supplement 6. Type 1 diabetes

Students should prepare a set of questions about Type 1 diabetes, its cause, its symptoms and its treatment. If a health professional is willing, ask them to visit the class to answer the questions. Alternatively, students should research the answers on the internet or in books. They should use what they find to prepare a brief health leaflet to help those who have just been diagnosed with the condition.

7. Consolidation: life in the desert

Tell students that they have been asked to identify the equipment needed for an expedition across a very hot, dry place such as the Sahara Desert. Remind them that as the expedition will constantly be on the move, equipment must be limited to only what is essential for survival.

Give them three minutes in small groups to write down an equipment list and then take examples from each group asking them to justify their choices. Make sure the 'essential' items are associated with the need to control body temperature and body water content.

Technician's notes

Be sure to check the latest safety notes on these resources before proceeding.

The following resources are needed for the temperature control demonstration:

temperature sensor or temperature strip thermometer
large bowl of cold water (about 10 °C)

The following resources are needed for the class practical B14.4, per group:

2 boiling tubes; warm water (about 50 °C)
2 paper towels
2 elastic bands
test-tube rack or 2 small beakers
large pipette (optional)
2 thermometers or temperature sensors
stopwatch or stop clock

Answers

Pages 267–268 Developing Practical Skills

1. The hot water in the tube is similar to the warm conditions inside the human body. The wet towel models sweating skin on the surface of the hot body.
2. Another tube set up identically, with hot water, etc. but with a dry paper towel surrounding it.
3. The graph should have time on the x-axis and temperature on the y-axis, with both sets of points drawn on the same axes and clearly labelled. Each set of points should be joined by suitable lines. There should be a comment about the outlier at 10 minutes of the dry towel.
4. The water in the tube with the wet towel cooled more rapidly than the water in the tube with the dry towel. This means heat energy was being transferred to the air more rapidly from the tube with the wet towel. Heat energy transferred from the water in the tube to the water in the towel would allow those water molecules to evaporate and move into the air, taking the heat energy with them.

Page 269

1. The maintenance of a constant internal environment, i.e. keeping the conditions inside the body within limits that allow cells to work efficiently.
2. Control of core body temperature (other answers possible).
3. **Supplement** Skin blood vessels (arterioles) dilate when the core body temperature is too high. This allows blood to reach the skin surface more easily and the heat energy it carries be transferred to the environment more rapidly.

   These arterioles constrict when the core body temperature falls too low. This reduces blood flow near the skin's surface, so heat energy cannot be transferred as easily to the skin surface and so cannot be transferred to the environment as quickly. This keeps more heat energy within the body.
4. **Supplement** When a change in a stimulus causes a control centre to trigger the opposite change in response, so keeping a condition within limits (returning it to its set point).
5. **Supplement**

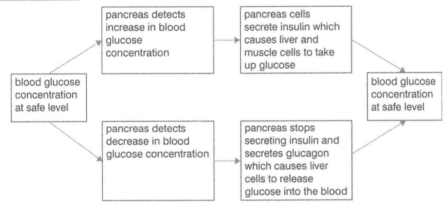

6. **Supplement** Regular injections of insulin into the fat beneath the skin.

Supplement Worksheet B14.4 Investigating a model for sweating

In this investigation you will use a model for sweating to investigate the effect of evaporation on heat loss from a surface.

Apparatus

2 boiling tubes	warm water (about 50 °C)
2 paper towels	2 elastic bands
test-tube rack or 2 small beakers	large pipette (optional)
2 thermometers or temperature sensors	stopwatch or clock

SAFETY INFORMATION
Take care handling warm water.

Method

1. Wrap one boiling tube in a dry paper towel, leaving the top open, and fix it in place with an elastic band.
2. Wet the other paper towel with water and then wrap it around the other tube and fix it in place with the other elastic band.
3. Without spilling any water (such as by using a large pipette), carefully fill each tube with the warm water and place it in the rack or in a beaker.
4. Place the thermometers or sensors into the water in each tube and leave for one minute.
5. Take the temperature of both tubes and set this as time 0. Record your measurements.
6. Take the temperature of both tubes every two minutes for the next 20 minutes and record your measurements.
7. Draw a suitable table to display your results.

Handling experimental observations and data

8. Use your results to draw a suitable graph, displaying both sets of results on the same graph.
9. Draw a curve through each set of points and identify any anomalous values. Try to explain any anomalous values.
10. Describe the curves on your graph and try to explain them using your knowledge of heat transfer and evaporation.
11. Explain the relevance of your results to the effect of sweating on body temperature regulation.

Planning and evaluating investigations

12. Explain why you did not take measurements from the time that the thermometers or sensors were placed into the water in the tubes.
13. Describe any problems that you had with this practical and suggest how you would change the method to produce more reliable results.

Learning episode B14.5 Tropic responses

Learning objectives

- Describe gravitropism as a response in which parts of a plant grow towards or away from gravity
- Describe phototropism as a response in which parts of a plant grow towards or away from the direction of the light source
- Investigate and describe gravitropism and phototropism in shoots and roots
- Supplement Explain phototropism and gravitropism of a shoot as examples of the chemical control of plant growth
- Supplement Explain the role of auxin in controlling shoot growth, limited to:
  (a) auxin is made in the shoot tip
  (b) auxin diffuses through the plant from the shoot tip
  (c) auxin is unequally distributed in response to light and gravity
  (d) auxin stimulates cell elongation

Common misconceptions

Many people say that 'plants do not move'. What they mean is that plants do not move from place to place as animals may do. There are many examples of plant movement (see Introduction: plant responses below) as well as the growth responses that cause parts of the plant to change position.

Note: the term *geotropism* is sometimes used to mean *gravitropism*.

Resources

Student Book pages 270–274

Worksheet B14.5a Investigating gravitropism

B14.5a Technician's notes

Worksheet B14.5b Investigating phototropism

B14.5b Technician's notes

Resources for demonstration and class practicals (see Technician's notes)

Approach

1. Introduction: plant responses

Ask students whether plants move. The answer should be 'yes' from their work on Section 1: *Characteristics and classification of living organisms*. Then give them two minutes to work in pairs to write down as many examples of plant movement as they can think of. Take examples from around the class.

If possible, play one or more video clips showing plant movement. There are many examples on the internet, such as: the closing of a Venus flytrap in response to touch by an insect; leaves tracking the Sun's position; flowers opening and closing in response to light intensity; and the growth of shoots such as *Rubus fructicosus* or climbing plants. These should quickly dispel any ideas that 'plants do not move'!

Introduce the idea that plants move in response to changes in the environment. Ask students to use the examples they have just seen, and those they listed, to identify the stimulus (the change in the environment) and response.

2. Tropisms

Help students to distinguish between plant movements that are the result of growth and those that are not. Explain that movements as a result of growth are called *tropisms*.

If possible, a few days before the lesson, place a potted plant in a dark area with the pot lying on its side, as shown in the diagram. Show students the plant with the pot base on the bench and ask them what may have happened to the plant for it to grow out to the side.

After discussion, explain what you did, and tell students that plant growth in response to gravity is called *gravitropism* (from *gravi*, 'relating to gravity'). You might also wish to introduce the idea of a *positive* response, towards the stimulus, and a *negative* response, away from the stimulus.

3. Gravitropism

Worksheet B14.5a gives students the opportunity to investigate the gravitropic response of the shoot and root of a seedling. Enough growth of the root to see the response will take several days; for the shoot, it will take one–two weeks. If there is not enough time, tell students that they will investigate the root response only.

Students should find that the shoot has a negative gravitropic response and the root has a positive gravitropic response. They should interpret these responses in terms of improved chances of survival.

SAFETY INFORMATION: Wash hands after handling seeds. Take care when using pins to stick seeds onto cotton wool or paper.

4. Phototropism

Worksheet B14.5b lets students carry out the investigation presented in the developing practical skills box on page 271 of the Student Book. Note that if you do not have enough boxes for all students as described in the instructions, you can simplify the set-up by putting one dish of seedlings in conditions of unidirectional light (for example, on a bright windowsill) and the other in the dark (for example, in a dark cupboard). The investigation will need about a week to produce sufficient growth for measuring results.

Students should find that shoots have a positive phototropic response, growing towards unidirectional light. Again, they should interpret these responses in terms of improved chances of survival.

SAFETY INFORMATION: Wash hands after handling seeds.

Supplement 5. Auxins

Ask students to read pages 271–273 of the Student Book on the plant chemical, auxin, and its control of phototropism and gravitropism. Ask them to summarise the information in a flow chart or sequence of drawings, to make sure they understand the effect of light or gravity on auxin and how this then leads to differential growth.

6. Consolidation: zig-zag plant

Ask students to imagine a plant with a zig-zag (Z-shaped) shoot, and to write bullet point notes describing how they could create such a plant without using wires to hold the shoot. They should explain why their method would produce the required result. They should compare their method with another student and work together to produce an improved method from their comparison.

Technician's notes

Be sure to check the latest safety notes on these resources before proceeding.

The following resources are needed for the demonstration of gravitropism:

a potted plant that has been put in a dark area, with the pot lying on its side for at least two days before the lesson

The following resources are needed for the class practical B14.5a, per group:

3 germinating seeds with root about 1 cm long (see note below)
Petri dish with lid
3 pins, long mounting pins are best
cotton wool or absorbent paper
scissors; water
marker pen
sticky tape
modelling clay
or other method for stabilising the Petri dish on its side

You may use any suitable seeds that germinate easily and are easy to handle, such as wheat, pea or bean. Soak the seeds for 24 hours in water and then leave to germinate in a warm, dark place for several days until the developing roots are about 1 cm long.

If signs of decay in the seeds are noted, the seeds and the cotton wool or paper should be disposed of as soon as possible and the container disinfected.

The following resources are needed for the class practical B14.5b, per group:

2 small boxes with lids
sharp knife or scissors
black paint and paintbrush or black paper and sticky tape
2 Petri dishes
absorbent paper
water
germinating seeds (see note above)
access to light, such as windowsill or light bank

You may use any suitable seeds that germinate easily and are easy to handle, such as wheat, pea or bean. Soak the seeds for 24 hours in water and then leave them to germinate in a warm dark place for several days until the shoots are just beginning to develop.

If the container is set in front of a light bank, it must be one that is designed for continuous illumination. Ordinary bench lamps (that could be knocked over) should not be left on for extended periods of time.

Answers

Page 271 Developing Practical Skills

1. The box should be set up near to a strong light source, so that the light entering the box is only coming from one direction. The seeds should be kept moist, but not wet, during germination and as the seedlings grow. Growth could be measured as angle between tip of shoot and base for quantitative data, or using photographic evidence for qualitative data.
2. The control needs to be set up in exactly the same conditions as the test apparatus, but with light reaching the shoots equally from all directions – for example, with the top of the box removed and an overhead light. (An alternative would be to have the shoots in a completely dark box.)
3. The seedlings in the windowed box should have grown so that their tips are heading towards the light, i.e. grown at an angle towards the light, whereas the control shoots should grow straight up.
4. Plants will need to be grown in water so that the roots can be exposed to light (for example, large bulbs such as hyacinth, grown suspended over water). Then two similar plants will need setting up so one has roots exposed to light from one direction only and the other to light all around the roots. Growth over a period of several weeks may be needed to show whether the roots with light from one direction grow in a particular direction in relation to the light, compared with the roots which receive light from all directions.

Page 272 Science in context: Gardener's tip

For example, if a plant needs to grow tall to reach more light because it is shaded by others. In this case, growing side shoots that would also be shaded will not help the plant and will be a waste of its resources.

Page 272 Science in context: Using auxins as weedkillers

The weedkiller is absorbed through the surface of leaves. Grasses have narrow leaves with a relatively small surface and so absorb little of the weedkiller.

Page 274

1. A growth response of a plant to a stimulus.
2. a) Shoots grow towards light.
   b) Roots grow in the direction of the force of gravity.
3. Supplement Auxin is produced in the tip of the growing shoot and diffuses down the shoot. Auxin on the bright/light side of the shoot moves across the shoot to the darker/shaded side as it diffuses down the shoot. Cells on the dark side of the shoot elongate more than the cells on the light side of the shoot, so the shoot starts to bend as it grows so that the tip is pointing towards the light.

Worksheet B14.5a Investigating gravitropism

In this investigation you will investigate the gravitropic response of developing roots, and if there is time, also of shoots.

Apparatus

3 germinating seeds with root about 1 cm long

Petri dish with lid

3 pins

cotton wool or absorbent paper

scissors

water

marker pen

sticky tape

modelling clay

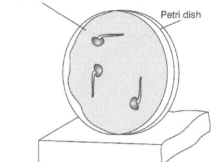

moist cotton wool or absorbent paper (top layer removed for this diagram)

Petri dish

SAFETY INFORMATION

Wash hands after handling seeds.

Take care when using pins to stick seeds onto cotton wool or paper.

Method

1. Use the pen to mark a point on the edge of the Petri dish. This is the reference point.

2. Cut two discs of cotton wool or several discs of absorbent paper to fit snugly within the dish.

3. Fill half of the base of the Petri dish with cotton wool or discs of absorbent paper. Pour on enough water to make the paper or cotton wool wet, leave for a minute then pour off any excess water.

4. Arrange three germinating seeds in the dish so that their roots are pointing in different directions. Carefully and gently attach each germinating seed to the cotton wool/paper using a pin.

5. Gently place another layer of cotton wool, or sufficient layers of absorbent paper, over the seeds to fill the depth of the dish. Replace the lid of the dish.

6. Stand the dish on its edge, with the marked point at the bottom, and use modelling clay to hold the dish in place so that it does not wobble.

7. After a few days, carefully remove the lid and the top layer of cotton wool or layers of paper to expose the seeds.

8. Draw the seedlings, making sure you record the relative position of the marked point that indicated the 'down' position.

9. If you are allowed to continue the practical for longer, carefully pour in a little more water, and then pour off the excess without disturbing the seeds. Replace the top layers and lid and stand the dish on its edge again, with the marked point down.

10. Repeat steps 8 and 9 every few days until the shoot has grown at least 1 cm.

Page 1 of 2

Handling experimental observations and data

11. Describe your results a) for the roots and b) for the shoots if you continued the practical.

12. Explain your results, using your knowledge of tropisms.

13. Explain the survival advantage to a plant of responding to gravity like this.

Planning and evaluating investigations

14. Covering the seeds above and below helped to exclude light. Explain why this is important in this investigation.

Worksheet B14.5b Investigating phototropism

In this investigation you will investigate the phototropic response of growing shoots.

Apparatus

two small boxes with lids

sharp knife or scissors

black paint and paintbrush, or black paper and sticky tape

2 Petri dishes

absorbent paper

water

germinating seeds

access to light

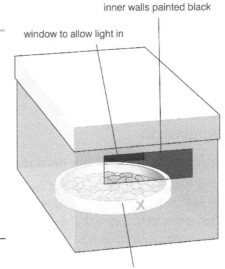

inner walls painted black

window to allow light in

Petri dish with damp
paper towel and seeds

SAFETY INFORMATION
Wash hands after handling seeds.

Method

1. If available, prepare one box by cutting a window out of one side as shown in the diagram. Either paint the inside of this box with black paint and leave to dry, or line it with black paper.
2. Line the base of each Petri dish with absorbent paper and pour in enough water to wet the paper. Leave the dish for about a minute, then pour off any excess water.
3. Place several seeds gently on the paper in each dish and then place one dish in each box and replace the lids.
4. Place the windowed box near a source of bright light, such as a windowsill or bank of lights.
5. Every couple of days, check that the absorbent paper in both dishes is still moist, but not wet. Add a little more water if needed. The paper should not be wet. If the paper is wet, the seeds may rot.
6. After a week, check the shoots of the seeds to see in which direction they are growing. If there has been insufficient growth for obvious results, continue the investigation for up to another week.
7. Draw the results from both dishes, remembering to indicate the direction of light for those in the windowed box.

Handling experimental observations and data

8. Describe your results a) for the dark box and b) for the windowed box.
9. Explain your results, using your knowledge of tropisms.
10. Explain the survival advantage to a plant of responding to light like this.

Planning and evaluating investigations

11. Explain how making the inside of the windowed box black may have improved the response of the seedlings in that box.

Learning episode B14.6 Consolidation and summary

Learning aims

- Review the learning points of the topic summarised in the end of topic checklist
- Test understanding of the topic content by answering the end of topic questions

Resources

Student Book pages 275–279

Approach

Introduce the learning episode

Ask students to work with a partner to make a list of key words from this topic. They could then work together to produce a spider diagram showing how the different concepts are linked. They could compare their list with the list of key terms given on page 275 of the Student Book. Discuss the checklist on pages 275–276 and use questioning to see how much of the content they are comfortable with.

Students could make flashcards of the key content and then use the flashcards to quiz each other on the information.

Develop the learning episode

Ask students to work individually through the end of topic questions on pages 278–279 of the Student Book without looking at the text. As they work, walk around the classroom observing their answers and asking questions as necessary to find out which questions are causing difficulties.

Finish the learning episode

After a set period, ask the students to stop working and discuss any areas of difficulty you observed as you walked round the class. Students should complete any unanswered questions for homework, but you should stress that they should try to answer the questions without looking at the text, so that they can see how much they have remembered.

Answers

End of topic questions mark scheme

The marks available for a question can indicate the level of detail you need to provide in your answer.

Question	Correct answer	Marks
1	D	1 mark
2	The *stimulus* is the heat detected by the heat sensors in the finger. This causes an electrical impulse to be triggered in the sensory *neurone*, which passes to the central nervous system (spinal cord). The nerve impulse passes to a relay *neurone* in the spinal cord, which passes the nerve impulse to a motor *neurone*. The nerve impulse passes along the motor neurone to muscles, which are *effectors*, in the arm. This causes the muscles to contract and produce the *response*, which is to draw the finger away from the heat. This pathway is an example of a *reflex arc*.	1 mark 1 mark 1 mark 1 mark 1 mark
3 a)	Answers may vary, but should include the following: eyes sense the direction and speed of the ball electrical impulses via nerve cells from the eye to the central nervous system including the brain brain identifies best place to stand and how to move the racket in order to hit the ball electrical impulses to muscles in the arm move the racket into position and then to hit the ball.	1 mark 1 mark 1 mark 1 mark
3 b)	No, this is not a reflex action, because thought is involved, using memories of training in order to give the best response.	1 mark 1 mark
4 a)	Outside, the person's pupils are small to prevent too much bright light damaging their eyes, but inside the house, they then widen to allow enough light in to see clearly.	1 mark
4 b)	Only some of the light entering the eye is able to get through the lens and reach the light-sensitive cells in the retina, and what light does get through may not be focussed properly, so the image produced will not be very clear.	1 mark 1 mark
4 c)	Some people's lenses cannot focus the light from something nearby clearly on the retina, and needs additional lenses in their spectacles to produce proper focusing for a clear image.	1 mark 1 mark
5 a)	Positive phototropism means growing towards light. If shoots grow towards the light, there is a better chance of the plant receiving more light on its leaves, and so it will be able to carry out more photosynthesis and produce more food.	1 mark 1 mark
5 b)	Positive gravitropism means growing in the direction of the force of gravity, i.e. down into the earth. Roots that grow downwards are more likely to find soil water (and dissolved mineral ions) and so be able to supply the substances that the plant needs for healthy and rapid growth.	1 mark 1 mark
Supplement 6 a)	Receptors: heat sensors in the skin and receptor cells in the hypothalamus that detect temperature of the blood flowing through them Monitoring area: hypothalamus in the brain	1 mark 1 mark

Question	Correct answer	Marks
	Effectors: sweat glands, blood vessel walls, muscles, including hair erector muscles	1 mark
Supplement 6 b)	Any one from: sweat glands produce sweat which evaporates from the skin surface; vasodilation	1 mark
Supplement 6 c)	Any two from: muscles start to shiver; body hair is raised; vasoconstriction	2 marks
Supplement 6 d)	Evaporation requires energy, so the evaporation of sweat takes heat energy from the body.	1 mark
	Increased blood flow near to the surface of the skin allows heat energy from the body to be transferred to the air faster than if the blood flow is reduced.	1 mark
Supplement 7	Negative feedback helps to keep a condition, such as blood glucose concentration, within narrow limits.	1 mark
	This means that even if blood glucose rises quickly after a meal, it will be brought back down, or if it falls during exercise, it will be brought back up.	1 mark
	This protects the body from damage that may be caused by extremes of blood glucose concentration.	1 mark
Supplement 8 a)	The curve for person A shows a small increase in blood glucose concentration for about half an hour after drinking the glucose, and then it drops again to its original level.	1 mark
	The curve for person B increases more rapidly than for person A, and for longer and only starts to decrease after about 90 minutes.	1 mark
Supplement 8 b)	Blood glucose concentration rises as glucose is absorbed into the blood from the alimentary canal.	1 mark
	Blood glucose concentration falls again as insulin is secreted and glucose is taken into muscle and liver cells from the blood.	1 mark
Supplement 8 c)	Person B has diabetes and does not secrete insulin,	1 mark
	so the concentration of glucose in the blood increases to very high levels.	1 mark
	Total:	38 marks

Introduction

This section covers drugs with a specific focus on antibiotics.

Links to other topics

Topics	Essential background knowledge	Useful links
1 Characteristics and classification of living organisms	1.1 Characteristics of living organisms 1.3 Features of organisms	
2 Organisation of the organism	2.1 Cell structure and size of specimens	
10 Diseases and immunity	10.1 Diseases and immunity	
16 Reproduction		16.1 Asexual reproduction 16.6 Sexually transmitted infections
18 Variation and selection		18.2 Selection

Topic overview

B15.1	Drugs and medicinal drugs
	This learning episode introduces the term *drug*, and then discusses antibiotics as examples of medicinal drugs.
	Supplement Students also look at how the development of antibiotic-resistant bacteria can be minimised.
	It may be helpful to arrange a visit from a health professional to discuss the correct use of antibiotics.
B15.2	Consolidation and summary
	This learning episode provides an opportunity for a quick recap on the ideas encountered in the section, as well as time for the students to answer the end of topic questions in the Student Book.

Careers links

These are some scientific careers that focus on this area of biology but careers in many other fields use the knowledge and skills gained studying science. **Pharmacologists** study the effect of drugs on animal and humans. **Pharmacists** dispense drugs and provide people with treatment advice for some medical conditions.

Starting points

The Student Book section opener puts the ideas in the section into context and sets the scene. It also allows students to acknowledge and value their prior learning, and provides a benchmark against which future learning can be compared.

- The questions provide a structure for introducing the section and can be used in a number of different ways:
- You could ask students to consider the questions as an introductory homework task.
- You could put students into groups to share their own ideas and understanding and then to report back to the whole class.
- Students could be given access to the Internet, preferably with a tight timescale, to find out the information required.

You could then use a spider chart or other form of wall chart to summarise everybody's ideas.

Recording these initial ideas allows you to retain them for reference as the individual topics are developed. In this way, your students' progress in learning can be readily acknowledged.

Learning episode B15.1 Drugs and medicinal drugs

Learning objectives

- Describe a drug as any substance taken into the body that modifies or affects chemical reactions in the body
- Describe the use of antibiotics for the treatment of bacterial infections
- State that some bacteria are resistant to antibiotics which reduces the effectiveness of antibiotics
- State that antibiotics kill bacteria but do not affect viruses
- Supplement Explain how using antibiotics only when essential can limit the development of resistant bacteria such as MRSA

Common misconceptions

The common understanding of the term *drug* is often restricted to those drugs that are illegal or harmful. It is important that students understand the full meaning of the term, to cover legal drugs such as nicotine and caffeine as well as medicines.

Resources

Student Book pages 282–285

Approach

1. Introduction

Introduce the term *drug* to students and give them a few minutes to jot down any ideas related to the term. Take a few examples from the class, and use them to help draw together a definition of the word. Compare this definition with the one given in the Student Book on page 282, and ask students to comment on any similarities and differences. Students may hold a limited view of what a drug is, so use this opportunity to broaden it to the definition given in the Student Book.

Note: If you consider any comment from a student of concern in relation to safeguarding, do not respond directly but take your concern to a member of staff who has responsibility for matters of child protection.

2. Antibiotics

Use the example of antibiotics to introduce medicinal drugs – those used by doctors to treat the symptoms and causes of illness. Many students will have been given antibiotics at some time in their lives. Take a few minutes to collect examples of illnesses that have been treated. If necessary add other examples, such as throat infection, ear infection, infected skin wounds, kidney infection. Although students do not need to know the names of particular antibiotics, it may help some students if you mention some common types such as penicillin, ampicillin or erythromycin.

Ask students what these illnesses have in common. It is possible that they will not know, but lead them to the answer that these are all caused by bacteria. Remind them that antibiotics cannot be used to kill viruses and therefore are of no use against illnesses such as flu or HIV.

Supplement Remind students of their work in Section 1 on the classification of organisms. Ask them to explain the differences between bacteria and viruses, and to use this to suggest possible reasons why antibiotics are not effective against viruses.

3. Problems with antibiotics

Ask students to make a list of questions about the use and misuse of antibiotics. These should include the problem of developing resistance, the need to complete a prescribed course of antibiotics, and their ineffectiveness against viruses.

Supplement Students should also look at the importance of only using antibiotics when absolutely essential – they may find it useful to read the information on pages 283–284 of the Student Book. (Note – the role of natural selection in the development of antibiotic-resistant bacterial strains will be looked at in Section 18: *Variation and selection*.)

If possible, arrange a visit from a health professional to answer the questions. Alternatively, students could carry out research in books or on the internet to find the answers.

Students could use their findings to produce a simple health leaflet about what antibiotics are and how they should be used.

4. Consolidation

Ask students to work in pairs to produce a brief answer for a web forum to the question, 'What is an antibiotic?' or 'Why should I follow what my doctor says about taking antibiotics?' Give them a few minutes to complete this and then ask them to exchange their answer with that of another pair and to identify one good point and one weakness in that answer. They should then return the answer, and consider how to improve the weakness identified in their own answer.

Answers

Page 285

1. a) A substance that, when taken into the body, affects chemical reactions in the body.
    b) A chemical that kills bacteria or prevents them from growing.
    c) When one type of bacteria is no longer killed or affected by an antibiotic.
2. If the infections are caused by bacteria, they can be treated with antibiotics, but if they are caused by viruses, they cannot be treated with antibiotics.
3. Supplement Using antibiotics only when essential so that bacteria are not exposed to antibiotics more often than necessary, as that is what increases the likelihood of the development of resistance.

Learning episode B15.2 Consolidation and summary

Learning aims

- Review the learning points of the topic summarised in the end of topic checklist
- Test understanding of the topic content by answering the end of topic questions

Resources

Student Book pages 287–288

Approach

Introduce the learning episode

Ask students to work with a partner to make a list of key words from this topic. They could then work together to produce a spider diagram showing how the different concepts are linked. They could compare their list with the list of key terms given on page 287 of the Student Book. Discuss the checklist on page 287 and use questioning to see how much of the content they are comfortable with.

Students could make flashcards of the key content and then use the flashcards to quiz each other on the information.

Develop the learning episode

Ask students to work individually through the end of topic questions on page 288 of the Student Book without looking at the text. As they work, walk around the classroom observing their answers and asking questions as necessary to find out which questions are causing difficulties.

Finish the learning episode

After a set period, ask the students to stop working and discuss any areas of difficulty you observed as you walked round the class. Students should complete any unanswered questions for homework, but you should stress that they should try to answer the questions without looking at the text, so that they can see how much they have remembered.

Answers

End of topic questions mark scheme

The marks available for a question can indicate the level of detail you need to provide in your answer.

Question	Correct answer	Marks
1	A	1 mark
2	Influenza is caused by a virus so cannot be treated by antibiotics.	1 mark
	Normally the body can overcome the virus after a few days of rest.	1 mark
	Strep throat is caused by a bacterium so antibiotics can be used to treat it.	1 mark
	Usually only doctors can prescribe antibiotics.	1 mark
3 a)	The bacteria cannot be killed by some/many antibiotics.	1 mark
3 b) Supplement	Some diseases are not caused by bacteria so antibiotics will not treat them.	1 mark
	Using antibiotics less frequently, only when essential, will limit the development of antibiotic-resistant strains of bacteria.	1 mark
	Total:	8 marks

Introduction

This section looks at the differences between asexual and sexual reproduction, and then covers reproduction in flowering plants and human reproduction.

Links to other topics

Topics	Essential background knowledge	Useful links
1 Characteristics and classification of living organisms	1.1 Characteristics of living organisms 1.3 Features of organisms	
2 Organisation of the organism	2.1 Cell structure and size of specimens	
3 Movement into and out of cells	3.1 Diffusion 3.2 Osmosis	
7 Human nutrition	7.1 Diet	
9 Transport in animals	9.3 Blood vessels 9.4 Blood	
10 Diseases and immunity	10.1 Diseases and immunity	
11 Gas exchange in humans	11.1 Gas exchange in humans	
12 Respiration	12.2 Aerobic respiration	
14 Coordination and response	14.3 Hormones in humans	
17 Inheritance		17.1 Chromosomes, genes and proteins 17.2 Mitosis 17.3 Meiosis 17.4 Monohybrid inheritance
18 Variation and selection		18.3 Selection

Topic overview

B16.1	**Sexual and asexual reproduction**
	This short learning episode looks at the differences between sexual and asexual reproduction, and sets the context for the rest of the work in this section. It provides practical work on taking cuttings of plants. (Note: The differences between asexual and sexual reproduction are covered further in Section 17: *Inheritance*, in the discussion of mitosis and meiosis.)
	Supplement Students will also discuss the advantages and disadvantages of the two kinds of reproduction in specific situations.
B16.2	**Sexual reproduction in plants**
	In this learning episode, students will learn how the structure of a flower is adapted to the method of pollination, by wind or by insects. This includes practical work looking at the structures in a flower. They will also study what happens during the fertilisation of a flower, and investigate the conditions that affect germination.
	Supplement Students will compare self-pollination and cross-pollination, and describe the details of fertilisation.

B16.3	**Sexual reproduction in humans**
	This learning episode introduces the structures and function of the organs of the human male and female reproductive systems Students will also compare the adaptive features of sperm and egg cells. Students will learn about the development of the embryo and fetus.
	Supplement Students will learn in more detail about the functions of the placenta.
B16.4	**Sexual hormones in humans**
	This learning episode describes the role of hormones in the development and regulation of secondary sexual characteristics, and the changes that occur during the menstrual cycle.
	Supplement Students learn about the role of the hormones controlling the menstrual cycle and pregnancy.
B16.5	**Sexually transmitted infections**
	This learning episode covers some sexually transmitted infections, particularly HIV and how it may lead to AIDS.
B16.6	**Consolidation and summary**
	This learning episode provides an opportunity for a quick recap on the ideas encountered in the section, as well as time for the students to answer the end of topic questions in the Student Book.

Careers links

These are some scientific careers that focus on this area of biology but careers in many other fields use the knowledge and skills gained studying science. **Horticultural technicians** grow, maintain and monitor plants for companies and research organisations. **Sonographers** use ultrasound to image and monitor fetus growth.

Starting points

The Student Book section opener puts the ideas in the section into context and sets the scene. It also allows students to acknowledge and value their prior learning, and provides a benchmark against which future learning can be compared.

- The questions provide a structure for introducing the section and can be used in a number of different ways:
- You could ask students to consider the questions as an introductory homework task.
- You could put students into groups to share their own ideas and understanding and then to report back to the whole class.
- Students could be given access to the Internet, preferably with a tight timescale, to find out the information required.

You could then use a spider chart or other form of wall chart to summarise everybody's ideas.

Recording these initial ideas allows you to retain them for reference as the individual topics are developed. In this way, your students' progress in learning can be readily acknowledged.

Learning episode B16.1 Sexual and asexual reproduction

Learning objectives

- Describe asexual reproduction as a process resulting in the production of genetically identical offspring from one parent
- Identify examples of asexual reproduction in diagrams, images and information provided
- Describe sexual reproduction as a process involving the fusion of the nuclei of two gametes to form a zygote and the production of offspring that are genetically different from each other
- Describe fertilisation as the fusion of the nuclei of gametes
- Supplement Discuss the advantages and disadvantages of asexual reproduction:
  (a) to a population of a species in the wild
  (b) to crop production
- Supplement State that nuclei of gametes are haploid and that the nucleus of a zygote is diploid
- Supplement Discuss the advantages and disadvantages of sexual reproduction:
  (a) to a population of a species in the wild
  (b) to crop production

Common misconceptions

Some students confuse the terms *reproduction* and *fertilisation*, which makes it difficult for them to understand asexual reproduction. It is important to define each term correctly at the start.

Resources

Student Book pages 292–296

Worksheet B16.1 Taking plant cuttings

B16.1 Technician's notes

Resources for demonstration and class practical (see Technician's notes)

Approach

1. Introduction

Write the word *reproduction* in the middle of the board and ask students to suggest related words. Ask how the words should be placed to build a concept map. If students are uncertain or confused, encourage discussion. Consider marking words that are not well understood with a question mark, so that you can use the map again at the end of the learning episode to identify what students have learned.

Supplement Students should include the terms *haploid* and *diploid* in relation to gametes and zygotes.

2. Differences

Ask students to use pages 292–295 of the Student Book to make notes about the two kinds of reproduction. If available, show students a potato plant with tubers, and explain that potato plants can reproduce asexually using tubers but that they can also reproduce sexually using flowers. Students could then use their notes to produce questions to test each other.

Supplement Students should also compare the advantages and disadvantages of the two forms of reproduction in relation to a population of a wild species, and to crop production, such as by drawing up a table to record their notes.

If there is time, and access to books or the internet, students could research other examples of species that use both sexual and asexual reproduction, and try to explain the advantages to them of doing so.

3. Taking plant cuttings

Worksheet B16.1 provides an opportunity for students to take cuttings of a plant. Although this is beyond the requirements of the syllabus, it will show students clear evidence of the ability of plants to reproduce asexually. Cuttings will need at least three weeks to develop roots, and longer if characteristics are to be measured for comparison. However, these measurements would be useful later for the discussion of the effects of genes and environment on variation in a species for Section 18: *Variation and selection*.

You could also use potato tubers and grow these over a whole season.

SAFETY INFORMATION: Remind students to take care when using sharp knives.

4. Consolidation: concept map

Return to the concept map started in the introduction. Ask students to resolve any question marks on the map. They should also suggest any additional words and how to add them to the concept map.

Technician's notes

Be sure to check the latest safety notes on these resources before proceeding.

The following resources are needed for the demonstration on differences:

tubers, such as potatoes, ideally on a plant that also has flowers

The following resources are needed for the class practical B16.1, per group:

healthy mature plant
sharp knife
cutting board
rooting hormone
small pot with drainage holes, e.g. yoghurt pot with 3 or 4 holes in the base
compost: use well-draining compost suitable for cuttings
pencil
water
large transparent plastic bag and tie

Many plants are suitable for this investigation, but choose something that grows quickly, such as busy Lizzie (*Impatiens*) or pot geranium (*Pelargonium*). You may prefer to give students shoots that have already been cut from the plant for step 3 on the worksheet. If so, these must be cut as late as possible before the lesson and kept with their cut ends in water until the lesson.

You could also use bulbs or potato tubers.

Answers

Page 295

1. Reproduction without the fusion of gametes, resulting in the production of genetically identical offspring from one parent.
2. Binary fission is where the genetic material is copied and the cell splits in half. Only one parent cell is involved and there is no fusion of gametes before division.
3. Supplement Advantages: no need for fertilisation, so reproduction faster and easier; if conditions remain stable, all new individuals will grow as well as the parent plant.
   Disadvantages: no genetic variation between plants, so if conditions change/the parent plant is susceptible to a particular disease, then all plants will do badly; this increases the risk that the plants in that area will all die.

Page 296

1. a) The fusion of the nuclei of a male gamete and a female gamete to produce a zygote.
   b) The production of genetically different offspring from two parents as a result of fertilisation.
2. Supplement a) a sex cell or gamete such as a sperm, male gamete in a pollen grain, or an egg cell
   b) a zygote, or fertilised egg cell
3. Supplement Advantage: produces individuals with new variations of genetic material that increase the chance of survival when conditions change.
   Disadvantage: variation in offspring may also result in many offspring being less well adapted to environmental conditions than parent plants and so producing a lower yield.

Worksheet B16.1 Taking plant cuttings

Plant growers use the natural ability of plants to produce new plants from parts of an old one by taking cuttings. Cuttings can be taken from stems, leaves or roots.

The new plants are produced as a result of asexual reproduction, as they grow from body cells of the parent plant. So their cells contain the same genetic information as the parent.

Apparatus

healthy mature plant

cutting board

small pot with drainage holes

pencil

large transparent plastic bag and tie

sharp knife

rooting hormone

compost

water

SAFETY INFORMATION
Take care when using sharp knives.
Wash your hands after completing the activity.

Method

1. Fill the pot with compost and gently firm the compost so that it does not completely fill the pot.
2. Water the compost until there is water coming through the drainage holes at the bottom of the pot. Then leave the pot to drain, so that the compost is moist but not waterlogged.
3. Cut a shoot from the plant that has at least four or five leaves. Remove the bottom few leaves to leave at least two leaves. Then cut across the shoot with a sharp knife at an angle to the shoot.
4. Holding the shoot gently but firmly, dip the cut end into the rooting hormone.
5. Use a pencil to make a hole in the compost and place the shoot into the hole so that the leaves are clear of the compost. Firm the compost around the shoot to hold it in place.
6. Place the pot in the large plastic bag and use the tie to close the bag.
7. Leave the cuttings in a bright place, such as a windowsill, for several weeks until the roots have started to develop. Then remove the plastic bag and water the compost to keep it moist but not waterlogged.

Handling experimental observations and data

8. Compare the characteristics of the new plant grown from the cutting with those of the parent plant. Explain any similarities and any differences.

Learning episode B16.2 Sexual reproduction in plants

Learning objectives

- Identify in diagrams and images and draw the following parts of an insect-pollinated flower: sepals, petals, stamens, filaments, anthers, carpels, style, stigma, ovary and ovules
- State the functions of the structures listed above
- Identify in diagrams and images and describe the anthers and stigmas of a wind-pollinated flower
- Distinguish between the pollen grains of insect-pollinated and wind-pollinated flowers
- Describe pollination as the transfer of pollen grains from an anther to a stigma
- State that fertilisation occurs when a pollen nucleus fuses with a nucleus in an ovule
- Describe the structural adaptations of insect-pollinated and wind-pollinated flowers
- Investigate and describe the environmental conditions that affect germination of seeds, limited to the requirement for: water, oxygen and a suitable temperature
- Supplement Describe self-pollination as the transfer of pollen grains from the anther of a flower to the stigma of the same flower or a different flower on the same plant
- Supplement Describe cross-pollination as the transfer of pollen grains from the anther of a flower to the stigma of a flower on a different plant of the same species
- Supplement Discuss the potential effects of self-pollination and cross-pollination on a population, in terms of variation, capacity to respond to changes in the environment and reliance on pollinators
- Supplement Describe the growth of the pollen tube and its entry into the ovule followed by fertilisation (details of production of endosperm and development are **not** required)

Common misconceptions

Many people confuse *pollination* and *fertilisation*. It is important for students to be able to distinguish properly between them. There can also be confusion between the dispersal of pollen and seed dispersal.

Resources

Student Book pages 297–307

Worksheet B16.2a Flower dissection

B16.2a Technician's notes

Worksheet B16.2b Investigating conditions for germination

B16.2b Technician's notes

Resources for demonstration and class practicals (see Technician's notes)

Approach

1. Introduction

Give students one minute to write down three important facts about flowers. They should discuss their facts with another student, and choose together their three most important facts. Take examples from the class to cover the widest possible range of facts. Make sure at least one fact is related to pollination.

Use the opportunity to ask students to distinguish between pollination and fertilisation in plants. If they are not sure, leave this for now but make sure you return to it at the end of the learning episode.

2. The structure of flowers

If possible, provide flowers from wind-pollinated and insect-pollinated plants for students to dissect. Worksheet B16.2a supports this task. Students will also need the diagram of flower structure (Fig.16.10) from page 297 of the Student Book. Consider providing a photocopy of this diagram to avoid damage to the books.

If there is not enough time for students to dissect a flower of each type, then give one half of the class wind-pollinated flowers and the other half insect-pollinated flowers. Allow time for students to present their results to the other half at the end of the practical work.

If practical work is not possible, ask students to research examples on the internet or from books, and to draw one example of each, clearly labelling each diagram to show how it is adapted for its particular mode of pollination.

Students could also research the shapes of pollen grains from different plants, to identify whether those from wind-pollinated plants differ from those of insect-pollinated plants.

SAFETY INFORMATION: Remind students to take care when using sharp knives. Wash hands thoroughly after the practical. Pollen from some plants can cause an allergic reaction in some students.

Supplement 3. The growth of pollen tubes

Look for a sequence of timed photographs or a video clip of pollen tube growth – ideally one that shows time passing.
Students can use the photographs or video to investigate how quickly the tubes grow and how this might relate to the time that elapses between pollinations and fertilisation.

The Practical Biology website (www.nuffieldfoundation.org/practical-biology) provides a sequence of timed photographs to show how they develop.

Supplement 4.Self-pollination and cross-pollination

Explain to students that some plants have flowers that contain both male and female reproductive structures, some have male-only and female-only flowers on the same plant, and others have male flowers and female flowers on separate plants.

They should consider each of these cases in terms of which may result in self-pollination (gametes from the same plant) and which may result in cross-pollination. They should bear in mind that the male and female reproductive structures do not necessarily ripen at the same time.

Ask them to suggest the advantages and disadvantages of self-pollination and cross-pollination, which they should be able to do from a similar comparison of sexual and asexual reproduction. From that they should be able to discuss the implications to a species of each type of pollination.

5. Investigating conditions for germination

Worksheet B16.2b provides a method for investigating the conditions needed for germination, and supports the developing practical skills box on page 306 of the Student Book. If students have suggested other possible conditions, add these to the investigation.

Note that class results will be analysed, taking averages of repeat tests from groups investigating the same set of conditions. So the more conditions investigated at the same time, the fewer groups and fewer repeats of a set of conditions will be available for analysis. Seeds do not always germinate successfully, so at least one repeat for each condition (light intensity, moisture level and temperature) is advisable.

It is important to control other variables as far as possible during the investigation. This will mean trying to provide a light place and a dark place at similar temperature, and a warm place and a cool place with similar light intensity. If any of these are not possible, remove that condition from the investigation.

Students should find that seeds need moisture and warmth to germinate successfully. Seeds in the dark should germinate as successfully as seeds in the light in similar moisture and temperature, particularly if they are large seeds. (Note: some very small seeds, such as poppy, need light for germination as they will only germinate when on or near the surface of soil.)

SAFETY INFORMATION: Remind students to wash hands thoroughly after handling seeds. Only keep germinating seeds for one week and dispose after this.

6. Consolidation: drawings

Ask students to sketch a flower with typical features of one type of pollination. They should exchange their sketch with a neighbour. Ask them to identify which type of pollination is involved and then annotate the features that are adaptations for this type of pollination. They should then discuss the sketch with the student who drew it to identify any other features to annotate.

Technician's notes

Be sure to check the latest safety notes on these resources before proceeding.

The following resources are needed for the class practical B16.2a, per group:

flowers from plants of different species
sharp knife or scalpel (only use scalpels when necessary) and cutting board
hand lens or magnifying glass
pencil and paper
optional: photocopy of Fig. 16.6 from page 293 of the Student Book

Offer a selection of flowers, some of which are wind-pollinated and some of which are insect-pollinated, but avoid complex 'flowers', such as daises and dandelions, which consist of a large number of true flowers within one flower head. Wind-pollinated flowers may be difficult to obtain, as they are very seasonal. It is easy to confuse the fruit of wind-pollinated plants with the flower.

Only use scalpels where necessary.

The following resources are needed for the demonstration on the growth of pollen tubes:

video clip of pollen tubes growing

The following resources are needed for the class practical B16.2b, per group:

20 seeds
2 Petri dishes and 2 paper towels
light sensor or temperature sensor or thermometer
access to light
access to dark
access to bright, cold place
marker pen

Use seeds that quickly germinate such as *Arabidopsis* or cress.

Note that some seeds sold for growing in gardens or at home are treated with fungicide that may cause an allergic response.

Answers

Page 298

1. The carpel made up of: the stigma, where pollen grains attach; style, which supports the stigma; ovary, which surrounds and protects the ovule, inside which is the female gamete.
2. The stamen made up of: an anther that contains pollen grains, inside which are the male gametes; filament, which holds the anther above the flower to help with shedding of pollen.

Page 301 Science in context: Flowers and pollinators

Successful pollination can only happen between flowers of the same plant species. So it benefits the plant, if after collecting pollen grains, the animal only visits other flowers of the same plant species to pollinate them. If the animals took the pollen to flowers of a different plant species, fertilisation could not take place.

Page 302 Science in context: The problem with bees

Bees rely on flowers for food, so they will have an easily accessible and plentiful source.

Page 303

1. Pollination is the transfer of pollen from a stamen to a stigma. Fertilisation is the fusion of the male gamete with the female gamete to form a zygote.
2. Any three differences from: wind-pollinated flowers are usually small, no colour (white), make large amounts of lightweight pollen; insect-pollinated flowers are usually large, may be brightly coloured, produce nectar and sometimes scent, make (relatively) small amounts of larger pollen grains.
3. Supplement a) Can make less pollen OR less waste of pollen as insects more likely to deliver pollen to flower than random distribution in wind.
   b) If the insect species die out, the plant will not get pollinated.

Page 305 Science in context: Conditions for germination

If the weather turns cold later, this may harm or kill the plants. If they have insect-pollinated flowers, the insects may not be active yet and the flowers may not get pollinated.

Page 306 Developing Practical Skills

1. Plan should include apparatus shown in diagram (Petri dishes, paper towel, water, seed). For each condition, there should be at least one set of control apparatus set up identically to the test apparatus, e.g. (a) one set in light, one in dark, (b) one set with water, one set without, (c) one set at warm temperature, one set at cool temperature. Ranges of conditions would also be appropriate, e.g. a range of light intensities, water regimes, or temperatures. Repeats of each set would be useful for averaging natural variation between seed responses.
2. A graph showing number of seeds germinated on y-axis against day on x-axis, with each set of points carefully marked, joined by a suitable line and clearly labelled.
3. Most germination of seeds at 10 °C happened between the 4th and 6th day. At 20 °C, most seeds germinated between the 3rd and 5th day.
4. The results show that these seeds germinate faster at 20 °C than at 10 °C.
5. At a warmer temperature, enzymes in the seeds work faster, making more new substances for building new cells. The growing plant inside the seed will cause the seed coat to split. So the warmer seeds split open faster than the cooler seeds.

Page 307

1. When the embryo in a seed starts to grow, splitting the seed coat and increasing in size and complexity.
2. a) Seeds need a supply of oxygen for aerobic respiration to release energy needed for growth, although they may be able to start germination using anaerobic respiration.
   b) Seeds need water for germination and will not germinate in dry soil – water is needed, for example, to transport substances needed by the growing embryo plant.
   c) Seeds need warmth for germination because this allows the chemical reactions involved in growth to happen at a fast enough rate, although the amount of warmth they need may depend on where they naturally grow. Seeds from plants that live in colder areas may need a period of deep cold before they will germinate. Seeds from plants that live in areas prone to fire may not germinate until after a fire.

Worksheet B16.2a Flower dissection

Flowers are adapted in different ways to increase the chance of pollination. Some depend on wind and others depend on an animal, such as an insect, to bring pollen to the stigma. In this practical you will dissect flowers to investigate how they are adapted to their method of pollination.

Apparatus

flowers from plants of different species

cutting board

pencil and paper

sharp knife

hand lens or magnifying glass

SAFETY INFORMATION
Take care when using sharp knives.
Wash your hands thoroughly after the practical.

Method

For each flower that you study, use the following method.

1. Use the hand lens or magnifying glass to look closely at the flower. Using the diagram from page 297 of the Student Book, try to identify the sepals, petals, male parts and female parts of the flower. Note that in some species the flowers only contain one sex.
2. Make annotated outline drawings of your flower to show:

    *   sepals – number and colour
    *   petals – number, colour and relative size (large, medium or small)

    Also note whether the flower has scent and record this as none, a little or strongly scented.

3. Use the knife to remove the sepals and petals, so that you can see the reproductive parts more clearly. Make annotated drawings of the following:

    *   stamens – number, relative size and any other distinctive features
    *   stigmas – number, relative size and any other distinctive features

4. Remove the stamens and look carefully at the female parts of the flower. Draw and label one carpel. If possible, note the position of the nectaries.

Handling experimental observations and data

5. For each flower, identify how it is pollinated and describe how each type of structure in the flower is adapted to improve the chances of it being pollinated. You should include the sepals, petals, anthers and stigmas in your descriptions.
6. Use your results to compare the adaptations of insect-pollinated and wind-pollinated flowers.

Worksheet B16.2b Investigating conditions for germination

In this practical you will investigate the conditions that seeds need for germination. You will test one set of conditions and your results will contribute to a class analysis. Your teacher will tell you whether you are testing light intensity, temperature or moisture level.

Apparatus

20 seeds

2 Petri dishes

2 paper towels

marker pen

light sensor, or temperature sensor or thermometer

access to light, access to dark, access to bright, cold place

SAFETY INFORMATION
Wash hands thoroughly after handling seeds.
Only keep germinating seeds for one week and dispose after this.

Method

a) Use the pen to mark the two dishes on the edge with your initials.
b) Fold a paper towel to fit it neatly in the bottom of a Petri dish. Repeat with the other towel and dish.
c) a) Light intensity or temperature: pour sufficient water into each dish to wet the paper towel, and pour off any excess water.
   b) Moisture level: wet the paper towel in only one dish.
d) Scatter ten seeds evenly across each dish.
e) a) Temperature: place one dish in a bright, warm place and one in a bright, cold place. Measure the light intensity to make sure the dishes are receiving a similar amount of light.
   b) Moisture level: place both dishes in the same light place.
   c) Light intensity, place one dish in a bright, warm place and the other in a dark, warm place. Measure the temperature to make sure the dishes are receiving a similar amount of heat.
f) Leave the dishes for up to a week in these positions. Every day, check and record how many seeds have germinated. Also check that the moist paper towels remain moist, adding a little water if needed.

Handling experimental observations and data

g) Gather all the data from the class into a table or spreadsheet, showing the results for each group separately.
h) Compare the repeat tests for each condition and identify any anomalies. Ignoring any anomalies, calculate an average number of germinated seedlings for each day in repeat dishes.
i) Use your results to draw separate graphs for temperature, moisture and light with days along the x-axis and number of seedlings along the y-axis. Each graph will have two lines.
j) Describe the results shown in each graph, and use your scientific knowledge to explain them.

Planning and evaluating investigations

k) Describe any problems with this investigation, and explain how the method could be improved to reduce their effect.

Learning episode B16.3 Sexual reproduction in humans

Learning objectives

- Identify on diagrams and state the functions of the following parts of the male reproductive system: testes, scrotum, sperm ducts, prostate gland, urethra and penis
- Identify on diagrams and state the functions of the following parts of the female reproductive system: ovaries, oviducts, uterus, cervix and vagina
- Describe fertilisation as the fusion of the nuclei from a male gamete (sperm) and a female gamete (egg cell)
- Explain the adaptive features of sperm, limited to: flagellum, mitochondria and enzymes in the acrosome
- Explain the adaptive features of egg cells, limited to: energy stores and the jelly coat that changes at fertilisation
- Compare male and female gametes in terms of: size, structure, motility and numbers
- State that in early development, the zygote forms an embryo which is a ball of cells that implants into the lining of the uterus
- Identify on diagrams and state the functions of the following in the development of the fetus: umbilical cord, placenta, amniotic sac and amniotic fluid
- Supplement Describe the function of the placenta and umbilical cord in relation to the exchange of dissolved nutrients, gases and excretory products between the blood of the mother and the blood of the fetus
- Supplement State that some pathogens and toxins can pass across the placenta and affect the fetus

Common misconceptions

Some parts of the male reproductive system and the urinary system are shared – some diagrams can be confusing because of this. Also, some diagrams of the female reproductive system may show the bladder misplaced to one side – students must realise that this has been done to make the diagram more intelligible and does not reflect the bladder's true position.

Some students think that a mother and her developing baby share their blood supply – they do not. Although the maternal blood and fetal blood are in close proximity in the placenta, they cannot exchange all materials freely – only what can diffuse from the mother to the baby and from the baby to the mother. This prevents some substances, for example, large blood proteins (such as the Rhesus factor), from passing from one to the other.

Resources

Student Book pages 307–311

B16.3 Technician's notes

Worksheet B16.3a Male reproductive system

Worksheet B16.3b Female reproductive system

Worksheet B16.3c The placenta

Resources for demonstration (see Technician's notes, following)

Approach

1. Introduction

Ask students to give the scientific terms for the organs of the human male reproductive system, and then for the human female reproductive system.

Be aware of continuing embarrassment of any student and manage the class if needed by allowing students to work more individually than in pairs or groups.

2. Human male reproductive system

Ask students to read page 307 of the Student Book on the human male reproductive system. They should make notes on the different organs in the system, and identify as many tissues and cells as they can, explaining the role they play in reproduction.

Worksheet B16.3a gives an outline of the male reproductive system that students can use to answer question 1 on page 309.

3. Human female reproductive system

Ask students to read pages 307–308 of the Student Book on the human female reproductive system. They should make notes on the different organs in the system, and identify as many tissues and cells as they can, explaining the role they play in reproduction.

Worksheet B16.3b gives an outline of the female reproductive system that students can use to answer question 2 on page 309.

4. Human gametes question

Ask students to read pages 49–50 of Section 2 in the Student Book, as well as pages 308–309, about the adaptive features of sperm and egg cells. Give them a few minutes to write a six mark question covering this content, and a mark scheme to go with their question. They should then exchange questions with a neighbour, try to answer the one they receive and then check the mark scheme. They should identify one good point about the question and mark scheme and also something that could be improved. They should return the question and mark scheme back to the other student, and use what they receive to improve their own work.

5. Fertilisation and development of the fetus

Remind students that sexual reproduction involves the fusion of the nuclei of male and female gametes to form a zygote (fertilisation). Explain that the zygote divides to form a ball of cells called the embryo which in humans implants into the uterus lining where it develops into a fetus. They can also read about fertilisation and the development of the fetus on pages 309–311 of the Student Book.

If possible, show students a video of the development of a fetus in the uterus. Note that many videos available on the internet are aimed at women who want to know what happens during pregnancy, so they discuss the development of the fetus in general terms.

Before the video, ask students to suggest a range of questions that they would like to answer from the video, such as when the limbs develop, or when the fetus responds to sound. Make sure the key requirements of the syllabus are included, such as the functions of the umbilical cord, placenta, amniotic fluid and sac. Discuss the best questions and pause the video at appropriate points for students to write down the answers. If any questions are not answered by the video, or a video is not available, students could carry out further research to find the answers.

6. The placenta

Worksheet B16.3c provides a diagram of a developing fetus in the uterus. Students can label and annotate this to explain the role of the placenta in supporting the developing fetus. If students are unsure of the needs of the fetus, remind them that it is a living organism and therefore needs to carry out many of the characteristics of living organisms discussed in Section 1: *Characteristics and classification of living organisms.*

Supplement Make sure students include the function of the placenta in relation to the exchange of dissolved materials, as well as acting as a barrier, for example, to some toxins and pathogens, but not to others such as nicotine from tobacco smoke or the rubella virus.

7. Consolidation: crossword clues

Ask students to write clues for a crossword on the human reproductive system and development of the zygote, embryo and fetus. They should write clues for at least ten of the scientific terms that have been used in the lesson. If there is time, they could test their clues out on another student (there are several good programs available on the internet for generating crossword puzzles).

Technician's notes

The following resources for are needed for the demonstration of the developing fetus:

video clip of the development of a fetus in the uterus

Answers

Page 309

1. a), b) and c) Sketch should be similar to Fig. 16.24. Labels and annotations as follows:
    *   testes, where sperm (male gametes) are produced
    *   sperm duct, which carries sperm to urethra
    *   prostate gland and seminal vesicles, which produce liquid in which sperm swim
    *   penis, which when erect delivers sperm into vagina of female
    *   urethra, the tube that carries sperm from sperm ducts to outside the body.
2. Sketch should be similar to Fig. 16.25. Labels and annotations as follows:
    *   ovaries, where egg cells form
    *   oviducts, which carry the eggs to the uterus and where fertilisation by sperm takes place
    *   uterus, where embryo implants into lining and fetus develops
    *   cervix, base of uterus where sperm are deposited during sexual intercourse
    *   vagina, where penis is inserted during sexual intercourse.
3. Supplement

	Egg cell	Sperm cell
Size	very large, 0.2 mm diameter	very small, 45 μm long
Numbers	thousands in ovary but usually only one released each month	>100 million produced each day
Motility	unable to move on its own	self-propelling with tail

Page 310 Science in context: Ultrasound scans

Most importantly, X-rays could harm the fetus. Also, X-ray images may show clear images of hard tissue such as bone, but not soft tissues such as in the developing fetus.

Page 311

1. a) The cell produced by fusion of a male gamete and female gamete.
    b) Formed from the division of cells in the zygote – until distinctive structures are obvious, such as limbs, when it becomes a fetus.
    c) Developing baby in the uterus (womb), from about 3 months after fertilisation.
2. In an oviduct.
3. The amniotic sac is made up layers of membrane and contains the fetus and the amniotic fluid. The amniotic fluid cushions the fetus to protect it from mechanical damage as well as helping to maintain the fetus at a constant temperature.
4. Supplement The placenta provides nutrients to the fetus from the mother's blood and carries waste from the fetus to the mother's blood to be excreted.

Worksheet B16.3a Male reproductive system

Add annotated labels to the diagram below to explain the function of each of the organs in the human male reproductive system.

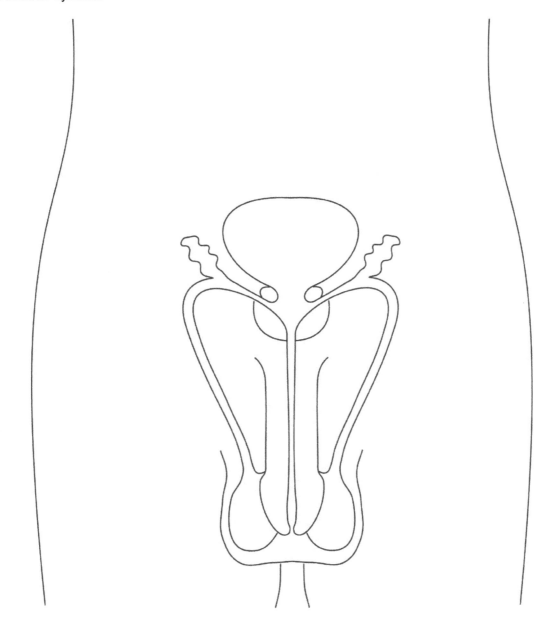

Worksheet B16.3b Female reproductive system

Add annotated labels to the diagram below to explain the function of each of the organs in the human female reproductive system.

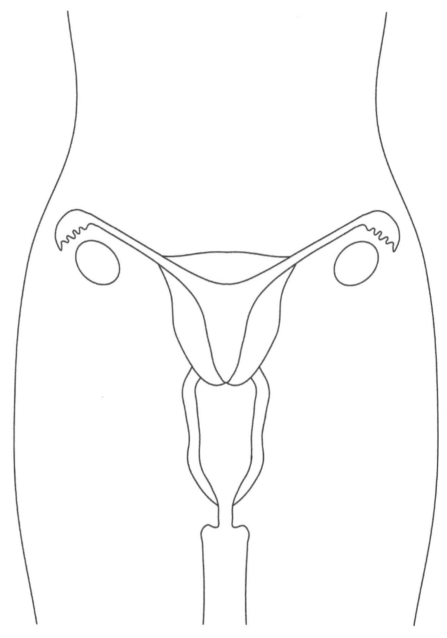

304

Worksheet B16.3c The placenta

The diagram shows a developing fetus in the uterus.

1. Label the following in the diagram:

    a) fetus   b) placenta   c) mother's body   d) umbilical cord   e) amniotic fluid.

2. Add notes to the labels for placenta, umbilical cord and amniotic fluid to explain what they do.
3. a) Use a coloured pencil to draw arrows and write notes to explain how food molecules and oxygen from the environment reach the cells in the fetus for respiration.
    b) Use a different colour of pencil to explain how waste products of metabolism are excreted from the fetus.

Learning episode B16.4 Sexual hormones in humans

Learning objectives

- Describe the roles of testosterone and oestrogen in the development and regulation of secondary sexual characteristics during puberty
- Describe the menstrual cycle in terms of changes in the ovaries and in the lining of the uterus
- `Supplement` Describe the sites of production of oestrogen and progesterone in the menstrual cycle and in pregnancy
- `Supplement` Explain the role of hormones in controlling the menstrual cycle and pregnancy, limited to FSH, LH, progesterone and oestrogen

Common misconceptions

Students should be reminded that male sexual hormones are also found in females, and female hormones are also found in males. It is the different relative concentrations in the body that produce different effects in males and females.

All materials about the menstrual cycle refer to the 28-day length of the average cycle and a period (bleed) or menstruation of about five days. It is important to make clear that many girls and women vary from this, with longer or shorter cycles and longer or shorter menstruation, and that this is usually completely normal. If any girl seems concerned about this, she should be advised to speak to see the school nurse or a doctor.

Resources

Student Book pages 312–314

Approach

1. Introduction

Revise what students already know about the menstrual cycle by giving them one or two minutes to write down three facts about it. Take examples from around the class and encourage discussion if some of the facts seem confused. Be sensitive to the responses of students and allow any students who seem embarrassed to stay out of the discussion unless they wish to join in.

2. Describing the menstrual cycle

Ask students to read pages 312–314 of the Student Book on the menstrual cycle and to make notes about the changes that take place during the cycle.

`Supplement` Students should identify the sites of production of the sexual hormones oestrogen, progesterone, LH and FSH, and explain their roles in controlling the menstrual cycle.

Explain that many of the videos on the internet describing the control of the menstrual cycle go beyond what is required for Cambridge IGCSE, so students should use their notes to make a storyboard for a video that is more suitable. They could work individually or in pairs to do this. They should research what diagram(s) to use as a basis for this, and write the text that should be voiced at each stage of the video. In the final part of the video, they should explain what happens if the egg is fertilised during a cycle.

3. Secondary sexual characteristics

Ask students to suggest some changes that happen in the body during puberty. Identify those that are usually termed *secondary sexual characteristics*. Ask them to suggest why they are called 'secondary' characteristics. They should be able to point out that they are not present from birth.

Explain that secretion of the sexual hormones increases at puberty. This includes oestrogen in women and testosterone in men, which control the development of secondary sexual characteristics.

Ask why these characteristics do not develop until puberty. Students should be able to explain that after puberty the body is mature enough for sexual reproduction, and that these characteristics either make sexual reproduction possible or increase attractiveness to the opposite sex so that sexual reproduction is more likely to take place.

4. Consolidation: red, yellow and green

Give students three cards, red, yellow and green. Then read out questions about the menstrual cycle, explaining that they should hold up the red card if they are uncertain of the answer, the yellow card if they think they know the answer and the green card if they are certain of the answer. Questions could include: Which hormone causes ovulation? Which hormone maintains the thickness of the uterus wall?

Answers

Page 314

1. Testosterone
2. Oestrogen OR progesterone
3. To make sexual reproduction possible, and to show that the individual is sexually mature.
4. a) The release of an egg from an ovary.
   b) The changes that happen in the female reproductive system over about 28 days, including the development and breakdown of the uterus lining and ovulation.
5. Supplement oestrogen: causes uterus wall to thicken and stimulates pituitary to secrete more LH.
   progesterone: maintains uterus lining and inhibits secretion of LH and FSH from pituitary.
   LH: causes ovulation/release of an egg from an ovary.
   FSH: stimulates development of egg in ovary.

Learning episode B16.5 Sexually transmitted infections

Learning objectives

- Describe a sexually transmitted infection (STI) as an infection that is transmitted through sexual contact
- State that human immunodeficiency virus (HIV) is a pathogen that causes an STI
- State that HIV infection may lead to AIDS
- Describe the methods of transmission of HIV
- Explain how the spread of STIs is controlled

Common misconceptions

Despite the large amount of scientific research into the causes and spread of sexually transmitted infections, there is a lot of misinformation available, particularly on ways of protecting against transmission. It is essential that students realise that only information that has been tested in a rigorous and scientific manner should be trusted, as the problems caused by an infection can be devastating.

Resources

Student Book pages 314–315

Approach

1. Introduction

Ask students if they know any examples of sexually transmitted infections (STIs). Note the syllabus only requires specific knowledge of HIV, but there are other STIs that can also be harmful if left untreated.

Make sure students appreciate that infections are caused by pathogens that must be passed to the uninfected person. In the case of STIs, this is usually through the exchange of body fluids during sexual intercourse. Ask students if they also know of ways that help to protect against transmission of STIs. It is possible that they are aware of different methods, although they may not know the full details. Take examples from around the class and note any areas of confusion or error. Make sure these are covered properly in the learning episode.

2. Sexually transmitted infections

Ideally, ask a health professional to visit and talk to the students about STIs, including discussion of HIV/AIDS. The discussion should include how HIV is transmitted, that it may lead to the condition known as AIDS, and should cover methods for the control of transmission. Students should prepare questions before the visit to make sure they learn what is required in the syllabus.

Also note student embarrassment and peer pressure may be factors in class activities surrounding this section. Consider offering students the option to submit questions anonymously and to talk after class alone with the teacher.

Ask students to use what they have learned to produce a poster or presentation on HIV, covering at least what they need to know for the syllabus.

3. Consolidation

Give students a few minutes to jot down three key facts to include in a health campaign for teenagers on HIV and how its spread can be controlled. Take examples from around the class, and encourage discussion to select the most important three facts to include.

Answers

Page 315

1. An infection that is transmitted through sexual contact.
2. Through sexual contact; via blood through cuts or sharing of needles for injecting drugs; across the placenta from mother to fetus; through milk from mother to baby when breast-feeding.
3. The virus attacks the immune system, reducing the ability of the body to fight off other infections. This leads to AIDS.
4. Do not have sexual contact with a partner infected with HIV; use barrier methods such as condoms or femidoms during intercourse.

Learning episode B16.6 Consolidation and summary

Learning aims

- Review the learning points of the topic summarised in the end of topic checklist
- Test understanding of the topic content by answering the end of topic questions

Resources

Student Book pages 316–321

Approach

Introduce the learning episode

Ask students to work with a partner to make a list of key words from this topic. They could then work together to produce a spider diagram showing how the different concepts are linked. They could compare their list with the list of key terms given on page 316 of the Student Book. Discuss the checklist on pages 316–318 and use questioning to see how much of the content they are comfortable with.

Students could make flashcards of the key content and then use the flashcards to quiz each other on the information.

Develop the learning episode

Ask students to work individually through the end of topic questions on page 319–321 of the Student Book without looking at the text. As they work, walk around the classroom observing their answers and asking questions as necessary to find out which questions are causing difficulties.

Finish the learning episode

After a set period, ask the students to stop working and discuss any areas of difficulty you observed as you walked round the class. Students should complete any unanswered questions for homework, but you should stress that they should try to answer the questions without looking at the text, so that they can see how much they have remembered.

Answers

End of topic questions mark scheme

The marks available for a question can indicate the level of detail you need to provide in your answer.

Question	Correct answer	Marks
1	D	1 mark
2 a)	Reproduction	1 mark
2 b)	Anthers	1 mark
2 c)	They contain/release the male gametes in the pollen grains.	1 mark
2 d)	Wind, because there are no bright petals or colour to attract insects,	1 mark
	and the anthers are held outside the flower	1 mark
	so that they are easily blown away by the wind.	1 mark
3 a)	Make less pollen, less waste of pollen as insects more likely to deliver pollen to flower than random distribution in wind.	1 mark
3 b)	If the insect species die out, the plant will not get pollinated.	1 mark

Question	Correct answer	Marks
Supplement 4 a)	This prevents self-pollination, which would reduce the genetic variation of the offspring.	1 mark
	Greater genetic variation in the offspring means that if environmental conditions change, there is a greater chance that some of the offspring will be adapted to growing well in the new conditions.	1 mark
Supplement 4 b)	Cross-pollination requires another plant to transfer pollen,	1 mark
	so if there is no other plant of the same species nearby the flowers will not be pollinated and the plant will not produce seed.	1 mark
Supplement 5 a)	They can reproduce faster as they do not need a mate,	1 mark
	so numbers increase more rapidly while food is available.	1 mark
	Offspring are genetically identical,	1 mark
	so if the parent aphid is well adapted to the conditions, the offspring will also be well adapted.	1 mark
Supplement 5 b)	Genetic variation in the offspring	1 mark
	will mean that if conditions are different in the following year, there are more likely to be some offspring that are well adapted to the new conditions.	1 mark
6 a)	When the seed begins to swell and break open the seed coat so that the embryo plant can grow.	1 mark
6 b)	Moisture is needed because the chemical reactions that happen during growth occur in solution.	1 mark
	Warmth is needed so that chemical reactions/enzyme action is as fast as possible.	1 mark
6 c)	Seeds need water, but waterlogged soil excludes air and therefore oxygen.	1 mark
	If the developing embryo cannot get enough oxygen for respiration, it will be unable to release enough energy for all the processes needed for growth and development.	1 mark
6 d)	Small seeds contain fewer food reserves than large seeds.	1 mark
	So seedlings from small seeds need to start photosynthesising more quickly than those from larger seeds.	1 mark
	If small seeds are planted too deeply, they may run out of energy supplied from their food reserves if they cannot get their leaves to the surface in time.	1 mark
6 e)	If they germinated too late in the year, they would not be able to grow enough before winter,	1 mark
	to flower and reproduce before the winter killed them.	1 mark

Question	Correct answer	Marks
7 a)	In the testes.	1 mark
7 b)	In the ovaries.	1 mark
7 c)	Inside one of the oviducts.	1 mark
7 d)	The sperm travels down the sperm duct and mixes with liquids from the prostate gland (and seminal vesicles) to form semen.	1 mark
	During sexual intercourse sperm are deposited close to the cervix at the top of the vagina.	1 mark
	The egg cell is released from the ovary and travels along an oviduct.	1 mark
	Sperm swim along the oviduct to reach the egg cell.	1 mark
8 a) i)	Circle, with numbers 1 to 28 around the circle	1 mark
	ovulation c. day 14	
ii)	menstruation c. days 1–5	1 mark
Supplement iii)	increase in oestrogen c. days 8–12	1 mark
	decrease c. days 12–14	1 mark
Supplement iv)	increase in progesterone c. days 14–18	1 mark
v)	decrease c. days 23–28	1 mark
Supplement 8 b)	LH triggers ovulation.	1 mark
	FSH stimulates egg development in the ovary.	1 mark
9 a)	Attached to the lining of the uterus.	1 mark
9 b)	Substances such as oxygen, carbon dioxide and glucose are exchanged between the mother and the developing fetus across the placenta.	1 mark
Supplement 9 c)	Diffusion of small molecules.	1 mark
	Osmosis of water.	1 mark
Supplement 9 d)	It prevents many pathogens and toxins crossing from the mother's blood into the fetus.	1 mark
10 a)	It can be transmitted during sexual contact.	1 mark
10 b)	Across the placenta before birth.	1 mark
	In breast milk after birth.	1 mark
10 c)	Avoiding unprotected sexual intercourse.	1 mark
	Avoiding sharing needles.	1 mark
	Total:	54 marks

Introduction

This section begins with a reminder of the structure of DNA from earlier work, before looking at the role of DNA in protein synthesis, and then at the inheritance of sex in humans. This leads on to a description of mitosis and meiosis, their roles in growth and reproduction, and their impact on variation in offspring. Inheritance in monohybrid crosses follows, with an opportunity to practise calculating the outcomes of crosses.

Links to other topics

Topics	Essential background knowledge	Useful links
1 Characteristics and classification of living organisms	1.1 Characteristics of living organisms 1.2 Concept and uses of classification systems	
2 Organisation of the organism	2.1 Cell structure and size of specimens	
3 Movement into and out of cells	3.3 Active transport	
4 Biological molecules	4.1 Biological molecules	
5 Enzymes	5.1 Enzymes	
10 Diseases and immunity	10.1 Diseases and immunity	
14 Coordination and response	14.1 Coordination and response	
16 Reproduction	16.1 Asexual reproduction 16.2 Sexual reproduction 16.3 Sexual reproduction in plants 16.4 Sexual reproduction in humans	
18 Variation and selection		18.1 Variation 18.3 Selection
21 Biotechnology and genetic modification		21.1 Biotechnology and genetic modification 21.3 Genetic modification

Topic overview

B17.1	**Chromosomes, genes and proteins**
	This learning episode helps students to understand the relationship between DNA, chromosomes, genes and alleles.
	Supplement Students will also learn about how the sequence of bases on DNA is related to the protein that is coded by them, and about the process of protein synthesis. The terms haploid and diploid are explained.
B17.2	**Inheriting characteristics**
	In this learning episode, students explore the monohybrid inheritance of characteristics, and the inheritance of sex in humans, through practical work and the use of genetic diagrams.
	Supplement Students also look at test crosses, along with the inheritance of codominant and sex-linked characteristics.

Supplement B17.3	**Mitosis and meiosis**
	This learning episode helps students to understand how cell division by mitosis and meiosis have different outcomes, and how these different kinds of cell division have different purposes.
B17.4	**Consolidation and summary**
	This learning episode provides an opportunity for a quick recap on the ideas encountered in the section, as well as time for the students to answer the end of topic questions in the Student Book.

Note the inheritance of sex included in section 17.1 of the syllabus has been included in section B17.2: *Inheriting characteristics*, once students are more familiar with the processes of inheritance.

Careers links

These are some scientific careers that focus on this area of biology but careers in many other fields use the knowledge and skills gained studying science. **Bioinformaticians** are responsible for storing and analysing the huge amount of data which is often involved in genome research and treatment.

Genetic (or genomic) counsellors work with patients and families to provide information and support to help them make decisions about their health based on genetic conditions.

Starting points

The Student Book section opener puts the ideas in the section into context and sets the scene. It also allows students to acknowledge and value their prior learning, and provides a benchmark against which future learning can be compared.

- The questions provide a structure for introducing the section and can be used in a number of different ways:
- You could ask students to consider the questions as an introductory homework task.
- You could put students into groups to share their own ideas and understanding and then to report back to the whole class.
- Students could be given access to the Internet, preferably with a tight timescale, to find out the information required.

You could then use a spider chart or other form of wall chart to summarise everybody's ideas.

Recording these initial ideas allows you to retain them for reference as the individual topics are developed. In this way, your students' progress in learning can be readily acknowledged.

Learning episode B17.1 Chromosomes, genes and proteins

Learning objectives

- State that chromosomes are made of DNA, which contains genetic information in the form of genes
- Define a gene as a length of DNA that codes for a protein
- Define an allele as an alternative form of a gene
- Supplement State that the sequence of bases in a gene determines the sequence of amino acids used to make a specific protein (knowledge of the details of nucleotide structure is **not** required)
- Supplement Explain that different sequences of amino acids give different shapes to protein molecules
- Supplement Explain that DNA controls cell function by controlling the production of proteins, including enzymes, membrane carriers and receptors for neurotransmitters
- Supplement Explain how a protein is made, limited to:
  (a) the gene coding for the protein remains in the nucleus
  (b) messenger RNA (mRNA) is a copy of a gene
  (c) mRNA molecules are made in the nucleus and move to the cytoplasm
  (d) the mRNA passes through ribosomes
  (e) the ribosome assembles amino acids into protein molecules
  (f) the specific sequence of amino acids is determined by the sequence of bases in the mRNA
  (knowledge of the details of transcription or translation is **not** required)
- Supplement Explain that most body cells in an organism contain the same genes, but many genes in a particular cell are not expressed because the cell only makes the specific proteins it needs
- Supplement Describe a haploid nucleus as a nucleus containing a single set of chromosomes
- Supplement Describe a diploid nucleus as a nucleus containing two sets of chromosomes
- Supplement State that in a diploid cell, there is a pair of each type of chromosome and in a human diploid cell there are 23 pairs

Common misconceptions

Students can find it difficult to make the connection between the concept of genes as part of a DNA molecule and how genes control the development of physical characteristics. It will help if you talk about genes that code for proteins, which have noticeable effects on characteristics such as the colour of skin and hair (caused by amount of the protein melanin that is produced). Some students may remember that enzymes are proteins, and that enzymes control reactions that affect how cells in the body work.

Resources

Student Book pages 326–330

Worksheet B17.1a The structure of DNA

Supplement Worksheet B17.1b The structure of proteins

Resources for a class practical (see Technician's notes, following)

Approach

1. Introduction

Write the word *gene* in the middle of the board. Ask students to suggest related words and how to link them to create a concept map for this subject. They should be able to suggest a wide range of links from their previous learning, including cell structure (nucleus), sexual reproduction (inheritance of characteristics) and variation in characteristics.

You could add to the concept map at the end of other activities in this section to build a full map on inheritance. If so, leave plenty of room around the outside for adding more words.

2. Chromosomes, DNA and genes

Worksheet B17.1a provides a cut-and-paste activity to help students relate these structures to each other and to their position in the nucleus. This would be a useful opportunity to point out that chromosomes only show their X-shaped form during cell division, and that most of the time they are 'uncondensed' – meaning that they are not coiled tightly and cannot be distinguished from each other. The diagram in Fig. 17.2 on page 326 of the Student Book may help some students.

Students may have difficulty marking a 'gene' on their diagrams. This is to be expected. Take the opportunity to discuss what a gene is and why we need further information to decide how many bases within the DNA strand equate to 'one gene'. Explain that different genes are of different lengths, depending on which protein they are coding for.

Supplement Remind students of what they learnt about the structure of DNA and its bases in Section 4: *Biological molecules.*

3. Alleles

Ask students to identify some obvious physical characteristics in humans (not including gender). They should then identify variation in those characteristics. Then ask: since a gene codes for a characteristic, how is inherited variation produced? They should be able to suggest variations in the genes.

Introduce the term *allele* for these different forms of the same gene, and ask students to consider how many alleles there might be for the genes that code for eye colour or any other inherited characteristic that has been suggested. (Note that for some genes there are only two alleles, but for others there may be many. You may wish to tell some students that most characteristics are actually under the influence of multiple genes, each of which may have multiple alleles.)

Supplement 4. Making proteins – 1

Explain that the sequence of bases in a gene determines the sequence of amino acids in the protein it codes for and that this is how genes influence different characteristics. Remind students that the functions of many proteins rely on their shapes. Encourage them to recall examples such as active transport carrier molecules (Section 3), enzymes (Section 5), antibodies (Section 10) and receptors for neurotransmitter molecules (Section 14). Worksheet B17.1b provides an activity looking at the effect of amino acid sequence on the shape of proteins. Give students a selection of beads and a piece of thread or string. Ideally, students should have two or three of some types of bead, and at least six different types.

Supplement 5. Making proteins – 2

Ask students to read the information on pages 327–329 of the Student Book, and look at Fig. 17.5 on page 328, to learn how proteins are formed, starting with the DNA code in the nucleus. They should use what they have learned to script an animation that describes the stages from the copying of the DNA to the formation of the protein, explaining why a particular gene produces a particular protein. They should also explain why, despite most body cells in an organism containing the same genes, many genes in a particular cell are not expressed.

When they have completed their scripts, students could exchange them and compare the script they receive with the information in the Student Book to look for strengths and weaknesses. They should mark two good points and identify two points that could be improved. They should return the script to the student who wrote it and decide how they could improve the two points marked on their script. Take examples of areas for improvement from around the class and discuss those that occur most frequently, to make sure students fully understand the process.

You could further test understanding by asking questions such as:

- Where is mRNA made?
- What happens on a ribosome?
- Suggest how a cell controls which proteins are made.

Supplement 6. Haploid and diploid cells

Show students a photograph of the human karyotype from a book (e.g. Fig. 17.4 on page 327 of the Student Book) or the internet. Explain that it shows the chromosomes from the nucleus of one human body cell. Introduce the terms *diploid* for a nucleus containing two sets of chromosomes, and *haploid* for a nucleus of a cell such as a sperm or egg that contains only one set of chromosomes.

Give students a few minutes to work individually or in pairs to answer the following questions. Then take answers from around the class to check that students have the right answers (shown underlined):

How many chromosomes are there in a human body cell? [23, <u>46</u>, hundreds, thousands]

How many genes are there in a human body cell? [23, 46, hundreds, <u>thousands</u>]

What is the largest number of alleles that a gene can have? [1, 2, <u>many</u>]

What is the largest number of alleles that one cell can have for a gene? [1, <u>2</u>, many]

How many chromosomes are there in a human gamete (sex cell)? [<u>23</u>, 46, hundreds, thousands]

It is important for students to understand that although a gene may occur in many different forms (alleles), an individual diploid cell can only have a maximum of two different alleles because there are two copies of each gene in each cell.

7. Consolidation

Either ask students to suggest words to amend any errors and add any new learning to the concept map created in the introduction, or ask them to produce definitions for an online dictionary for the following terms: *DNA, chromosome, gene, allele.*

Technician's notes

Be sure to check the latest safety notes on these resources before proceeding.

The following resources are needed for task 4, per group:

a selection of beads, containing at least two or three of some types and ideally at least six types of bead
approximately 20 cm string, wool or thick thread
sticky tape

Answers

Page 327

1. gene, chromosome, nucleus, cell
2. A gene codes for a protein, and so a particular characteristic; an allele is one form of the gene, coding for a variation in the protein or characteristic. Any suitable example, e.g. gene for eye colour, allele for blue eye colour or brown eye colour.
3. The passing on of genetic information from one generation to the next.

Page 330

1. a) The sequence of bases on the DNA strand is copied to make the strand of mRNA.
   b) The mRNA strand moves from the DNA in the nucleus to the ribosome in the cytoplasm.
   c) The ribosome uses the sequence of bases on the mRNA strand to produce the correct sequence of amino acids to make the protein
2. Many of the genes are 'switched off so they are not expressed.
3. 24

Worksheet B17.1a The structure of DNA

Cut out the diagrams below. Arrange them in order of size, starting with the largest, and paste them into your workbook in this order.

Link the diagrams by indicating which part of the previous diagram the following diagram represents, such as by circling the part it shows and adding an arrow to link them.

Add the following labels at the correct points on the diagrams. You may need to use some of the labels more than once: chromosome   gene   DNA   nucleus   cell.

Add any other notes to your diagram to help you remember all the details of the structure of DNA.

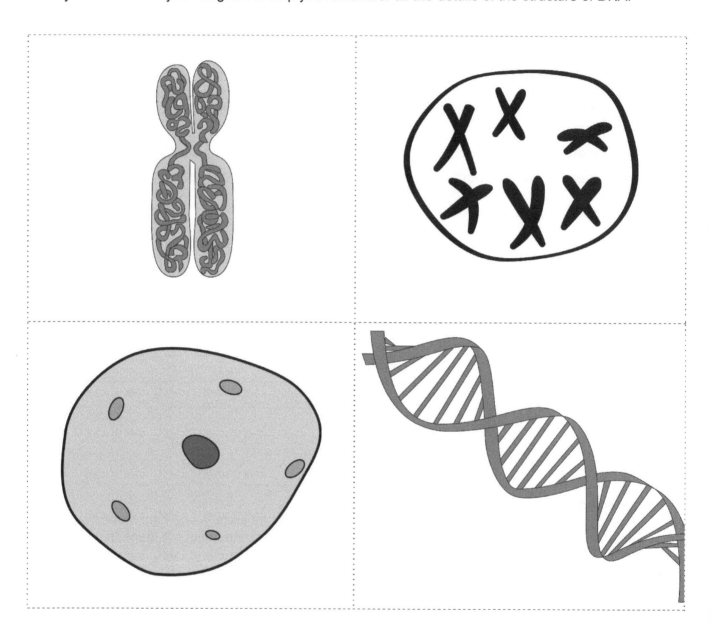

318

Supplement Worksheet B17.1b The structure of proteins

In this activity you will model the effect of amino acid sequence on the shape of a protein. You will use beads to represent amino acids – each different type of bead is a different amino acid. The string of beads will represent a protein.

Apparatus

a selection of beads

string or wool

sticky tape

Method

1. Tie a knot in one end of the piece of string.
2. Loosely thread a mixture of beads onto a piece of string or thread and knot the other end. Leave plenty of space on the string so you can easily fold the string into different shapes.

Handling experimental observations and data

Some amino acids form bonds with any identical amino acids in the same protein.

3. Choose one bead type in your string where you have more than one of that type, and fold the string in a way that brings those beads close together. (It may help to use a bit of sticky tape to hold them in position.)

   a) What effect does this have on the shape of the string?

   b) What effect would this have had on the shape of the string if the beads were in different positions?

Some amino acids repel other amino acids in the same protein, pushing them as far apart as possible.

4. Choose two different types of bead in your string, and push them as far apart as possible.

   a) What effect does this have on the shape of the string?

   b) What effect would this have had on the shape of the string if the beads were in different positions?

Look at the diagram of an enzyme in Fig. 5.2 on page 91 of the Student Book.

5. Use the findings from your modelling work to explain:

   a) why each enzyme has a particular shape

   b) why different enzymes have different shapes.

Learning episode B17.2 Inheriting characteristics

Learning objectives

- Describe the inheritance of sex in humans with reference to X and Y chromosomes
- Describe inheritance as the transmission of genetic information from generation to generation
- Describe genotype as the genetic make-up of an organism and in terms of the alleles present
- Describe phenotype as the observable features of an organism
- Describe homozygous as having two identical alleles of a particular gene
- State that two identical homozygous individuals that breed together will be pure-breeding
- Describe heterozygous as having two different alleles of a particular gene
- State that a heterozygous individual will not be pure-breeding
- Describe a dominant allele as an allele that is expressed if it is present in the genotype
- Describe a recessive allele as an allele that is only expressed when there is no dominant allele of the gene present in the genotype
- Interpret pedigree diagrams for the inheritance of a given characteristic
- Use genetic diagrams to predict the results of monohybrid crosses and calculate phenotypic ratios, limited to 1:1 and 3:1 ratios
- Use Punnett squares in crosses which result in more than one genotype to work out and show the possible different genotypes
- Supplement Explain how to use a test cross to identify an unknown genotype
- Supplement Describe codominance as a situation in which both alleles in heterozygous organisms contribute to the phenotype
- Supplement Explain the inheritance of ABO blood groups: phenotypes are A, B, AB and O blood groups and alleles are I^A, I^B and I^o
- Supplement Describe a sex-linked characteristic as a feature in which the gene responsible is located on a sex chromosome and that this makes the characteristic more common in one sex than in the other
- Supplement Describe red-green colour blindness as an example of sex linkage
- Supplement Use genetic diagrams to predict the results of monohybrid crosses involving codominance or sex linkage and calculate phenotypic ratios

Common misconceptions

Students frequently get muddled about the terms *genotype* and *phenotype*. If this happens, encourage them to develop a mnemonic that will help them remember them correctly.

Some students think that because human females have two identical sex chromosomes, then all other animals are the same. This is not the case. In birds and some insects, for example, males have identical sex chromosomes and females have two different sex chromosomes.

Resources

Student Book pages 332–345

Worksheet B17.2 Investigating monohybrid inheritance

B17.2 Technician's notes

Resources for class practical (see Technician's notes, following)

Approach

1. Introduction

Ask students to suggest physical characteristics that are inherited from parents (other than those related to gender). When you have a small selection of suggestions, ask students how they know those characteristics are inherited. They should also consider how those characteristics are passed on, and whether we can predict which offspring may inherit a particular characteristic or not. This discussion should produce a range of ideas, although some may be quite vague. Tell students that the study of the inheritance of characteristics is called *genetics*, and that they will start by studying some simple examples.

2. Genetics terminology

Use the diagram of definitions in genetics (Fig. 17.7) on page 333 of the Student Book to introduce the terms *homozygous, heterozygous, dominant allele, recessive allele, genotype* and *phenotype*. Ask students to write their own definitions and to compare with another student's definitions to see if they could be improved.

Test their understanding by telling them that an imaginary animal has two alleles for the gene for hair colour. Capital B represents the allele for blue hair and small b represents the allele for white hair. Ask questions such as the following:

- Which letter represents the recessive allele (and how do you know)? [b, lower case/small letter]
- What is the genotype of a homozygous dominant individual? [BB]
- What is the phenotype of a homozygous recessive individual? [white hair]
- What is the genotype and phenotype of a heterozygous individual? [Bb, blue hair]
- Which allele(s) would be found in the gametes of a homozygous dominant individual? [only B]
- Which allele(s) would be found in the gametes of a heterozygous individual? [some B, some b]

Take answers from around the class to check understanding. Alternatively, do this as a red/yellow/green card activity to check how certain students are of the answers.

3. Investigating monohybrid inheritance

Worksheet B17.2 gives students the opportunity to model monohybrid inheritance using counters or other small objects of two colours to represent the alleles. This supports the Developing practical skills box on pages 336–337 of the Student Book. It offers four possible crosses for the inheritance for a gene with two alleles. It might help weaker students if you go through the questions as a group or class for the first cross of homozygous dominant and homozygous recessive.

For the practical, students should be able to draw genetic diagrams, either in layout form or as Punnett squares, so some practice in this with the first cross will be helpful. Alternatively, let students use the diagrams on pages 334 and 335 of the Student Book for guidance.

Using the genetic diagrams, students should get the following theoretical results:

- homozygous dominant × homozygous recessive: genotype all heterozygous, phenotype all of dominant colour
- heterozygous × heterozygous: genotypes 1 homozygous dominant: 2 heterozygous : 1 homozygous recessive; phenotypes 3 dominant colour : 1 recessive colour
- homozygous dominant × heterozygous: genotypes 1 homozygous dominant : 1 heterozygous; phenotypes all dominant colour
- homozygous recessive × heterozygous: genotypes 1 heterozygous : 1 homozygous recessive; phenotypes 1 dominant colour: 1 recessive colour.

Results can be expressed as ratios, percentages or probabilities.

Explain that pedigree diagrams are another type of diagram, but one that shows the inheritance of phenotypes over several generations. They don't show genotypes but can sometimes be interpreted to identify the genotypes of individuals. Ask students to read about pedigree diagrams on pages 344–345 of the Student Book and attempt the questions.

Supplement Ask students to use what they have learned to explain why the crossing of an individual of unknown alleles with the homozygous recessive is a 'test cross', and what that cross will show.

4. Mendel and investigating inheritance

Although knowledge of Gregor Mendel's investigations of inheritance in peas is beyond the requirements of the syllabus, it offers an interesting alternative or extension to the practical work. Students could research how he set up his investigations to make sure they would produce as reliable results as possible, and what results he got from his crosses. They should be able to explain Mendel's results.

Supplement ## 5. Investigating codominance

Students should repeat part of Worksheet B17.2, but with the understanding that the two colours of the counter represent codominant alleles. They should then compare the proportions of each genotype and phenotype in the offspring with those from the same crosses in the previous investigation and explain any differences.

Ask students to read pages 339–340 of the Student Book and explain how the different ABO blood groups are inherited.

6. Inheritance of sex in humans

Ask students to compare the photographs of the human X and Y chromosomes on page 341 of the Student Book. Tell them that these chromosomes are the sex chromosomes and link the shape of the chromosomes to the notation of XX for female and XY for male.

Ask students to identify what the gametes from a man and a woman will contain in terms of sex chromosomes. They should realise that the egg cells will all contain one X chromosome, while the sperm may contain one X or one Y chromosome. In fact, there should be equal numbers of X sperm and Y sperm cells produced (but they may not fully understand this until they cover meiosis later in this section).

As an extension, ask students to create their own genetic diagram to show the inheritance of sex in humans. Then ask: if a couple already have two boys, what is the chance that their next child is also a boy? Using their genetic diagrams, students should be able to confirm that this is always 50%.

Supplement ## 7. Sex linkage

Explain that, because the X chromosome is much bigger than the Y chromosome, many of the alleles on the X chromosome have no corresponding allele present on the Y chromosome. These genes are described as sex-linked, and this can cause characteristics controlled by these genes to be more common in males than in females. Use the example of red-green colour blindness to illustrate this. Students could use the information from pages 342–343 of the Student Book to help them produce Punnett squares for other possible crosses in addition to that shown on page 343.

8. Consolidation: key facts

Either ask students to suggest words to add any new learning to the concept map created in the Introduction of Learning episode B17.1, or ask them to write a list of the five most important facts that they have learned about inheritance. They should then compare their list with a neighbour, and agree the five most important facts from the two lists.

This could be repeated in fours, and then examples taken from each group to draw up a class list of the five key facts to remember.

Technician's notes

Be sure to check the latest safety notes on these resources before proceeding.

The following resources are needed for the class practical B17.2, per group:

10 counters each of 2 colours (other small items could be used, such as beads or buttons, as long as they are all identical except for the 2 colours)
2 small pots, such as yoghurt pots

Answers

Page 333

1. a) A dominant allele is an allele that is expressed in the phenotype if it is present at all in the genotype.
   b) A recessive allele is an allele that is only expressed in the phenotype when there is no dominant allele of the gene present in the genotype.
   c) Having two identical alleles of a particular gene.
   d) Having two different alleles of a particular gene.
2. a) 2 b) 1 c) 2

Page 336

1. The inheritance of a characteristic controlled by one gene.
2. genotype (the alleles in the chromosomes) BB, phenotype (what the organism looks like) brown; genotype Bb, phenotype brown (because the allele for brown colour is dominant); genotype bb, phenotype black (because the organism doesn't have the allele for brown colour).
3. a) The answer may be presented as a full layout diagram or a Punnett square, showing the adult genotypes and phenotypes (male BB brown and female bb black), the possible gametes produced (male B and B, female b and b), genotypes and phenotypes of possible offspring (all Bb brown).
   b) None of the offspring will be black, so the probability is 0.

Page 338 Science in context: Mendel's peas

1. So that, when the plants were bred together, each parent could only pass on one type of allele to their offspring, so there was no possible confusion as to which parent each allele was inherited from.
2. Random variation is possible in the results, but the larger the sample, the less effect any random (or 'fluke') results have on the overall pattern.
3. He removed the anthers from each flower, so no pollen could be transferred by insect. He also covered each flower after he had hand-pollinated it, so that other pollen could not get to the stigmas.
4. Any characteristic may be used, with alleles appropriately designated with capital letter for dominant and lower-case letter for recessive allele. Parents used should show one with phenotype of dominant allele, homozygous, e.g. BB, and one parent with phenotype of recessive allele, i.e. bb. First cross will produce all individuals with phenotype of dominant allele but heterozygous in genotype, i.e. Bb. Crossing of these individuals will produce characteristic 1 BB : 2 Bb : 1 bb in genotype and 3 dominant characteristic to 1 recessive characteristic in next generation.
5. If Mendel had not been as thorough about his method, then his results would not have been as clear and predictable. So he would not have been able to have drawn clear and repeatable conclusions about the way characteristics are inherited in pea plants.

Pages 336–337 Developing Practical Skills

1. Start with two beakers, containing the same number of beads, but half the beads in each beaker are red and half are blue (because half the gametes will receive the dominant allele during meiosis and the other half will receive the recessive allele). Start with many beads in each pot, well mixed.
2. For each 'fertilisation' take one bead (gamete) from one beaker, without looking because fertilisation is random. Then take one bead, without looking, from the other beaker. Place the two beads together to represent the genotype of the offspring.
3. There would be 20 red beads and 20 blue beads in each pot because (as a result of meiosis) half the gametes will receive one allele from the diploid parent cell and half will receive the other allele. Genetic diagram or Punnett square, using letters of own choice linked to red and blue beads, such as the following.

R is dominant allele, represented by red bead; r is recessive allele, represented by blue bead.

		gametes	
		R	r
gametes	R	RR	Rr
	r	Rr	rr

Predicted probabilities are 1 RR : 2 Rr : 1 rr for genotypes and 3 dominant : 1 recessive for phenotypes.

4. Actual results are 1 red/red : 2.4 red/blue : 0.6 blue/blue for genotypes and 5.7 dominant : 1 recessive for phenotypes.
5. The actual results vary quite a bit from the predicted results because only a small number of repeats was carried out.
6. If the number of repeats were increased, it is likely that the actual results would get closer to the predicted ones.

Science In Context, page 339

1. Using pure-breeding parent plants means that, when the plants were bred together, the results in the offspring were not confused by a mix of alleles in one or both of the parents.
2. Random variation is possible in the results. So the larger the sample, the more likely that any random variation will be averaged out.
3. He removed the anthers from every flower, so that pollen could not be transferred by insect. He also covered each flower after he had hand-pollinated it, so that other pollen could not get to the stigma.
4. Any characteristic may be used, with alleles appropriately designated with capital letter for dominant and lower case letter for recessive allele. Parents used should show one with phenotype of dominant allele, homozygous, e.g. BB, and one parent with phenotype of recessive allele, i.e. bb. First cross will produce all individuals with phenotype of dominant allele but heterozygous in genotype, i.e. Bb. Crossing of these individuals will produce 1 BB : 2 Bb : 1 bb in genotype and 3 dominant characteristic : 1 recessive characteristic in the next generation.
5. If Mendel had not been as thorough about his method, then his results would not have been as clear and predictable. So he would not have been able to draw clear and repeatable conclusions about the way characteristics are inherited in pea plants.

Supplement Page 340

1. An individual showing the dominant characteristic could either be homozygous dominant or heterozygous. Which it is can be revealed by a test cross with the homozygous recessive. If the individual with the dominant phenotype is heterozygous then some of the offspring produced will show the recessive phenotype, but if they are homozygous then there is no possibility that any offspring will show the recessive phenotype.
2. When both alleles in a heterozygous individual are expressed in the phenotype, and there is no dominance of one allele over the other.
3. Genetic layout diagram or Punnett square with following outcomes:

		father's gametes	
		I_A	I_B
mother's gametes	I_o	$I_A I_o$ blood group A	$I_B I_o$ blood group B
	I_o	$I_A I_o$ blood group A	$I_B I_o$ blood group B

Page 341

1. XX
2. XY
3. At each fertilisation there is a 0.5 probability that the X egg will be fertilised by an X sperm or a Y sperm. So the chance of the child being born male is 0.5, or 1 in 2, or 50%.

Supplement Page 343

1. A characteristic that is controlled by a gene on a sex chromosome, so that it is expressed more commonly in one sex than in the other.
2. If a girl has red-green colour blindness, then she must have inherited one recessive allele for the condition from her father and one from her mother. So her father must have had an X chromosome with the recessive allele, and would have had red-green colour blindness, as there is no allele for this characteristic on the Y chromosome. Her mother could either have been homozygous for the recessive allele, and so had red-green colour blindness, or heterozygous, with normal colour vision.

Page 345

1. three
2. two
3. two
4. The alleles for freckles must be dominant. This is because I has no freckles but both of her parents, C and D, do. This means I must be homozygous recessive and C and D must be both heterozygous. If having no freckles was dominant, then if I had no freckles, then at least one of C or D would also have had to have no freckles.

Worksheet B17.2 Investigating monohybrid inheritance

In this investigation you will model the inheritance of a characteristic controlled by one gene: monohybrid inheritance. You will use coloured counters (or other objects) to represent the alleles for the gene.

Apparatus

10 counters each of two colours

2 small pots

Method

1. Decide which of the colours will represent the dominant allele and which represents the recessive allele and write these down. Choose what the alleles code for, such as spotted coat colour and black coat colour in leopards, and write these down.
2. The pots represent the collection of gametes produced by each parent. To start, you will look at a cross between a homozygous dominant parent and a homozygous recessive parent. Since each parent is homozygous, they can only produce gametes containing one type of allele. Place ten counters of the dominant colour in one pot, and place ten counters of the recessive colour into the other pot.
3. To carry out a cross, take one counter from one pot and one from the other pot to produce the genotype of one offspring. Note down the genotype and the phenotype of this offspring.
4. Continue taking one counter from each pot, to represent more crosses from these parents, until you have used all the counters. In each case record the genotype and phenotype of the offspring.
5. Set up the pots again for different crosses. Remember that the gametes of a homozygous individual will all have the same allele. However, for a heterozygous individual half the counters will be of the dominant colour and half of the recessive colour. Try the following crosses, recording the genotypes and phenotypes of each offspring:

   heterozygous × heterozygous

   homozygous dominant × heterozygous

   homozygous recessive × heterozygous.

Handling experimental observations and data

6. Use your results to calculate the proportion of different genotypes for each cross, as a ratio, probability or percentage.
7. Use your results to calculate the proportion of different phenotypes for each cross, as a ratio, probability or percentage.
8. Draw a genetic diagram for each of the crosses you carried out. (Remember to identify which letter represents which allele in your diagram.)
9. Calculate the theoretical proportions for each genotype and phenotype for each of the crosses.
10. Compare your practical results with the theoretical values, and suggest an explanation for any differences.
11. Use your answer the previous question to explain why two heterozygous parents may produce all offspring with the recessive phenotype.

Supplement Learning episode B17.3 Mitosis and meiosis

Learning objectives

- Supplement Describe mitosis as nuclear division giving rise to genetically identical cells (details of the stages of mitosis are **not** required)
- Supplement State the role of mitosis in growth, repair of damaged tissues, replacement of cells and asexual reproduction
- Supplement State that the exact replication of chromosomes occurs before mitosis
- Supplement State that during mitosis, the copies of chromosomes separate, maintaining the chromosome number in each daughter cell
- Supplement Describe stem cells as unspecialised cells that divide by mitosis to produce daughter cells that can become specialised for specific functions
- Supplement State that meiosis is involved in the production of gametes
- Supplement Describe meiosis as a reduction division in which the chromosome number is halved from diploid to haploid resulting in genetically different cells (details of the stages of meiosis are **not** required)

Common misconceptions

Students commonly confuse mitosis and meiosis because of the similarity of the words. Help them produce their own mnemonics, such as 'miTosis produces Two cells, MEiosis produces gaMEtes'.

Resources

Student Book pages 330–331

B17.3 Technician's notes

Resources for activity (see Technician's notes, following)

Approach

1. Introduction

Remind students that one of the characteristics of living organisms is growth, and ask them what needs to happen in a body for it to grow. They should identify cell division as one of the processes needed.

Point out that all cells in a body contain the same features and the same genetic information. Give students a minute or so in pairs or small groups to consider what must happen in a body cell before it can divide to make more body cells. Then take suggestions from around the class. The key point to get from this is that everything in the cell must be duplicated before division, including the nucleus, so that the new cells are genetically identical.

2. The role of mitosis

If possible, show students a video of asexual reproduction, such as budding in *Hydra* or yeast, or binary fission in bacteria. Remind students of their work on asexual reproduction in Section 16: *Reproduction*, and to use that knowledge to explain what they are seeing. If needed, point out that all the cells that are formed in asexual reproduction are genetically identical. They should be able to draw parallels with the discussion on growth in the Introduction. You could also show a photograph of a mending wound, and ask students how the body is repairing itself. Again, they should draw parallels with growth and asexual reproduction in that identical cells are being produced by the division of existing cells.

Ask students to read pages 330–331 in the Student Book and to summarise the key points about mitosis.

3. Stem cells

Ask students to work as a group to think up some key questions to research on 'stem cells'. When they have done this, check that the questions include what the term means, what is special about stem cells, how they divide and what their purpose is. Students should then carry out research to answer their questions. They should use their research to write a short paragraph titled 'What's so special about stem cells?' Make sure they make the connection that stem cells are produced by mitosis, and that all the cells formed are genetically identical although the genes that are expressed in daughter cells change as they become specialised.

4. Meiosis

From earlier work, students should remember that gametes contain half the number of chromosomes of body cells – remind them of the terms *haploid* and *diploid*. Explain that they are formed by a different type of cell division: meiosis.

Ask students to draw up and complete a table to summarise the similarities and differences between the two forms of cell division. They should use information from pages 330–331 in the Student Book to do this.

5. Consolidation: quick questions

Ask students to write three quick questions on what they have learned in this section. They should then test out their questions on another student. Check to see whether there are any recurring areas of weakness, and make sure you address them by providing suitable support.

Technician's notes

Be sure to check the latest safety notes on these resources before proceeding.

The following resources are needed for the activity on the role of mitosis:

video showing asexual reproduction, such as budding in *Hydra* or yeast, or binary fission in bacteria

Answers

Supplement Page 331

1. a) Mitosis
    b) It produces cells that are genetically identical.
2.

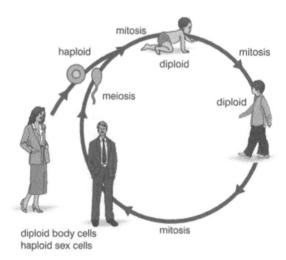

3. a) Meiosis

    b) Mitosis

4. Meiosis produces non-identical cells, so there is variety in the gamete cells. When the gamete cells fuse, this will mean that the offspring will vary from each other.

Learning episode B17.4 Consolidation and summary

Learning aims

- Review the learning points of the topic summarised in the end of topic checklist
- Test understanding of the topic content by answering the end of topic questions

Resources

Student Book pages 346–350

Approach

Introduce the learning episode

Ask students to work with a partner to make a list of key words from this topic. They could then work together to produce a spider diagram showing how the different concepts are linked. They could compare their list with the list of key terms given on page 346 of the Student Book. Discuss the checklist on pages 346–348 and use questioning to see how much of the content they are comfortable with.

Students could make flashcards of the key content and then use the flashcards to quiz each other on the information.

Develop the learning episode

Ask students to work individually through the end of topic questions on pages 349–350 of the Student Book without looking at the text. As they work, walk around the classroom observing their answers and asking questions as necessary to find out which questions are causing difficulties.

Finish the learning episode

After a set period, ask the students to stop working and discuss any areas of difficulty you observed as you walked round the class. Students should complete any unanswered questions for homework, but you should stress that they should try to answer the questions without looking at the text, so that they can see how much they have remembered.

Answers

End of topic questions mark scheme

The marks available for a question can indicate the level of detail you need to provide in your answer.

Question	Correct answer	Marks
1	B	1 mark
2	The nucleus in a cell contains the chromosomes, which are made from DNA.	1 mark
	A gene is a short piece of that DNA that codes for a protein or characteristic.	1 mark
3	All the eggs produced by a woman contain one X chromosome.	1 mark
	The sperm from a man contains either an X chromosome or a Y chromosome.	1 mark
	When the nucleus of a sperm fuses with the nucleus of an egg, the zygote can only be either XX (female) or XY (male),	1 mark
	so there are only two sexes.	1 mark

4 a)	Using any suitable symbols for wet and dry ear wax, such as E for wet and e for dry, either a layout or a Punnett square including the following information:	

		father's gametes	
		e	e
mother's gametes	E	Ee wet earwax	Ee wet earwax
	e	ee dry earwax	ee dry earwax

	suitable use of letters for alleles correct gametes for parents correct genotypes for offspring	1 mark 1 mark 1 mark
4 b)	Predicted genotypes 1 Ee: 1 ee, 50% or 1 in 2 chance of either outcome. Predicted phenotypes 1 wet: 1 dry, 50% or 1 in 2 chance of either outcome	1 mark 1 mark
4 c)	The chance of inheriting wet or dry earwax is the same at each fertilisation. So it is a 50% chance each time. Random fertilisation means it could give the same result each time.	1 mark 1 mark
Supplement 5	The cross will produce plants that are heterozygous, with one allele for red and one for white. If either allele was dominant, then all the offspring would have flowers of the colour that the dominant allele produces. Instead the flowers show splashes of each colour, suggesting that both alleles are being expressed, which is codominance.	1 mark 1 mark 1 mark
Supplement 6 a	It is a sex-linked characteristic that is more commonly seen in men than in women.	1 mark
Supplement 6 b)	The allele for colour blindness is recessive, so a girl must inherit one allele from each parent.	1 mark
Supplement 6 c)	From his mother, as the gene does not exist on the Y chromosome that he inherits from his father	1 mark

Supplement 6 d)		father's gametes	
		X^R	Y
mother's gametes	X^r	$X^R X^r$ (female, colour vision [carrier])	$X^r Y$ (male, colour blind)
	X^r	$X^R X^r$ (female, colour vision [carrier])	$X^r Y$ (male, colour blind)

	correct gametes for parents correct genotypes for offspring	1 mark 1 mark
Supplement 6 e)	On the X chromosome, in the part that is not matched by the Y chromosome, as it is a sex-linked characteristic.	1 mark

Supplement 7		Mitosis	Meiosis	8 marks
	Number of cells produced from one cell	2	4	
	Daughter cells genetically identical or different	identical	different	
	Daughter cells haploid or diploid	diploid	haploid	
	Purpose	growth and repair of body cells to produce offspring in asexual reproduction	to form gametes in humans and flowering plants	

Supplement 8	If there were no meiosis, the chromosome number of the zygote would double each time there was fertilisation.	1 mark
	Meiosis halves the number of chromosomes, and then the full number is restored during fertilisation.	1 mark
	Total:	33 marks

Exam-style questions and sample answers have been written by the authors. In examinations, the way marks are awarded may be different. References to assessment and/or assessment preparation are the publisher's interpretation of the syllabus requirements and may not fully reflect the approach of Cambridge Assessment International Education.

The marks available for a question can indicate the level of detail you need to provide in your answer.

Pages 351–361

Question	Correct answer	Marks
1	A	1 mark
2	D	1 mark
Supplement 3	B	1 mark
4 a)	A: right atrium	1 mark
	B: right ventricle	1 mark
	C: left atrium	1 mark
	D: left ventricle	1 mark
4 b)	<table><tr><th>Component of blood</th><th>Function</th></tr><tr><td>Red blood cells</td><td>transport oxygen</td></tr><tr><td>White blood cells</td><td>phagocytosis antibody production</td></tr><tr><td>Platelets</td><td>blood clotting</td></tr><tr><td>Plasma</td><td>Any four from: transport blood cells transport nutrients transport ions transport urea transport carbon dioxide transport hormones (allow distribute heat)</td></tr></table>	1 mark 2 marks 1 mark 4 marks
Supplement 5 a)	By the atrioventricular and semilunar valves.	1 mark
	When the atria/ventricles contract, blood is forced through the valves.	1 mark
	Any backflow of blood closes the valves.	1 mark
	Inversion of the valves is prevented by tendons.	1 mark
Supplement 5 b)	The left ventricle has thicker muscle,	1 mark
	as the blood it pumps has to reach the whole body/ needs to produce a higher pressure.	1 mark
	The blood the right ventricle pumps only has to reach the lungs/ is pulmonary circulation.	1 mark
	Low pressure pulmonary circulation prevents damage to lungs/capillaries.	1 mark
Supplement 5 c) i)	Antibodies are passed from one person to another, e.g. mother to child.	1 mark
Supplement 5 c) ii)	Vaccine contains dead or weakened form of the pathogens,	1 mark
	that have the same antigens as the active pathogen.	1 mark
	Lymphocytes produce antibodies against these antigens.	1 mark
	Memory cells are produced	1 mark
	Memory cells quickly respond so that antibodies are quickly produced if exposed to active pathogen.	1 mark 1 mark
6 a) i)	glucose → ethanol + carbon dioxide	2 marks

ii)	This releases the energy yeast cells need to carry out life processes.	1 mark
6 b)	The energy released from each molecule of glucose in anaerobic respiration is much less than that released in aerobic respiration.	1 mark
7 a)	A: (left) kidney B: ureter C: bladder D: urethra	1 mark 1 mark 1 mark 1 mark
7 b)	lungs carbon dioxide	1 mark 1 mark
Supplement 7 c) i)	In the liver	1 mark
Supplement ii)	From excess amino acids, by deamination, by removing the nitrogen-containing part of the molecule.	1 mark 1 mark
Supplement iii)	It is toxic	1 mark
8	Growth in response to the direction of light is called **phototropism**. If the growth is towards light, it is called **positive phototropism**, as shown by plant **shoots**. This response benefits plants because they receive more light for **photosynthesis**. Growth in response to gravity is called **gravitropism**. Plant roots show **positive gravitropism**. This response helps the plant roots to grow **downwards** so the plant can obtain the **water/mineral ions** it needs.	8 marks – 1 mark for each
9 a) i)	(stimulus/hot object) → receptor → sensory neurone → relay neurone in CNS/spinal cord → motor neurone → effector → (response/moves hand)	5 marks
9 a) ii)	Relay neurones in the spinal cord connect with other neurones, which pass impulses up the spinal cord to the brain.	1 mark 1 mark
9 b) i)	chemical substance, produced by (endocrine) gland, carried by blood, to target organ(s)	1 mark 1 mark 1 mark 1 mark
9 b) ii)	response is electrical in the nervous system, and chemical in the hormonal system, response carried by nerves/neurones in the nervous system, and in the blood in the hormonal system response is very quick in the nervous system, and slower in the hormonal system response is short-lived with the nervous system, but longer-lived with the hormone system	1 mark 1 mark 1 mark 1 mark
10 a)	To produce a new organism by sexual reproduction, two **gametes/sex cells** fuse. This process is known as **fertilisation**. Usually, sexual reproduction involves **two** parent organisms of the same species. The **offspring** formed are genetically different from each of the parents.	4 marks – 1 mark for each
Supplement 10 b) i)	Asexual reproduction: Advantage: only one parent required so large numbers can be produced quickly/ because genetically identical, can survive well in conditions suited to them. Disadvantage: if the environment changes, or there is disease, lack of variation means that they could all die out.	1 mark 1 mark
Supplement 10 b) ii)	Sexual reproduction: Advantage: genetic information from both parents leads to variety in the offspring, leading to better chances of some surviving in a changing/new environment. Disadvantage: two parents required, and in some habitats, for example, a desert, it may be difficult for two parents to meet/it takes longer to produce offspring.	1 mark 1 mark

10 c)	A: oviduct; carries egg/zygote to uterus (allow where fertilisation occurs) B: ovary; produces eggs C: uterus, where fetus develops D: cervix; produces mucus / keeps fetus secure in uterus	1 mark for each part, and 1 mark for each function – 8 marks total
11 a)	A: stigma B: style C: ovary D: carpel E: anther F: filament G: stamen H: petal I: ovule J: sepal	1 mark 1 mark 1 mark 1 mark 1 mark 1 mark 1 mark 1 mark 1 mark 1 mark
11 b i)	petals – small as no need to attract insects, green/inconspicuous, as no need to attract insects	1 mark 1 mark
11 b) ii)	stigma – large and feathery/large surface area to collect pollen, hang down outside the flower to collect pollen	1 mark 1 mark
11 b) iii)	stamens – large to produce huge numbers of pollen, hang down outside the flower to release pollen into wind	1 mark 1 mark
11 b) iv)	pollen grains – large numbers produced (as chances of reaching another flower when carried by the wind are low (lower than if carried by an insect), small/light, to be easily carried by wind	1 mark 1 mark
12 a)	flagellum for swimming (to egg) many mitochondria to supply the energy for swimming/movement/contraction of the flagellum acrosome at the tip of the sperm contains enzymes so that the sperm (nucleus) can penetrate the egg nucleus to fuse with/fertilise egg cell nucleus	1 mark 1 mark 1 mark 1 mark
Supplement 12 b)	oestrogen; stimulates repair and thickening of uterus lining in preparation for pregnancy progesterone; maintains uterus lining in preparation for pregnancy FSH; stimulates development of egg cell LH; stimulates release of egg/ovulation	2 marks 2 marks 2 marks 2 marks
13	Answer given in Student Book	
Supplement 14	1: (nitrogenous) bases 2: gene/double helix/DNA 3: chromosome 4: nucleus	1 mark 1 mark 1 mark 1 mark
15 a) i)	(Hint: In questions of this type, tell students that they should always have a good look through the family tree, and see what genotypes they can identify, before they even look at the questions.) Ff Father A does not have cystic fibrosis, so must have at least one dominant allele/F, but as his son C has cystic fibrosis, the father must be Ff (and not FF – if he were FF, there would be no possibility of producing a child with cystic fibrosis).	1 mark 1 mark
15 a) ii)	ff Son C suffers from cystic fibrosis, so must have two copies of the recessive allele/f	1 mark 1 mark

15 a) iii)	Both Ff	2 marks
	Neither have cystic fibrosis so each must have at least one dominant allele/F, but as daughter H has cystic fibrosis (so must be ff), they must also have one recessive allele/f each.	1 mark
15 b)	As neither have cystic fibrosis, I and J must each have at least one dominant allele/F.	1 mark
	None of their children have cystic fibrosis so they must all have inherited at least one dominant allele/F from their parents.	1 mark
	However, the grandson C has cystic fibrosis, so must be ff, and must have inherited one recessive allele/f from his father A, which he must have got from one of his parents, I or J.	1 mark
	So I is either FF or Ff, and J is either FF or Ff.	1 mark
Supplement 16 a) i)	In organisms heterozygous for codominant alleles both alleles contribute to the phenotype,	1 mark
	neither allele is completely dominant or recessive.	1 mark

Supplement
16 a) ii)

A red bull and a white cow:

		Female = C^WC^W (1 mark) possible alleles in eggs	6 marks
		C^W (1 mark)	
Male = C^RC^R (1 mark) possible alleles in sperm	C^R (1 mark)	C^RC^W (1 mark) roan cattle (1 mark)	

A roan bull and a white cow:

		Female = C^WC^W (1 mark) possible alleles in eggs	8 marks
		C^W (1 mark)	
Male = C^RC^W (1 mark) possible alleles in sperm	C^R	C^RC^W (1 mark) roan cattle (1 mark)	
	C^W (1 mark for both alleles)	C^WC^W (1 mark) white cattle (1 mark)	

Supplement
16 b)

(The father must be $I^A I^O$ and the mother $I^B I^O$.)

		Mother = $I^B I^O$ (1 mark for both parental genotypes) possible alleles in eggs		6 marks
		I^B	I^O (1 mark for all parental alleles)	
Father = $I^A I^O$ possible alleles in sperm	I^A	$I^A I^B$ Blood group AB (1 mark)	$I^A I^O$ Blood group A (1 mark)	
	I^O	$I^B I^O$ Blood group B (1 mark)	$I^O I^O$ Blood group O (1 mark)	

| 0.25 / 25% / probability of 1 in 4 of a child having blood group A | 1 mark |

Supplement 17 a)	Queen Victoria = X^HX^h	1 mark
	Prince Albert = X^HY	1 mark
	Prince Albert did not suffer from haemophilia	1 mark
	and as haemophilia is linked to the X chromosome, his X chromosome must contain a normal blood clotting allele,	1 mark
	Queen Victoria and Prince Albert had a son with haemophilia (Leopold) so Queen Victoria must have one haemophilia allele.	1 mark
	We know she has one allele for haemophilia rather than two, because she did not suffer from haemophilia.	1 mark 1 mark
Supplement 17 b)	The haemophilia allele is only found on the X chromosome, and there is no corresponding allele on the Y chromosome.	1 mark
	So if a haemophilia allele is present in the genotype, males are more likely to be affected as they have only one X chromosome.	1 mark
	It is possible for females to have haemophilia,	1 mark
	but as they have two X chromosomes, it is less likely that they will have two haemophilia alleles/ the chances of them having two haemophilia alleles is low.	1 mark
Supplement 17 c)	Edward VII (a son of Victoria and Albert) did not suffer from haemophilia/ he was X^HY,	1 mark
	so haemophilia would only have appeared in this line of the Royal Family if passed on by the person they had children with.	1 mark
Supplement 18 a)	i) mitosis ii) meiosis iii) mitosis iv) meiosis v) neither	5 marks
Supplement 18 b)	If the cells producing the gametes did not divide by meiosis,	1 mark
	the gametes would have the same number of chromosomes as body cells,	1 mark
	so at fertilisation the chromosome number of the zygote would be double that of the parent.	1 mark
	A reduction division/meiosis has to take place to produce gametes with half the number of chromosomes,	1 mark
	so that on fertilisation, the normal number of chromosomes is restored.	1 mark

Introduction

This section describes variation as a result of the effects of genes and environment, and leads on to how mutation is a source of new variation. This follows with a look at adaptive features of organisms, and how they link to selection, both artificial and natural.

Links to other topics

Topics	Essential background knowledge	Useful links
1 Characteristics and classification of living organisms	1.2 Concepts and uses of classification systems	
2 Organisation of the organism	2.1 Cell structure and size of specimens	
4 Biological molecules	4.1 Biological molecules	
15 Drugs	15.1 Drugs	
16 Reproduction	16.1 Asexual reproduction 16.2 Sexual reproduction 16.3 Sexual reproduction in plants 16.4 Sexual reproduction in humans	
17 Inheritance	17.1 Chromosomes, genes and proteins 17.2 Mitosis 17.3 Meiosis 17.4 Monohybrid inheritance	
19 Organisms and their environment		19.4 Populations
20 Human influences on ecosystems		20.1 Food supply 20.4 Conservation
21 Biotechnology and genetic modification		21.1 Biotechnology and genetic modification 21.3 Genetic modification

Topic overview

B18.1	Variation
	In this learning episode, students study the causes of variation in the phenotype as a result of variation in alleles and changes in the environment. This leads to a discussion of mutation as the cause of new alleles, and some of the factors that can increase the rate of mutation.
	Supplement Students will look in more detail at gene mutations and other sources of genetic variation.

B18.2	**Adaptive features**
	In this learning episode, students will learn to interpret information about an organism in order to describe its adaptive features.
	Supplement Students will also learn about the adaptive features of hydrophytes and xerophytes.
B18.3	**Selection**
	In this learning episode, students learn about natural selection and how it occurs. This leads on to selective breeding by humans of plants and animals for desirable features.
	Supplement Students will study how natural selection can lead to increased adaptation of a species to its environment, and also to evolution in a species using the example of antibiotic resistance in bacteria. Students will compare the processes of natural and artificial selection.
B18.4	**Consolidation and summary**
	This learning episode provides an opportunity for a quick recap on the ideas encountered in this section, as well as time for the students to answer the end of topic questions in the Student Book.

Careers links

These are some scientific careers that focus on this area of biology but careers in many other fields use the knowledge and skills gained studying science. **Plant breeders** use selective breeding to help improve the plants that they grow. **Artificial insemination technicians** support farmers to breed farm animals, often using carefully chosen livestock.

Starting points

The Student Book section opener puts the ideas in the section into context and sets the scene. It also allows students to acknowledge and value their prior learning, and provides a benchmark against which future learning can be compared.

- The questions provide a structure for introducing the section and can be used in a number of different ways:
- You could ask students to consider the questions as an introductory homework task.
- You could put students into groups to share their own ideas and understanding and then to report back to the whole class.
- Students could be given access to the Internet, preferably with a tight timescale, to find out the information required.

You could then use a spider chart or other form of wall chart to summarise everybody's ideas.

Recording these initial ideas allows you to retain them for reference as the individual topics are developed. In this way, your students' progress in learning can be readily acknowledged.

Learning episode B18.1 Variation

Learning objectives

- Describe variation as differences between individuals of the same species
- State that continuous variation results in a range of phenotypes between two extremes; examples include body length and body mass
- State that discontinuous variation results in a limited number of phenotypes with no intermediates; examples include ABO blood groups, seed shape in peas and seed colour in peas
- State that discontinuous variation is usually caused by genes only and continuous variation is caused by both genes and the environment
- Investigate and describe examples of continuous and discontinuous variation
- Describe mutation as genetic change
- State that mutation is the way in which new alleles are formed
- State that ionising radiation and some chemicals increase the rate of mutation
- Supplement Describe gene mutation as a random change in the base sequence of DNA
- Supplement State that mutation, meiosis, random mating and random fertilisation are sources of genetic variation in populations

Common misconceptions

The word *mutation* is generally associated with bad things such as cancer. Although a change in a gene is the cause of such illness, the term applies to any change in a gene. We never see the majority of such changes because they have no noticeable bad or good effect. As students will see in learning episode B18.1, mutation is the source of variation for natural selection, and so it is essential for evolution.

Resources

Student Book pages 365–369

Worksheet B18.1a Variation in humans

Worksheet B18.1b Skin melanoma

Approach

1. Introduction

Give students a few minutes to write down all the variation in phenotypic characteristics that they can see in the students in the class. Take examples, and each time ask what the cause of the variation is. Keep the discussion general and avoid any discussion that focuses on individuals. Try to elicit the two key sources of variation: from genes and from the environment. Use this to distinguish between phenotypic variation (the variation we see) and genetic variation (the variation in the alleles of genes, relating back to work in Section 17: *Inheritance*).

If students have done work in earlier years on the inherited and environmental causes of variation, you only need to summarise that work here. If they seem unsure of this work, give them the opportunity to investigate it further in the next task.

2. Variation in humans

Worksheet B18.1a provides an opportunity to investigate phenotypic variation in humans, including continuous and discontinuous characteristics. If possible, students should enter their data into a spreadsheet and analyse the results on computer. This will make it easier to gather a larger amount of data and so improve the reliability of conclusions. Some students may need help working out how to group the data and calculate proportions.

Warn students of the risks of drawing conclusions from differences in small amounts of data (due to the effect of random variation). For extra challenge, some students could be introduced to the idea of using standard errors on the data, in order to test the significance of any differences although it should be made clear to students that these aspects are beyond the requirements of the syllabus.

It is possible that students will not find any significant results, because of small sample sizes. However, the characteristics listed are all known to have some genetic component.

Although you may wish to refer to a variety of examples of variation, make sure that students also learn those examples listed in the syllabus: body length and body mass as examples of continuous variation, and ABO blood groups and seed shape and seed colour in peas as examples of discontinuous variation.

3. Mutations

Use page 368 of the Student Book to introduce mutations and their causes. Ask students to make notes of the key points and then to test each other on what they have learned by asking questions.

If there is access to the internet, students could then research examples of mutations and their effects. Alternatively they could research evidence for the effect of cancer-causing chemicals (mutagens) on the rate of mutation (for example, effects of smoking on lung, mouth and throat cancer).

4. Skin melanoma

Worksheet B18.1b provides a graph from Cancer Research UK on the increase in incidence of skin melanoma in the UK between 1975 and 2008. Questions on the sheet ask students to consider the genetic and environmental causes of this increase in mutation. The sheet introduces the terms *correlation* and *cause* to help students identify between the two.

Please note that the detail in this worksheet is beyond the requirements of the syllabus, but it provides opportunity to use knowledge of how ionising radiation can increase the rate of mutation.

Supplement 5. Sources of genetic variation

Ask students to read pages 368–369 of the Student Book about gene mutations and other sources of genetic variation and to summarise the key points. Explain that mutations are the ultimate source of variation, but that the random aspects of meiosis, mating and fertilisation serve to constantly bring alleles together in new combinations, leading to variation between individuals. If some students find this difficult to follow, ask them to start by thinking about how it is that two siblings (not identical twins) who share the same parents still have their own unique set of genetic characteristics.

6. Consolidation: true or false

Give students two pieces of paper and ask them to write *true* on one and *false* on the other. Then read out statements based on what they should have learned in this learning episode, some of which are true and some false, and ask them to hold up the correct answer.

Statements could include:

- Mutations always cause cancer. [false]
- Height is a characteristic that varies due to genes and the environment. [true]

Answers

Page 367

1. All the differences between individuals of the same species.
2. Discontinuous, e.g. ABO blood group or seed shape in peas; continuous, e.g. body length or body mass.
3. Discontinuous is usually caused by alleles in genes, because there are a limited number of alleles for a gene that produce each different variation of the characteristic. Continuous is caused by both genes and the environment because environmental factors affect the variation produced by genes, producing a range of variation.

Page 369

1. A genetic change that may produce a new allele that results in a different form of a characteristic.
2. Ionising radiation, such as ultraviolet radiation, X-rays or gamma rays; chemical mutagens such as the chemicals in tobacco smoke.
3. **Supplement** Mutation, meiosis, random mating, random fertilisation.

Worksheet B18.1a Variation in humans

Some of the characteristics in our phenotype are inherited from our parents, through our genes. Others are changed by conditions in the environment. Many are affected by both genes and environment. In this investigation you will collect data on a number of human characteristics to decide whether or not any variation is inherited, caused by the environment, or both.

Method

1. Choose some of the following characteristics, and draw up a table to record your results. You will need to ask a large number of people, including those who are related, so make sure there is room to record who is related to whom.
   Characteristics you could study are:

   - hair: naturally straight or curly
   - hair: natural colour
   - freckles: presence or absence
   - height
   - ear lobes: attached or unattached
   - eye colour: blue, brown or green
   - shoe size
   - longest finger on hand: first, second or third finger.

   There are other characteristics you could include, but remember that some, such as body size and proportion, change with age and so you should be aware of this if comparing adults and children. You should also consider how to record variation for each characteristic. For example, if you only have the categories of straight and curly hair, how will you classify hair that is a little wavy?
2. Gather data from as many people as you can, especially from members of the same family.
3. If possible, transfer your data to a spreadsheet for analysis.

Handling experimental observations and data

4. For each characteristic, identify whether it is continuous or discontinuous, and explain the effect of this on how easy it was to record variation in the characteristics.
5. For each characteristic, calculate the proportion of individuals in the whole group who have each variation that you recorded.
6. Taking one family at a time, calculate the proportion of individuals within the family who have each variation of a characteristic. Repeat this for all characteristics and all families.
7. For each characteristic, compare the proportions from families with the proportions of the whole group. Describe any differences.
8. If a characteristic is much more or less common in a family group than in the group as a whole, it is possible that the variation is inherited. Looking at your results, can you identify any characteristics that may be inherited?

Planning and evaluating investigations

9. Explain why you cannot say definitely that any particular characteristic is controlled only by genes or by the environment.
10. Explain why, when researching the cause of variation in a characteristic, scientists usually gather data from thousands of people.
11. In some studies of the causes of variation, scientists only compare data from identical twins. Explain why this helps them produce more reliable conclusions.

Worksheet B18.1b Skin melanoma

Skin melanoma is an aggressive form of skin cancer. In 2008 it was the sixth most common cause of cancer in men and women in the UK, although many people with this form of skin cancer are successfully treated and live for many years after treatment.

The graph shows the trend in number of people diagnosed with skin melanoma in the UK between 1975 and 2008.

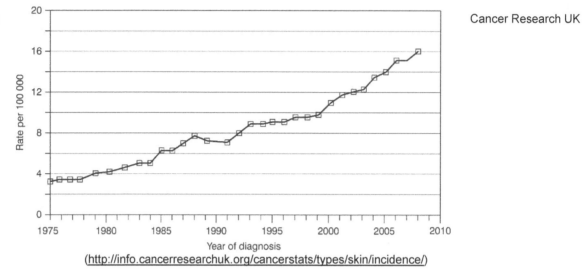

*Data source:

Cancer Research UK

Year of diagnosis
(http://info.cancerresearchuk.org/cancerstats/types/skin/incidence/)

1. Describe the trend in number of people diagnosed with skin melanoma between 1975 and 2008.
2. Some families show a higher than average rate of developing skin melanoma.

   a) What does this suggest?

   b) Explain, using the graph, why this probably is not the only cause of skin melanoma.

   Scientists studying the causes of skin melanoma have shown that the following groups are more likely to be diagnosed with skin melanoma than other groups:

   - people with fair skin who do not tan easily
   - women (compared with men)
   - people who have more money
   - people who spend more time in lower latitudes (nearer the Equator).

   The relationship between factors, such as skin type and occurrence of melanoma, is known as a *correlation*. This does not mean that one factor *causes* the other. To find a cause, we need to understand why the melanoma starts to form.

   It is known that UV radiation, from the Sun or from using sunbeds, is a mutagen that causes cells to change and become cancerous.

3. For each of the groups above, suggest as many reasons as you can why this cause of mutation could produce the correlation.
4. Take one example from your answer to question 3. Describe how you would investigate it to see whether it was a factor in the changing incidence of skin melanoma. Describe what data you would collect to test this and explain your choice.

Learning episode B18.2 Adaptive features

Learning objectives

- Describe an adaptive feature as an inherited feature that helps an organism to survive and reproduce in its environment
- Interpret images or other information about a species to describe its adaptive features
- Supplement Explain the adaptive features of hydrophytes and xerophytes to their environments

Common misconceptions

Students may be familiar with examples of the adaptations of organisms to their environment from earlier work. Understanding needs to be developed in this learning episode so that students relate adaptations not only to survival, but also to the chance to reproduce and pass on those adaptations to offspring.

Resources

Student Book pages 369–371

B18.2 Technician's notes

Resources for class activity (see Technician's notes)

Approach

1. Introduction

Give students a few minutes to jot down the names of some organisms and a particular feature that is associated with each organism – for example, cheetahs can run very fast, or whales have a streamlined body shape. Take a range of examples from round the class to write on the board, then ask students to work in pairs or small groups to identify why those particular features are useful to each organism. After a few minutes, take answers and lead to the idea that the features are related to the environment in which the organisms live.

2. Adaptive features

Introduce the term *adaptive feature*, and ask what it means or if students can suggest a meaning. They should recognise that this is what they described in the introduction. Then provide them with images of a range of organisms, or the names of a range, that live in different environments, such as polar regions, hot deserts, deep forest, open grassland, the bottom of a pond or surface waters. Ask students to select one or more organisms to research and to produce a list of adaptive features like those shown in Fig. 18.5 and Fig. 18.6 on pages 369 and 370 in the Student Book. They should not only identify the feature but also try to suggest how it improves the chances of survival.

Allow time to compare organisms that live in the same environments, to identify similarities and differences. This should help students draw general conclusions about which adaptations are most closely related to which environment.

Conclude by linking this work back to the work in Section 17: *Inheritance*, to point out that survival in an environment is not the only result of adaptations, but that reproduction passes adaptations to offspring.

Supplement 3. Hydrophytes and xerophytes

Introduce the term *hydrophyte* and *xerophyte* and then ask students to research examples of each. They should draw a labelled diagram for each plant that indicates how their features are adapted for their environments.

4. Consolidation: 20 questions

Give students a few minutes to think of an organism and all its adaptive features. Choose one student and ask others to ask questions in order to identify to which environment it is adapted. Questions can only be answered by 'yes' or 'no'. Repeat with several students to cover a range of different environments.

Technician's notes

Be sure to check the latest safety notes on these resources before proceeding.

The following resources are needed for the activity in task 2, per student or group:

photos or drawings of a range of organisms from different environments, such as:
polar regions: polar bear, walrus, Arctic fox
hot desert: cactus, desert rat, meerkat, thorny devil
deep forest: sloth, jaguar, liana, orchid
open grassland: ostrich, llama, lion, wildebeest
pond bottom: pond snail, dragonfly larva
surface waters: fish, mosquito larva

Answers

Page 371

1. An inherited feature that helps an organism to survive and reproduce in the environment in which it normally lives.
2. a) Any one from: white hair in winter – camouflage protects against being seen by predators; brown hair in summer – also camouflage protects against being seen by predators; thick fur in winter – for insulation against cold; large feet – for better grip on snow and loose rock for running away from predators.
    b) Any one from: fat in hump – provides energy and water from respiration when water and food are lacking; large feet – make it easier to walk/run on loose sand; large eyelashes/nostrils that close – prevent damage during sandstorms.
3. Supplement Sketch of cactus with labels to show the following:
    * extensive root systems to capture as much water as possible
    * reduced or no leaves to reduce water loss by transpiration
    * green stems for photosynthesis
    * stomata sunk deep in pits in the stem to reduce rate of transpiration
    * succulent stem (or leaves) to store water
    * hairy surface to reduce rate of transpiration from stomata.

Learning episode B18.3 Selection

Learning objectives

- Describe natural selection with reference to:
  - (a) genetic variation within populations
  - (b) production of many offspring
  - (c) struggle for survival, including competition for resources
  - (d) a greater chance of reproduction by individuals that are better adapted to the environment than others
  - (e) these individuals pass on their alleles to the next generation
- Describe selective breeding with reference to:
  - (a) selection by humans of individuals with desirable features
  - (b) crossing these individuals to produce the next generation
  - (c) selection of offspring showing the desirable features
- Outline how selective breeding by artificial selection is carried out over many generations to improve crop plants and domesticated animals and apply this to given contexts
- Supplement Describe adaptation as the process, resulting from natural selection, by which populations become more suited to their environment over many generations
- Supplement Describe the development of strains of antibiotic resistant bacteria as an example of natural selection
- Supplement Outline the differences between natural and artificial selection

Common misconceptions

One of the most common misconceptions about selection (natural or artificial) is that the selecting factor somehow causes the variation in the population – for example, that using an antibiotic suddenly makes some bacteria resistant to it. Explain that, before an antibiotic is used, there is no way of telling whether some bacteria are already resistant or not. As a result of natural variation caused by mutation, some of the bacteria are more resistant than others, but this only becomes apparent when the antibiotic is used.

Some people find it difficult to appreciate that evolution is an undirected process, caused by random changes in environmental conditions that change the impact of natural selection on the variation in a group of organisms. Instead, they think of evolution as a directed 'progress' toward something 'better'. It is important to emphasise the fundamental role of change in environmental conditions in natural selection, so that they realise that evolution is not directed toward a particular goal.

Resources

Student Book pages 372–377

Worksheet B18.3 A model of natural selection

B18.3 Technician's notes

Resources for a demonstration and class practical (see Technician's notes)

Approach

1. Introduction

Show students a range of pictures of animals of different breeds, such as cats or sheep, or plants of different varieties, such as varieties of *Brassica oleracea* (cabbage, kale, kohlrabi, cauliflower, etc.). (Make sure that there are images of different breeds/varieties of the same species of animal or plant, which should vary as much as possible in phenotypic characteristics.) Tell students that the pictures show varieties of the same species. Refer back to Section 1.2 *Concept and uses of classification systems*, to revise what *species* means if students are unsure.

Explain that originally there was one wild form of the species (e.g. wolf for dogs and wild cabbage for cabbage varieties). Give students a few minutes to work in pairs or small groups to consider how the different varieties were developed. Take suggestions from around the class and discuss any differences. Do not criticise any incorrect ideas at this point, but make sure they are tackled in the rest of the learning episode.

2. Natural selection

Worksheet B18.3 gives students an opportunity to model natural selection as a result of 'predation' of coloured sticks on two different colour backgrounds. Some students may be tempted to cheat in order to change the results, so be prepared to discount results that do not look reasonable.

Check that students have drawn up a suitable table for their results before they carry out the tests.

Students should find that the colour of stick that most closely matches the colour of the background increases in number over the few generations tested, because the other colour is most obvious. (At some point the 'rarity' value of the non-camouflaged colour will create a balance of the two colours in the population.)

Students could follow up this practical with research in books or the internet for real-life examples of natural selection, such as the predation of banded snails by thrushes.

Supplement Students should apply what they have learned to write a paragraph to answer the question, 'How does natural selection lead to populations of organisms becoming better adapted to their environment over many generations?'

3. Describing natural selection

Ask students to read pages 372–374 of the Student Book on natural selection, and to use the bullet lists in the text to create a flow chart that explains how natural selection leads to the passing on of more copies of their genes to the next generation by some individuals compared with others. They should be encouraged to write their flow chart in their own words, in brief notes, and not to copy out text. It might be helpful to give them a particular example to describe, such as the selection of more effective camouflage.

Students should include reference to competition and differential survival and reproduction in their flow chart.

Supplement ## 4. An example of evolution

Ask students to define the term *evolution*. Note – the syllabus doesn't actually use the term *evolution* in this section of the syllabus, but it does appear elsewhere (see Section 1.2: *Concept and uses of classification systems*) and students should understand it. The key idea is that it is a change in the characteristics of a species. Explain that unless there is a change in differential survival and reproduction (for example, due to a change in the environment or the appearance of a mutation that is beneficial), natural selection will not lead to evolution. Give them a few minutes to discuss this in pairs or small groups and then take any questions and sort out any confusion.

Ask students to read pages 373–374 of the Student Book on the increase of antibiotic resistance in bacteria. Then ask questions such as the following to test their understanding:

- Why is the increase in antibiotic resistance in bacteria an example of evolution? [because antibiotic use is resulting in a change in characteristics in the population]
- In terms of natural selection, why is antibiotic resistance increasing so quickly? (This needs an answer that is more than 'We are using lots of antibiotics'.) [because the antibiotic kills a large proportion of the population initially, so only those that are resistant may survive and reproduce]
- How could antibiotic resistance be reduced? [any suitable measures, such as: not using antibiotics, but this produces its own problems; killing bacteria by other means such as good hygiene, making sure people complete their course of antibiotics so that the risk of passing on resistant bacteria is minimised; only using antibiotics when really needed.]

5. Selective breeding

Ask students to research a range of animal and plant species that have been developed through selective breeding (also called *artificial selection*) to produce breeds and varieties of economic importance. Examples could include sheep for meat and wool, chickens for meat and egg-laying, varieties of wheat or rice.

For each example, they should identify two or more breeds/varieties and explain how they have economic importance, and how they have been developed by crossing individuals with the most desirable features. They should also try to identify an original 'wild' form of the species, to help show how the characteristics have changed as a result of selection.

Supplement Students should compare the processes of natural selection and artificial selection, for example by drawing flow charts to show what happens in each process. Ask them to identify similarities and differences in the two processes.

6. Consolidation

Ask students to suggest words to write at least ten clues for a crossword on key words from their work on variation. Give them three minutes to do this, then ask them to try their clues out on another student and work together to improve the clues. Take examples from around the class for each key word (there are many software packages available for creating crosswords – some of them are free).

Technician's notes

Be sure to check the latest safety notes on these resources before proceeding.

The following resources are needed for the demonstration in the introduction:

pictures of animals of different breeds, such as cats or sheep, and/or pictures of plants of different varieties, such as varieties of *Brassica oleracea* (cabbage, kale, cauliflower, etc.).

Make sure that there are images of different breeds/varieties of the same species of animal or plant, which should vary as much as possible in phenotypic characteristics.

The following resources are needed for the class practical B18.3, per group:

30 sticks each of 2 different colours
2 coloured backgrounds
large pot for mixing up sticks
blunt forceps
stopwatch or clock

The sticks could be toothpicks, or any similar short sticks. Alternatively, use rice grains or similar that will absorb colour well.

The sticks should be dyed with food colouring, and should match the colours of the two backgrounds as closely as possible. If possible, avoid using red with green, because of the possibility of red-green colour blindness in students.

Answers

Page 374

1. Answer along the lines of: the influence of the environment on the inheritance of a characteristic, such that some variations of the characteristic cause the individual organisms to be more successful at producing offspring than others and so pass on the alleles for that characteristic to their offspring.
2. a) If the individuals in a population are not all the same (there is variation), natural selection will favour some over others.
   b) If there is competition between variations, those that are better adapted will be more likely to survive to produce offspring and pass on their alleles.
   c) Individuals with particular variations of adaptive features be more likely to survive and reproduce than individuals with other variations. So they are more likely to produce offspring that carry their alleles, and so those alleles will become more common in the next generation.
3. Supplement As the better-adapted individuals contribute more offspring to the next generation than those that are less well adapted, more individuals will have the alleles for the better adaptations. This means that more individuals in the next generation will be better suited to their environment.
4. Supplement Diagram should show the following:
   person infected with bacteria > bacteria grow in number inside patient > treatment of patient with antibiotics kills off least-resistant bacteria but most-resistant bacteria survive > some of these bacteria escape into the environment from the patient and infect another person > the same antibiotic cannot be used on that patient as the bacteria are resistant.

Page 376 Science in context: Tulip mania

Selective breeding can only work with alleles that already exist. It can bring together, in new combinations, alleles that were already present in different individuals, but it does not produce new alleles. If the alleles necessary for a perfectly black flower do not already exist, then selective breeding will not be able to create them.

Page 377

1. Any two suitable examples, e.g. large eggs produced by chickens, unusual flower colours, crops that produce large amounts of grain.
2. Supplement Parent organisms with desirable characteristics are selected and bred together. Offspring with the best combination of characteristics are selected and then bred together. The process is repeated over many generations to produce a number of individuals that all exhibit the desired characteristics.
3. Supplement Natural selection is the selection by the environment for features that are best adapted to the conditions in that environment. Selective breeding is the selection of features by people of plants and animals that the people think are most useful or most attractive

Worksheet B18.3 A model of natural selection

Natural selection occurs when some individuals in a group are better able to survive and produce more offspring than others. The factor that makes this possible is called the selection factor. One obvious selection factor is predation, for example predation of snails by birds. Snails that have a similar colour to their background are better camouflaged and are less likely to be seen and eaten by a bird.

In this practical you are going to investigate the effect of 'predation' on a population of coloured sticks. You will work in pairs, with one student acting as the 'predator' and one as the experimenter.

Apparatus

30 sticks each of 2 different colours

2 coloured backgrounds

large pot for mixing up sticks

blunt forceps

stopwatch or clock

Method

1. The experimenter should place 20 sticks of each colour into the pot and mix them thoroughly.
2. Without the predator looking, the experimenter should scatter the sticks on one of the coloured backgrounds, making sure the sticks do not overlap. They should then start the stopwatch.
3. The predator has 30 seconds to collect as many sticks as possible, one at a time, with the forceps and place them in the empty pot.
4. At the end of 1 minute, count how many sticks there are of each colour still on the background. Record the number of each colour, and the colour of the background.
5. Empty the pot of sticks that were 'caught' and return the sticks still left on the background to the pot, two of the same colour at a time. For each pair of sticks returned to the pot, add another stick of the same colour. (This models the survivors breeding and producing one 'offspring' of the same colour.)
6. Repeat steps 2–5 twice more, recording the number of each colour of sticks that are left on the background each time.
7. Repeat the investigation with the other colour background.
8. If there is time, swap roles as predator and experimenter.
9. Draw a suitable table to display your results.

Handling experimental observations and data

10. Describe your results.
11. Explain your results using what you know about predators and camouflage.
12. Explain what your results suggest about natural selection.

Planning and evaluating investigations

13. Describe any problems you had with this investigation. How do you think they affected your results?
14. Suggest how the method could be adjusted to avoid these problems.

Learning episode B18.4 Consolidation and summary

Learning aims

- Review the learning points of the topic summarised in the end of topic checklist
- Test understanding of the topic content by answering the end of topic questions

Resources

Student Book pages 378–381

Approach

Introduce the learning episode

Ask students to work with a partner to make a list of key words from this topic. They could then work together to produce a spider diagram showing how the different concepts are linked. They could compare their list with the list of key terms given on page 378 of the Student Book. Discuss the checklist on pages 378–379 and use questioning to see how much of the content they are comfortable with.

Students could make flashcards of the key content and then use the flashcards to quiz each other on the information.

Develop the learning episode

Ask students to work individually through the end of topic questions on pages 380–381 of the Student Book without looking at the text. As they work, walk around the classroom observing their answers and asking questions as necessary to find out which questions are causing difficulties.

Finish the learning episode

After a set period, ask the students to stop working and discuss any areas of difficulty you observed as you walked round the class. Students should complete any unanswered questions for homework, but you should stress that they should try to answer the questions without looking at the text, so that they can see how much they have remembered..

Answers

End of topic questions mark scheme

The marks available for a question can indicate the level of detail you need to provide in your answer.

Question	Correct answer	Marks
1	C	1 mark
2 a)	Discontinuous variation,	1 mark
	because there are only two possible states, and no intermediates.	1 mark
2 b)	Continuous variation,	1 mark
	An example such as weight or height could be given.	1 mark
2 c)	There is a range of variation between two extremes.	1 mark
3 a)	Random change in an allele.	1 mark
3 b)	ionising radiation (or example)	1 mark
	mutagenic chemicals (or example)	1 mark
3 c)	Ionising radiation from UV light as a result of over-exposure of skin to sunlight,	1 mark
	because skin is more exposed to this form of radiation than other parts of the body.	1 mark
3 d) i)	Both curves increased,	1 mark
		1 mark

Question	Correct answer	Marks
	from about 500 new cases to over 2000 new cases for men and from about 300 to around 1500 new cases for women.	
3 d) ii)	Any suitable explanation that refers to increased exposure to stronger sunlight, e.g. more holidays in tropical regions, people trying to get darker tans.	1 mark
	This increases the risk of over-exposure to UV light, which is ionising radiation that can cause cells to turn cancerous.	1 mark
4 a)	White colour for camouflage in the snow,	1 mark
	makes it easier for the polar bear to get nearer to the seals without being seen and therefore more likely to catch seals for food.	1 mark
	(allow other examples with suitable explanations)	
4 b)	*Either*:	1 mark for adaptive feature
	thick fur or fat, for insulation against the cold	
	or:	1 mark for suitable explanation
	large feet, for better grip on snow and ice.	
Supplement 5 a)	Annotations should include:	
	• green photosynthetic cells around the outside of the stem, not in leaves	1 mark
	• no leaves to reduce water loss through transpiration	
	• thick body of succulent water-storing tissue	1 mark
	• spines to deter herbivores from eating stem.	1 mark
	Also, not visible:	1 mark
	• extensive root system to gather as much water as possible	
	• stomata sunk in pits in stem to reduce rate of transpiration.	1 mark
		1 mark
Supplement 5 b)	Xerophyte,	1 mark
	because the features it has helps it to survive in dry conditions.	1 mark
6	Any suitable example that illustrates the following:	
	Individuals in a population show variation in their characteristics.	1 mark
	If some variations are better adapted to the environment, those individuals will be healthier and produce more offspring.	1 mark
	So those individuals will pass on more copies of their genes to the next generation than individuals that are not as well adapted.	1 mark
Supplement 7	Treatment of person 1 with antibiotic A kills off the less-resistant bacteria, but more-resistant bacteria survive; the more-resistant bacteria escape to the environment and infect person 2.	1 mark
	Person 2 falls ill, antibiotic A will not control the infection, so person 2 is given a different antibiotic (antibiotic B).	1 mark
	The less-resistant bacteria are killed by antibiotic B but the more-resistant bacteria survive; these bacteria escape to the environment and infect person 3.	1 mark
	Person 3 falls ill and antibiotics A and B will not have an effect, so they are treated with antibiotic C … and so on.	1 mark
	Total:	34 marks

Introduction

This section introduces students to ecology, and gives them opportunities to study the flow of energy through food chains and food webs. This is followed by work on nutrient cycles, and on population studies.

Links to other topics

Topics	Essential background knowledge	Useful links
1 Characteristics and classification of living organisms	1.1 Characteristics of living organisms 1.2 Concept and uses of classification systems 1.3 Features of organisms	
2 Organisation of the organism	2.1 Cell structure and size of specimens	
3 Movement into and out of cells	3.3 Active transport	
4 Biological molecules	4.1 Biological molecules	
6 Plant nutrition	6.1 Photosynthesis	
7 Human nutrition	7.1 Diet 7.2 Digestive system 7.5 Absorption	
8 Transport in plants	8.1 Xylem and phloem 8.2 Water uptake	
12 Respiration	12.1 Respiration 12.2 Aerobic respiration 12.3 Anaerobic respiration	
13 Excretion in humans	13.1 Excretion in humans	
16 Reproduction	16.1 Asexual reproduction 16.2 Sexual reproduction	
20 Human influences on ecosystems		20.1 Food supply 20.2 Habitat destruction 20.3 Pollution 20.4 Conservation

Topic overview

B19.1	**Energy flow, food chains and food webs**
	This learning episode introduces students to some of the terminology used in ecological studies, principally concerning food chains and webs. It also gives students an opportunity to draw and interpret pyramids of numbers and biomass.
	Supplement Students will consolidate their understanding of the transfer of energy through food chains and webs. They will also look at pyramids of energy.
B19.2	**Nutrient cycles**
	Students will study the carbon cycle with particular attention to the roles of living (and once-living) organisms.
	Supplement Students will also study the nitrogen cycle, and the importance of microorganisms in this cycle.
B19.3	**Populations**
	Students will study the factors that affect population growth and interpret information about population growth.
	Supplement Students will also learn to interpret sigmoid growth curves.
B19.4	**Consolidation and summary**
	This learning episode provides an opportunity for a quick recap on the ideas encountered in this section, as well as time for the students to answer the end of topic questions in the Student Book.

Careers links

These are some scientific careers that focus on this area of biology but careers in many other fields use the knowledge and skills gained studying science. **Ornithologists** study birds and may be involved in classification and conservation work. **Mycologists** and **plant pathologists** study fungi and the diseases that they can cause in plants. **Countryside officers and rangers** manage open spaces, the organisms that live there and the people that use the spaces.

Starting points

The Student Book section opener puts the ideas in the section into context and sets the scene. It also allows students to acknowledge and value their prior learning, and provides a benchmark against which future learning can be compared.

- The questions provide a structure for introducing the section and can be used in a number of different ways:
- You could ask students to consider the questions as an introductory homework task.
- You could put students into groups to share their own ideas and understanding and then to report back to the whole class.
- Students could be given access to the Internet, preferably with a tight timescale, to find out the information required.

You could then use a spider chart or other form of wall chart to summarise everybody's ideas.

Recording these initial ideas allows you to retain them for reference as the individual topics are developed. In this way, your students' progress in learning can be readily acknowledged.

Learning episode B19.1 Energy flow, food chains and food webs

Learning objectives

- State that the Sun is the principal source of energy input to biological systems
- Describe the flow of energy through living organisms, including light energy from the Sun and chemical energy in organisms, and its eventual transfer to the environment
- Describe a food chain as showing the transfer of energy from one organism to the next, beginning with a producer
- Construct and interpret simple food chains
- Describe a food web as a network of interconnected food chains and interpret food webs
- Describe a producer as an organism that makes its own organic nutrients, usually using energy from sunlight, through photosynthesis
- Describe a consumer as an organism that gets its energy by feeding on other organisms
- State that consumers may be classed as primary, secondary, tertiary and quaternary according to their position in a food chain
- Describe a herbivore as an animal that gets its energy by eating plants
- Describe a carnivore as an animal that gets its energy by eating other animals
- Describe a decomposer as an organism that gets its energy from dead or waste organic material
- Use food chains and food webs to describe the impact humans have through overharvesting of food species and through introducing foreign species to a habitat
- Draw, describe and interpret pyramids of numbers and pyramids of biomass
- Discuss the advantages of using a pyramid of biomass rather than a pyramid of numbers to represent a food chain
- Describe a trophic level as the position of an organism in a food chain, food web or ecological pyramid
- Identify the following as the trophic levels in food webs, food chains and ecological pyramids: producers, primary consumers, secondary consumers, tertiary consumers and quaternary consumers
- Supplement Draw, describe and interpret pyramids of energy
- Supplement Discuss the advantages of using a pyramid of energy rather than pyramids of numbers or biomass to represent a food chain
  Supplement Explain why the transfer of energy from one trophic level to another is often not efficient
- Supplement Explain, in terms of energy loss, why food chains usually have fewer than five trophic levels
- Supplement Explain why it is more energy efficient for humans to eat crop plants than to eat livestock that have been fed on crop plants

Common misconceptions

If students carry out research on food webs and food chains, they may meet statements such as 'All food webs begin with a plant', and 'All energy for life on Earth comes from the Sun'. Neither of these is completely true. There are many producers other than photosynthesising plants, most of which are bacteria and too small to be easily noticed, and/or they live in environments that are difficult to explore, such as in mid-ocean ridges or in deep caves. Some of these have unique biochemistry that takes energy from the breakdown of inorganic chemicals to make 'food' rather than using light energy from the Sun. However, it is fair to say that they contribute little to the food chains and food webs that we normally see.

Resources

Student Book pages 386–399

Worksheet B19.1 Drawing pyramids of numbers

Approach

1. Introduction

Students should have some understanding of ecological studies from earlier work. Give them a minute or so to write down what they remember about the terms *food chain* and *food web*. Ask can they remember any other terms, such as *habitat*, and, if so, also write down what they mean. Then take examples from around the class and encourage discussion to come up with class definitions for the terms. This should help you identify how much they remember from earlier work.

The following tasks assume that students have remembered little from previous work. The tasks may need adjusting if they have a lot of previous knowledge.

2. Energy inputs

Ask students what all organisms need in order to survive, grow and reproduce. Refer them back to their earlier work on the key characteristics of organisms from Section 1.1: *Characteristics of living organisms.* They should be able to respond with 'food', and to describe sources of this for both animals and plants.

Ask them what the food is used for. Explain that in animals it is a source of energy as well as providing essential raw materials, while in plants 'food' is created in photosynthesis as their source of energy with additional raw materials (mineral ions) being absorbed from the ground with water. Make sure students make the link with respiration, the process for releasing energy from 'food molecules' for use in other life processes.

Reinforce the link between 'food', which they will have studied in earlier work, and 'energy stored in food molecules' to help them develop their understanding for this course.

3. Food chains and webs

Ask students to work in pairs and to write down a food chain that they remember. (If they are struggling, give them an example of organisms that could be used to make a food chain, such as grass, rabbit, fox.) They should use their food chain to write a set of 'rules' for drawing a food chain. Take examples from around the class to draw up a set of class rules. They should include the following:

- the chain is set out with the organisms in order of who eats what, placing the 'eater' on the right of what they eat
- an arrow should point from the organism being eaten to the organism that is the 'eater'.

Note the rules above are written in the simplest terms. Students may have a greater understanding and be able to give names for the feeding (trophic) levels in the chain (if not, do not introduce them yet). Extend their understanding to realise that chemical energy stored in the tissues of the organism being eaten is transferred to the animal eating it.

Ask students what information they would need to develop their food chain into a food web. If there is time and access to the internet, they could look for examples of food webs, then choose one and identify as many food chains as possible within it.

4. Feeding levels

Ask students to read pages 386–389 of the Student Book about food chains and food webs and the naming of the different levels at which organisms feed. They should then annotate one food chain and food web used in the task above, or the diagram of a food web on page 389 of the Student Book, to name the levels.

Make sure students can identify *producers*, *consumers* (*primary*, *secondary*, *tertiary* and *quaternary*), *herbivores* and *carnivores*.

Alternatively, they could test each other by asking questions, such as: 'Name one primary consumer in the food web', or 'Which secondary consumer eats xxx and xxx?'

Introduce the term *trophic level* as the scientific name for a feeding level and expect students to use it in the remainder of their work in this section.

Note, this is also a good opportunity to introduce the role of *decomposers*, which are traditionally left out of food chains, although they may be included in some food webs. Students may like to suggest why this is. The role of decomposers will be covered in more detail in work on the carbon and nitrogen cycles later in this section.

Supplement 5. Energy gains and losses

Ask students to draw a simple sketch of a plant. They should annotate the sketch to show the energy transfers to and from the plant. It might help to remind them of their work on respiration and photosynthesis in earlier chapters, and that living tissue and waste materials are stores of energy locked up in the molecules within them. They may have difficulty working out the transfer of energy to the environment as a result of respiration that results in heating. They need to understand that the energy transferred to the environment as heat is no longer available to organisms.

Then ask them to repeat the activity using a simple sketch of an animal, such as a human. They should compare both annotated sketches with Fig. 19.8 and Fig. 19.9 on page 393 of the Student Book, to identify anything they have missed or are not clear about.

Give them the following questions and time to discuss the answers in small groups:

- Why is energy flow through a food chain always one-way and not recycled?
- Why is the energy in one trophic level always less than that in the trophic level that feeds on it?
- What impact does this have on the length of a food chain?
- Why is it more efficient in terms of energy transfers for humans to feed on plants than to feed on animals that fed on the plants?

Take suggested answers from groups, encouraging others to add to or amend the suggestions in order to form as clear an answer as possible for each question.

6. Interpreting human impacts

Ask students to read the section 'Interpreting human impacts on food chains and food webs' on page 390 of the Student Book. They could then use one of the examples given, or research their own example of overharvesting of a food species or introduction of a foreign species into a habitat. They should use what they learn to try to describe the changes in terms of food chain or food web diagrams.

If students carry out research on different examples, allow opportunity for them to present their results to the rest of the class, or to produce a poster that describes what has happened in text and diagrams.

7. Pyramids of numbers

Ask students to take a simple food chain from the last activity, such as grass → rabbit → fox, and to think about the number of organisms that would be found in each trophic level in a community of organisms. (For now, take care to choose a chain of organisms that will produce a traditional shape in a pyramid of numbers: that is, avoid something where there are fewer individuals in a lower trophic level than the one above, like lettuce → caterpillars → insect-eating bird.)

Then show them a picture of the plants and animals in an ecosystem, such as the African savannah or a lake or pond, and ask them to identify the organisms found in this ecosystem (including the plants), and to arrange them into a food web. Again, they should consider the number of individuals in each level.

Ask students to comment on what they found in both cases. They should suggest that generally the number of individuals in each level, for a single food chain or for a community, decreases as you go from producer to top consumer level.

8. Drawing pyramids of numbers

Worksheet B19.1 explains how to construct pyramids of numbers. It also gives students the opportunity to construct their own pyramid of numbers from data. Alternatively, provide an opportunity for students to research numbers for their own food chain. Give them the rules on the worksheet to construct their pyramids. Note that the example on the worksheet produces a 'pyramid' that is not a 'traditional' pyramid, and the final question asks students why this happens. Encourage discussion to consider the value of pyramids of numbers.

9. Pyramids of biomass

Introduce students to the term *biomass* as the dry mass of tissue in an organism, and ask students how they might gather data to change a pyramid of number for a food chain into a pyramid of biomass. As an extension to worksheet B19.1, provide the following data:

- the average mass of 1 lettuce was 92 g
- the average mass of 1 caterpillar was 3.9 g
- the average mass of 1 thrush was 83 g.

Ask students to use these values to convert the pyramid of numbers on the worksheet to a pyramid of biomass. Make sure they consider the total numbers as well as the average biomass for each type of organism. They should compare the shapes of the two pyramids and discuss the reasons for any similarities/differences.

Then ask students to use pages 394–396 of the Student Book, on pyramids of numbers and biomass, to draw up a table to show the advantages and disadvantages of each.

Supplement 10. Pyramids of energy

Following on from the previous task, explain to students that even pyramids of biomass are not always a true 'pyramid' shape as they are a 'snapshot' of the biomass at just one moment in time. Introduce them to pyramids of energy which, because they are measured over a period of time, are always a 'pyramid' shape. You could ask students to research to find out suitable data (or provide it for them) to enable them to plot pyramids of energy for particular food chains or webs.

Ask students to use pages 397–399 of the Student Book, or their own research, to draw up a table to show the advantages and disadvantages of using pyramids of energy, compared with pyramids of numbers and biomass.

11. Consolidation: definitions

Give students a few minutes to write notes for a web dictionary to explain food webs, feeding (trophic) levels and energy flow through them. They should include definitions of the key words (in bold) in the Student Book. They should swap their notes with a neighbour to check them and see how they could be improved. They should then swap them back again and make any changes that are needed to improve their notes.

 © HarperCollins*Publishers* Ltd 2021

Answers

Page 386 Science in context: Energy input from the Sun

In winter, animals migrate to sunnier, warmer areas where food is more readily available. Plant seeds germinate in spring and summer when it is warm enough and light enough for efficient photosynthesis and so growth.

Page 388 Science in context: Other producers

Jupiter is much further from the Sun than we are and so gets much less light. Also, ice would further reduce any light reaching the ocean below. This means there wouldn't be enough light energy for photosynthesis.

Page 391 Science in context: Extension in Hawaii

On the one hand, when new species have arrived in the past, before humans, some existing species may always have gone extinct. On the other hand, animals that have arrived with humans may never have naturally got there otherwise, and so would never have caused the damage they have.

Page 392

1. Producer: an organism that makes its own organic nutrients, usually using energy from sunlight, through photosynthesis.
   Consumer: an organism that gets its energy by feeding on other organisms.
   Herbivore: an animal that gets its energy by eating plants.
   Carnivore: an animal that gets its energy by eating other animals.
   Decomposer: an organism that gets its energy from dead or waste organic material.
   Trophic level: the position of an organism within a food chain, food web or ecological pyramid.
2. The Sun provides light energy, transferred as chemical energy to build plant tissue, which is then transferred as chemical energy through all other organisms in the ecosystem.
3. A food chain only shows the relationship between one producer, one herbivore that eats the producer, one carnivore that eats one herbivore, and so on. A food web shows the feeding relationships between all, or many of, the organisms living in an area.
4. a) Food webs help us to understand the relationship between organisms in an area, and can help us predict what might happen to the organisms as a result of a change to the ecosystem.
   b) It can be difficult to organise the information in a food web because some organisms feed at many trophic levels, and it may not be possible to include all organisms (e.g. decomposers) on a food web because of space for the drawing.

Supplement Page 394

1. Energy in light from Sun (gain) > some reflected, some passes straight through, some wrong wavelength (losses) > energy from light transferred to chemical substances during photosynthesis > energy transferred to environment from photosynthetic reactions and from respiration as heat (losses) > energy stored in plant biomass.
2. Energy stored in food (gain) > energy in undigested food lost transferred to environment in faeces (loss) > energy stored in absorbed food molecules transferred to energy in waste products such as urea in urine and transferred to the environment (loss) > energy released in respiration transferred as heat to environment (loss) > energy stored in animal biomass.
3. Not all the energy gained is stored in new tissue in the organism because a lot of the energy is lost from the organism in waste or as heat from respiration

Page 396 Developing Practical Skills

1. Plants → snails; plants are producers and snails are primary consumers

2. Plan should include the following:

   - method for sampling standard size areas
   - counting all individual plants and snails in each area
   - repeat samples so that means can be calculated to average out variability.

3. Take several typical plants. Dry out in warm oven overnight. Measure mass. Repeat drying for a few more hours and measure mass. Repeat until two consecutive masses are the same. Calculate an average mass per plant. Then do the same for several snails and calculate the average mass for a snail. Multiply the average biomass of each type of organism by the number of each type.
4. Plants 42, snails 4.
5. Pyramid with two bars centred one on top of other, bottom bar 42 units wide and top bar 4 units wide.
6. Pyramid with two bars centred one on top of the other, bottom bar 1596 units wide and top bar 24 units wide.
7. The pyramid has the typical shape of the top trophic level being smaller than the lower level. This is because there are more plants than snails, with each snail feeding off several plants.
8. The pyramid has the typical shape of the top trophic level being smaller than the lower level. This is because the total biomass of all the plants is greater than that of all the snails. This can be explained because not all the biomass in the plants is passed on to the snails – i.e. the snails do not eat all of the plants.

Page 399

1. Pyramid of numbers: a diagram showing the numbers of organisms at each trophic level in a food chain or food web in an area. Pyramid of biomass: a diagram showing the total biomass of the organisms at each trophic level in a food chain or food web in an area.
2. Any suitable example that includes producers, primary consumers and secondary consumers from a reasonable food chain. Count the number of individuals feeding at each level within the area. Draw a pyramid of three layers, starting with producers at the bottom and ending with secondary consumers at the top, with the bar for each level drawn to scale.
3. Supplement A diagram showing the total amount of energy in all the organisms at each trophic level in a food chain or food web in an area.
4. Supplement Although pyramids of numbers and biomass are often pyramid-shaped, this may not be true in the case of pyramids of numbers if there are many, very small organisms feeding on a few, very large organisms, or in the case of pyramids of biomass if organisms in one trophic level have much shorter life-spans than those in another. Pyramids of energy are always pyramid-shaped because it is always true that only a relatively small proportion of the energy at one trophic level is passed onto the next.
5. Supplement The pyramid for plant > animal > human should show three layers, widest at the bottom for plant and shortest at the top for human. The pyramid for plant > human should show two layers, the bottom one the same width as in the other pyramid, but the top one for humans should be wider than in the other. Explanation should indicate that eating the plants ourselves means more food is available, as energy is not lost to the environment from an intermediary animal level.

Worksheet B19.1

1. Correctly plotted pyramid of numbers drawn to a suitable size, with clearly labelled scale and bars.
2. This is an 'inverted' pyramid with the bar for the producers (lettuces) not being the largest.
3. This is because the producers (lettuces) are much larger in size than the primary consumers/herbivores (caterpillars).

Worksheet B19.1 Drawing pyramids of numbers

Pyramids of numbers are drawn according to a particular set of rules. Use the rules to answer the questions below.

The pyramids are drawn to scale, so graph or squared paper will help, and you will need a calculator.

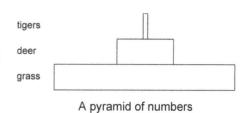

tigers

deer

grass

A pyramid of numbers

Rules for drawing pyramids of numbers

1. Pyramids of numbers are always drawn with the producer level at the bottom, and they work upwards through the food chain to the top consumer at the top.
2. Start with the largest value for a feeding level (often, but not always, the producer level). Work out a scale that you can use to fit this number within the width of your paper. For example, if the value is 2000 and your diagram can be 10 cm wide, then 1 cm on your scale will represent the value 200.
3. Use this scale to calculate the relative sizes of the values for the other feeding levels.
4. Mark the maximum line width you can use along the bottom of your paper, and mark the midpoint.
5. Starting with the producer level, mark the width of the scaled value at the bottom of the page. Centre the line on the midpoint of your original line. From this width, create a rectangle to form the first step of the pyramid as shown above. It doesn't matter how high this step is, because the height doesn't represent anything – it just needs to look right on your paper.
6. On the top line of the producer step, mark the width that represents the value for the primary consumer/herbivore level. Make sure this width is centred on the midpoint of the producer step.
7. Draw in a step of the same height as the producer level but using the width you marked out for the primary/consumer herbivore level, so that it sits on top of the producer step.
8. Repeat steps 6 and 7 for each successive level. Label each step with the feeding level or organism that it represents.

Questions

A student was studying the food chain: lettuce → caterpillar → thrush in the school vegetable garden. She watched the garden for 2 hours one day and counted the following numbers of organisms: 56 caterpillars, 4 lettuces, 2 thrushes.

1. Draw a pyramid of numbers for this food chain, using these values.
2. What do you notice about the shape of this pyramid?
3. Suggest why the shape is like this.

Learning episode B19.2 Nutrient cycles

Learning objectives

- Describe the carbon cycle, limited to: photosynthesis, respiration, feeding, decomposition, formation of fossil fuels and combustion
- **Supplement** Describe the nitrogen cycle with reference to:
    - (a) decomposition of plant and animal protein to ammonium ions
    - (b) nitrification
    - (c) nitrogen fixation by lightning and bacteria
    - (d) absorption of nitrate ions by plants
    - (e) production of amino acids and proteins
    - (f) feeding and digestion of proteins
    - (g) deamination
    - (h) denitrification
- **Supplement** State the roles of microorganisms in the nitrogen cycle, limited to: decomposition, nitrification, nitrogen fixation and denitrification (generic names of individual bacteria, e.g. *Rhizobium*, are **not** required)

Common misconceptions

Combustion of fossil fuels rapidly contributes carbon to the atmosphere. This is easier to understand if students understand clearly that the fuels are formed from the remains of huge quantities of dead plant or animal material deposited over millions of years and changed by geological processes over this time. Since plant and animal tissue is rich in carbon, so are fossil fuels.

Although nitrogen gas forms the majority of air, we tend to ignore it because it is not obviously relevant to living organisms. It is important to emphasise the essential role of nitrogen-fixing bacteria in converting nitrogen gas to forms that plants can absorb, so that all organisms can get the nitrogen-containing compounds they need for life.

Resources

Student Book pages 399–405

Worksheet B19.2a Constructing the carbon cycle

Supplement Worksheet B19.2b The nitrogen cycle

B19.2 Technician's notes

Resources for class practical (see Technician's notes)

Approach

1. Introduction

Revise work from Section 6: *Plant nutrition*, and Section 12: *Respiration*, by asking students to write down equations for photosynthesis and respiration.

If they are not sure, remind them of the practical work they carried out to investigate photosynthesis and respiration. Focus on the measurement of the net release of carbon dioxide using hydrogencarbonate indicator. Ask them to write down (or give them) an equation for the burning of fuel (combustion).

Give them a minute or so to compare the equations and write down three key points from the comparison. Take examples from around the class to make sure that students grasp that combustion and respiration are similar in releasing carbon dioxide, and photosynthesis is opposite in taking in carbon dioxide.

2. Constructing the carbon cycle

Worksheet B19.2a provides cards that students can cut out and use to construct a diagram of the carbon cycle. Explain that the term *stores of carbon* on the worksheet means 'something in which carbon is found'. Also make sure that students understand the terms *combustion* (burning, in this case of organic material such as fossil fuels) and *decomposition* (decay of dead organisms and animal wastes).

Encourage students to make the link between the carbon dioxide released during decay and the respiration of the decomposers as they use the food materials to release energy for growth and reproduction.

Students should lay out the cards first in what they think is the right arrangement, and then compare their layout with the diagram of the carbon cycle on page 401 of the Student Book before sticking them into their workbook.

You may wish to point out to some students that they will need several copies of the word *respiration.* For extra challenge, students could be encouraged to identify the form of carbon in each of the stores shown in their diagram.

If there is time, students could research in books or on the internet how much carbon there is in each carbon store given on the worksheet.

Supplement 3. The nitrogen cycle

Worksheet B19.2b contains a partially-completed diagram of the nitrogen cycle. Explain that students should complete the missing labels for the diagram. Before looking at the Student Book, they could try to add any labels they can in pencil.

Using a large version of the diagram on the board, take suggestions from the class for all labels that do not relate to bacteria. Ask students to suggest what might fill the remaining gaps. If needed, hint that it is a kind of organism.

Then ask students to read pages 401–405 of the Student Book on the nitrogen cycle and to complete the labelling. They should write a sentence for each type of bacterium in the cycle to explain its role.

Supplement 4. Root nodule bacteria

If available, give students the opportunity to look at a prepared slide of nitrogen-fixing bacteria (*Rhizobium*) in a nodule of a root of a legume plant. Alternatively, show them the image on page 402 of the Student Book. Explain that the relationship between the bacteria and the plant is a specialised one called a mutualistic relationship in which each partner benefits.

Ask students to suggest how the bacteria and plant benefit from such a close relationship. [Bacteria get a protected environment with nutrients in which to grow, the plant gets a good supply of nitrates directly from the bacteria.]

If you have the resources available, you could grow legumes in tall glass jars for several weeks. Students can then see the root nodules through the sides of the glass jars.

If there is time, students could then research the use of legume crops in plant rotation to maintain soil fertility as an alternative to using fertiliser.

5. Consolidation: true or false

Give students two pieces of paper and ask them to write *true* on one and *false* on the other. Then read out some statements about the carbon cycle, some of which are true and some false, and ask students to hold up the correct answer each time. Suitable statements include:

- Fossil fuels are formed from dead plant or animal material. [true]
- Carbon is released from living organisms to the air by photosynthesis. [false]

Supplement Repeat this with appropriate statements for the nitrogen cycle.

Technician's notes

Be sure to check the latest safety notes on these resources before proceeding.

The following resources are needed for the root nodule bacteria activity, per student or group:

prepared slide of *Rhizobium* bacteria in the root nodule of a legume

Answers

Page 401

1. a) Respiration releases carbon dioxide into the atmosphere from the breakdown of complex carbon compounds inside organisms.
    b) Photosynthesis fixes/converts carbon dioxide from the atmosphere into complex carbon compounds in plant tissue.
    c) Decomposition decays/breaks down dead plant and animal tissue by decomposers, releasing carbon dioxide into the atmosphere during their respiration.
2. a) carbon dioxide, b) complex carbon compounds, c) complex carbon compounds

Page 403 Science in context: Maintaining soil fertility

The plants would have their own supply of nitrates which would mean that they would need no nitrogen-containing fertiliser. This would save money, energy and reduce pollution.

Supplement Page 404

1. a) Bacteria that increase the amount of nitrate ions in the soil by converting ammonium ions to nitrite ions and then to nitrate ions.
    b) Bacteria that convert atmospheric nitrogen gas directly a form that plants can use.
    c) Bacteria that reduce the amount of nitrate ions in soil by converting them to nitrogen gas.
2. For soil to be fertile, it must contain not only the elements that plants need for growth, but these elements must also be in a form that plants can take in, for example, nitrogen needs to be in the form of nitrate ions. Without this, plants will not grow well and become stunted.
3. Decomposers break down complex nitrogen compounds in dead plant and animal tissues and animal waste. This releases ammonium ions that nitrifying bacteria convert to nitrate ions that plants need. Without decomposers, the bacteria would have nothing to work on, and the concentration of nitrate ions in the soil would decrease.

Pages 404–405 Developing Practical Skills

1. Plan should include:

    growing one plant of each kind with each type of mineral ion solution
    keeping all plants in identical conditions, i.e. same light, temperature and amount/frequency of watering, so that other factors that can affect the rate of growth are controlled
    after several days/weeks measure the growth of each plant, using some suitable measure of growth, e.g. increase in height.

2. Prediction should suggest that the wheat plant with the nitrate-containing solution will grow better than the one without, but that there is unlikely to be any difference in the two legume plants.
    Wheat/all mineral ions: $(20.6 - 5.2) / 5.2 \times 100\% = 296\%$
    Wheat/without nitrates: $(13.6 - 4.8) / 4.8 \times 100\% = 183\%$
    Legume/all mineral ions: $(18.1 - 3.6) / 3.6 \times 100\% = 403\%$
    Legume/without nitrates: $(19.3 - 4.1) / 4.1 \times 100\% = 371\%$
3. The wheat plant with nitrates grew much more than the wheat plant without. There is no great difference in growth between the two legume plants.

4. The difference in the wheat plants is because the plant in the solution lacking nitrates cannot get the nitrogen it needs to make proteins, etc. for healthy growth.

5. The lack of difference between the legume plants is because the plant in solution lacking nitrates can get the nitrates it needs for healthy growth from the nitrogen-fixing bacteria in its roots.

6. **Supplement** Plants that contain nitrogen-fixing bacteria can grow as well in conditions when nitrates are limited as they can when nitrates are available, but plants without these bacteria grow less well when nitrates are lacking.

7. This experiment compares two different kinds of plants, so there may be something about the species that causes the difference. It also only compares one plant in each condition, which will not allow for natural variability between individuals. So, some of the results may be affected by chance.

8. Growing legumes of the same species, some that are inoculated with bacteria and some without, in conditions with and without nitrates, and growing a large number in each situation, would get rid of variation between species and allow for averaging to reduce the effect of random variation between individual plants.

Page 410 Science in context: Human population growth

The populations in both types of country have increased, but there has been a much greater increase in developing countries. The overall increases, particularly rapid in the 20th and 21st centuries, can be explained by an increase in birth rate, due to a greater abundance of food as a result of improved technology, and a decrease in death rate, as a result of improved medicine, hygiene and health care. One reason for the lesser increases in developed countries, compared with developing countries, is families limiting the number of children they have.

Worksheet B19.2a Constructing the carbon cycle

The boxes below show processes in the carbon cycle in **bold** text, and stores of carbon in normal text.

1. Cut out the boxes and arrange them to form a diagram that describes the carbon cycle. You may need to repeat some of the process names to complete your diagram.
2. When you are sure of the position, stick them into your workbook, and complete the diagram by adding arrows to link the boxes correctly.
3. Add pictures and other annotations to help explain more clearly what is happening in the diagram.

respiration	**feeding**	animal material
photosynthesis	**formation of fossil fuels**	dead plant and animal material
combustion	carbon dioxide in the atmosphere	fossil fuels
decomposition	plant material	

Supplement Worksheet B19.2b The nitrogen cycle

Complete the missing labels in this diagram of the nitrogen cycle.

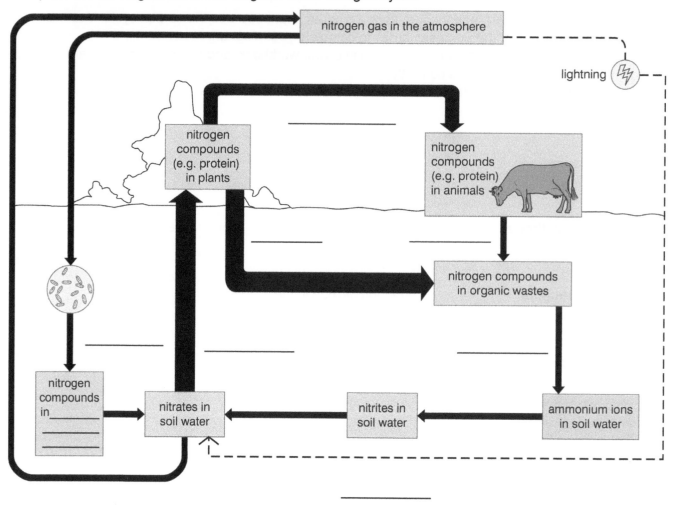

Learning episode B19.3 Populations

Learning objectives

- Describe a population as a group of organisms of one species, living in the same area, at the same time
- Describe a community as all of the populations of different species in an ecosystem
- Describe an ecosystem as a unit containing the community of organisms and their environment, interacting together
- Identify and state the factors affecting the rate of population growth for a population of an organism, limited to food supply, competition, predation and disease
- Identify the lag, exponential (log), stationary and death phases in the sigmoid curve of population growth for a population growing in an environment with limited resources
- Interpret graphs and diagrams of population growth
- Supplement Explain the factors that lead to each phase in the sigmoid curve of population growth, making reference, where appropriate, to the role of limiting factors

Common misconceptions

As the word *population* is in common usage, students need to become aware of its more specific, scientific definition.

Some students do not realise that an area can only sustain a certain size of population.

Resources

Student Book pages 405–410

Worksheet B19.3a Population change factors

Worksheet B19.3b Investigating change in population size

Worksheet B19.3c Explaining sigmoid curves

B19.3 Technician's notes

Resources for the class practical (see Technician's notes)

Approach

1. Introduction

Ask students to suggest suitable phrases that contain the word *population*. Once you have a few suggestions, give students a few minutes to work alone or in pairs and to use the phrases to produce a definition for the word. Take examples from around the class and discuss any variations.

Then ask students to read the scientific definition of the word on page 405 of the Student Book. Ask them how their general definition varies from this, and what they need to consider when talking about populations of living organisms. The Student Book gives one example of a plant population and one of an animal population. Ask students to write down one more example of each. Take examples from around the class and discuss whether these are reasonable examples.

Repeat the exercise using the terms *ecosystem* and *community*. The latter is likely to cause the most difficulty as it is commonly used to describe a group of people living in the same place. Students need to understand that the scientific use of the word means all the organisms of different species in a particular area. This may also be an appropriate place to introduce the term *habitat* (if it hasn't been used earlier) which appears more specifically in the next section, Section 20: *Human influences on ecosystems*.

2. Population change

Worksheet B19.3a provides some cards with words related to population change. Students could use these in a variety of ways, for example:

- to create word equations that explain why a population increases or decreases in size
- to discuss what affects the rate of change of a factor
- to discuss the effect of environmental conditions on the factors that affect population size.

You may need to discuss and explain the meaning of the words *immigration* and *emigration* before students begin this learning episode. Remind students that the scientific definition of *population* includes a restriction on the area in which the organisms live. Therefore, immigration and emigration may cause as much, or more, population change than births and deaths.

3. Investigating change in population size

Worksheet B19.3b provides a method for investigating the change in population size of duckweed (*Lemna*, a floating plant) in a limited environment. To provide clean duckweed, your technician can cultivate a few generations in regularly-changed clean, nutrient-balanced, aerated water to ensure a pure, uncontaminated patch of duckweed which will give a bit of experience and predictability in cultivating the plants for student use. Dispose of plant material safely to ensure non-native plants cannot contaminate the environment.

A suitable animal alternative is the brine shrimp (*Artemia*), which can be bought from biological suppliers. Follow the supplier's instructions carefully on the setting up and care of the shrimp. Counting shrimp may be more difficult than counting duckweed plants because the animals are often moving, so students may need instruction on estimating numbers.

This investigation needs to run for a few weeks (depending on conditions) in order to reach a stationary or death phase. If there are time and resources, this investigation could be extended to look at the effect of differences in environmental conditions: for example, set up different pots in different light intensities or with different concentrations of nitrate fertiliser (avoid high concentrations, which may kill the plants).

If necessary, discuss with students how they are going to record their results and what is the most suitable graph for them to draw. For Worksheet B19.3b, a suitable table would have time (in days) in one column and number of plants in another. A suitable graph would be time (in days) on the *x*-axis and number of plants on the *y*-axis.

SAFETY INFORMATION: Wash hands thoroughly after handling duckweed or pond water. Do not remove the covering from the jars. Dispose of plant material safely to ensure non-native plants cannot contaminate the environment.

4. Sigmoid curves

Ask students to compare the growth curve that they produced from their investigation using Worksheet B19.3b with the sigmoid growth curve on page 408 of the Student Book. They should identify similarities and differences. Explain that the curve shown in the book is one based on many experiments with different kinds of organisms, and so may not closely match their own results.

If there are similarities, particularly in the earlier stages of the curve, encourage students to describe them using the correct terminology for each phase as given in the Student Book. Ask them to suggest a description for each phase and to think of a way of linking the shape of the curve at that point to the name. This will help them remember the terms in the future. It might help if students find the definition of the term *exponential* or *log* for the second phase.

Supplement 5. Explaining sigmoid curves

Worksheet B19.3c has questions on explaining the phases of a sigmoid curve to help students understand what is happening at each phase.

The explanations given for the change in phases of the curve are described in terms of nutrient availability. Remind students that some factors in the environment may be limited, for example, nothing can be brought in to replenish nutrients that are used up. You could link this back to Learning episode 19.2 *Nutrient cycles*, to remind students of the importance of decomposers in the return of nutrients to the environment from dead organisms.

Another possible reason for the shape of the curve is damage to the environment, such as producing toxic waste. The amount of waste produced will increase as population size increases, which will then limit the birth rate and increase the death rate. Eventually conditions will become too toxic for organisms to live in. It is worth introducing this idea briefly now, to link to work in Section 20: *Human influences on ecosystems.*

6. Consolidation

Ask students to write a short description of population growth, and the factors that can cause it to change, for an online encyclopaedia. They should include examples in their description. They should then exchange their description with a neighbour and suggest how their neighbour's description could be improved. Encourage the pairs to discuss their suggestions and work together to complete their descriptions.

Technician's notes

Be sure to check the latest safety notes on these resources before proceeding.

The following resources are needed for the class practical B19.3b, per group:

5 duckweed plants
tap water left overnight to allow chlorine to evaporate
beakers or glass jars
plastic film (clingfilm)
access to light
marker pen

Duckweed (*Lemna*), or any other small floating aquatic plant, is suitable for this investigation. Be aware that duckweed is not available from ponds over winter. Select healthy plants for the experiment.

Use certified clean duckweed, if available. If you are unable to source a supplier of certified clean duckweed, cultivate a few generations in regularly-changed clean, nutrient-balanced, aerated water to ensure a pure, uncontaminated patch of duckweed which will give a bit of experience and predictability in cultivating the plants for student use.

There are safety issues if the plants die and start to decompose. You should be alert to this possibility and dispose of decomposing cultures as soon as you notice this.

Dispose of plant material safely to ensure non-native plants cannot contaminate the environment.

Clean all containers with 1% Virkon solution for 10 minutes.

Answers

Page 406

1. Population: a group of organisms of the same species living in the same area at the same time. Community: all of the populations of different species in an ecosystem.
2. All the organisms and the environmental factors that interact within in an area; examples include a lake, desert, tropical rainforest, coral reef, or anything similarly large-scale that has reasonably definable boundaries.
3. Populations of different species that live in different habitats form the community of organisms that live in an ecosystem.

Page 408

1. Food supply can increase population growth because it can increase birth rate and survival, reducing death rate. It can also cause an increase in immigration and decrease in emigration.
2. Predation and disease can decrease population growth because they increase the death rate.

Page 409

1. Lag phase, exponential (log) phase, stationary phase, death phase.
2. Microorganisms in a fermenter (or other suitable example) because conditions for growth are ideal and there is no other organism to predate on the populations.
3. Supplement Lag phase is when the individuals in the population are preparing for growth and reproduction but population size is increasing very slowly. Exponential (log) phase is where growth in population size is rapid because birth rate is fast due to ideal conditions. Stationary phase is where growth levels off and birth rate and death rate are equal, due to a limiting factor such as limited nutrients. Death phase is where population size falls because death rate is greater than birth rate, due to lack of a nutrient or increase in toxic conditions.

Worksheet B19.3c

1. Birth rate is low because the population hardly increases in size. It could be that the population is young and not reproducing.
2. Birth rate is much greater than death rate because the population size is increasing rapidly.
3. Supplement The nutrients are being absorbed from the environment into the organisms for making new materials in their bodies.
4. Birth rate and death rate are equal because population size is not changing.
5. Supplement Nutrient availability is likely to be limiting because the nutrients are removed from the environment by the organisms as they grow. This leaves fewer nutrients in the environment for new organisms.
6. Death rate is greater than birth rate because population size is decreasing.
7. Supplement There are probably no or very few nutrients left in the environment, so no or very few new organisms can grow. It could also be caused by a build-up of toxic waste or a lack of space.

Worksheet B19.3a Population change factors

The cards below show words related to population change and some of the factors that can affect whether a population increases or decreases in size.

Your teacher will explain how you are to use these cards.

births	deaths
birth rate	death rate
immigration	emigration
immigration rate	emigration rate
food supply	competition
predation	disease
population size	change in population size

Worksheet B19.3b Investigating change in population size

In this practical you will measure the change in population size of an organism. Duckweed plants (*Lemna*) are suitable for this as they are easy to grow and to count.

Apparatus

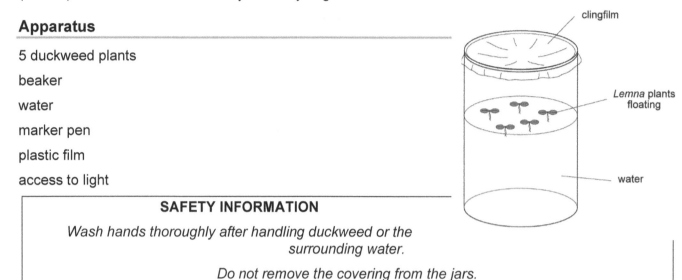

5 duckweed plants

beaker

water

marker pen

plastic film

access to light

SAFETY INFORMATION

Wash hands thoroughly after handling duckweed or the surrounding water.

Do not remove the covering from the jars.

Method

1. Use the marker pen to label the beaker with your name.
2. Half fill the beaker with water.
3. Select five healthy plants and float them on the top of the water.
4. Cover the top of the beaker with plastic film and use a sharp point, such as that of a pencil, to make a few air holes in the plastic.
5. Leave the beaker in a bright place.
6. Every day, count how many separate plants there are. Note that each plant may have several leaves, so make sure you count plants, not leaves. Once the surface of the water is almost fully covered with plants, you may have to estimate the number of plants. At this point you need only take a count every few days.
7. Record all your results in a suitable table.

Handling experimental observations and data

8. Use your results table to draw a suitable chart or graph.
9. Describe the shape of your graph.
10. Explain how population growth changes over the time that you observed the plants.

Planning and evaluating investigations

11. Describe any problems you had with this investigation. How do you think they affected your results?
12. Suggest how the method could be adjusted to avoid these problems.

Worksheet B19.3c Explaining sigmoid curves

The graph shows a sigmoid growth curve for a population of organisms growing in an environment with limited resources, for example bacteria or yeast growing in a fermenter.

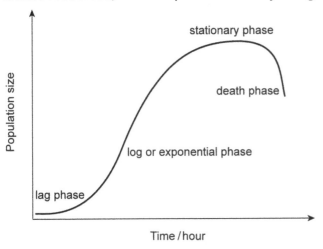

Lag phase

During this phase, the individuals are adjusting to the conditions. This is the 'lag' before there are any signs of population growth. In this phase the death rate is very low.

1. Using this information, what can you say about the birth rate? Explain your answer.

Exponential or log phase

During this phase, the population is doubling at a constant rate – this is what *exponential* means.

Start with a small number, for example, 2, and double it. Do this 15 times, then sketch a graph to plot the 'population size' against generation. The curve you have sketched is a logarithmic curve (hence 'log' phase). It is a particular mathematical shape.

2. What can you say about the birth rate and the death rate in this population during this phase? Explain your answer.
3. Supplement What is happening to the nutrients in the environment in which this population lives?

Stationary phase

During this phase population growth levels out – the population size stays the same (becomes stationary).

4. What can you say about the birth rate and death rate in this population during this phase? Explain your answer.
5. Supplement These changes in birth rate and death rate are caused by environmental factors that are now limited in availability. Suggest one factor that may now be limiting. Explain your answer.

Death phase

6. What can you say about the birth rate and death rate in this population during this phase? Explain your answer.
7. Supplement Suggest what has caused this change in birth rate and death rate since the stationary phase.

Learning episode B19.4 Consolidation and summary

Learning aims

- Review the learning points of the topic summarised in the end of topic checklist
- Test understanding of the topic content by answering the end of topic questions

Resources

Student Book pages 411–417

Approach

Introduce the learning episode

Ask students to work with a partner to make a list of key words from this topic. They could then work together to produce a spider diagram showing how the different concepts are linked. They could compare their list with the list of key terms given on pages 411 of the Student Book. Discuss the checklist on pages 411–413 and use questioning to see how much of the content they are comfortable with.

Students could make flashcards of the key content and then use the flashcards to quiz each other on the information.

Develop the learning episode

Ask students to work individually through the end of topic questions on pages 414–417 of the Student Book without looking at the text. As they work, walk around the classroom observing their answers and asking questions as necessary to find out which questions are causing difficulties.

Finish the learning episode

After a set period, ask the students to stop working and discuss any areas of difficulty you observed as you walked round the class. Students should complete any unanswered questions for homework, but you should stress that they should try to answer the questions without looking at the text, so that they can see how much they have remembered.

Answers

End of topic questions mark scheme

The marks available for a question can indicate the level of detail you need to provide in your answer.

Question	Correct answer	Marks
1	C	1 mark
2 a)	carnivore because it eats animals	1 mark
2 b)	primary consumer	1 mark
2 c)	grass → zebra → lion	
	organisms in correct order	1 mark
	arrows pointing in right direction.	1 mark
2 d)	food web showing grass at bottom	1 mark
	arrows to zebra, gazelle and wildebeest in middle layer	1 mark
	arrows from zebra, gazelle and wildebeest to lions in top layer.	1 mark

3 a)	They are broken down by the action of decomposers.	1 mark
	Some of the products of digestion are absorbed by the decomposers and some soak into the ground.	1 mark
3 b)	During the winter it is too cold for decomposer organisms to grow	1 mark
	so the leaf litter remains on the ground.	1 mark
	In spring, when it gets warm again, the decomposer organisms start to grow and break down the leaf litter.	1 mark
4 a)	pyramid of three bars drawn in proportion	1 mark
	centred one above the other with producer at bottom and secondary consumer at top	1 mark
	lowest bar 5 units wide labelled lettuce or producer, middle bar 40 units wide labelled caterpillar or primary consumer, top bar 2 units wide labelled thrush or secondary consumer	1 mark
4 b)	The pyramid has an inverted shape.	1 mark
	The data are very limited, for example the thrushes will go on to feed on other insects elsewhere, so the pyramid is not a reliable picture of what is happening in the community.	1 mark
4 c)	To measure its biomass, an organism needs to be dried,	1 mark
	which means it has to be killed.	1 mark
5 a)	The predatory insects would most likely decrease in number,	1 mark
	because there would be a lack of food.	1 mark
5 b)	The insectivorous birds would also decrease in number	1 mark
	because they feed on herbivorous insects/predatory insects/spiders which have all decreased in number, so there is a lack of food.	1 mark
5 c)	The mice might increase in number,	1 mark
	because there might be more food for them when there are no herbivorous insects.	1 mark
5 d)	The snakes may not change in number	1 mark
	as they could replace the animals that would be decreasing in number due to the loss of herbivorous insects with mice and seed-eating birds that might be increasing in number.	1 mark
6 a)	caterpillars,	1 mark
	because they eat the leaves on the tree	1 mark
6 b)	Although it is very large, the tree is only one organism, which means that its bar is much smaller than the caterpillars bar.	1 mark
6 c)	caterpillar bar would be much smaller,	1 mark
	because they would have changed into butterflies and flown away	1 mark

6 d)	typical pyramid shape, widest at the bottom	1 mark
	there is a lot of biomass in the tree so the bottom bar would be very wide	1 mark
Supplement 7	Of the energy that is available to a trophic level:	
	• in plants some is lost immediately because not all light is absorbed by the chlorophyll	1 mark
	• some is lost to the environment as heat as a result of respiration and other chemical reactions	1 mark
	• in animals, some is lost immediately as energy stored in the chemicals of undigested food as faeces	1 mark
	• some is lost to the environment as energy stored in the chemicals in the waste products of reactions such as urea	1 mark
	• and some is lost to the environment as heat as a result of respiration.	1 mark
	So always, less energy is available to the next trophic level than was available from the previous trophic level.	1 mark
Supplement 8	Regions near the Equator receive more light energy than at higher latitudes.	1 mark
	So more plant tissue can be produced and more animals can be produced from the chemical energy in that plant material.	1 mark
	Even though energy is lost at each trophic level in both food chains, because there is more energy for the plants in a tropical chain, there is more energy at each level up the pyramid,	1 mark
	meaning that there may be sufficient energy for more levels in that chain than in a high northern latitude chain.	1 mark
Supplement 9	When the farmer feeds the grain to the chickens, some of the energy in the grain is lost to the environment through respiration of the chickens, etc.	1 mark
	So there is less energy available in his food if the farmer eats the chickens than if the farmer ate the grain.	1 mark
10 a)	Light intensity increases through the morning as the Sun rises,	1 mark
	and decreases through the afternoon as the Sun sets.	1 mark
10 b)	As light intensity increases, rate of photosynthesis increases,	1 mark
	so more carbon dioxide is removed from the air due to photosynthesis than is returned by respiration of plants and animals in the forest.	1 mark
	As light intensity decreases, rate of photosynthesis decreases until it stops when it is dark,	1 mark
	so more carbon dioxide is given off into the atmosphere by respiration of all organisms in the forest, than is removed by photosynthesis.	1 mark

Supplement 11 a)	Nitrifying bacteria need aerobic conditions.	1 mark
	Denitrifying bacteria can grow in anaerobic conditions.	1 mark
	Denitrifying bacteria convert nitrogen compounds in soil to nitrogen gas, which escapes to the atmosphere.	1 mark
Supplement 11 b)	Plants need nitrogen for healthy growth.	1 mark
	Waterlogged soils have low concentrations of nitrates, so plants will grow poorly without another source of nitrogen.	1 mark
	Digesting animal tissue releases nitrogen compounds that the plants can use for growth.	1 mark
Supplement 12	Decomposer bacteria, nitrifying and nitrogen-fixing bacteria are all essential for adding nitrates to the soil.	1 mark
	Since plants can only take nitrogen in as nitrates from soil water, without these bacteria the nitrate concentration of soil water would drop,	1 mark
	which would reduce the rate of plant growth.	1 mark
	Less plant growth will provide less food for animals,	1 mark
	eventually leading to starvation and death in the community.	1 mark
13 a)	Population growth is likely to increase as adults have more food, so birth rate will increase	1 mark
	and death rate will decrease.	1 mark
	Immigration, as other mice are attracted to the area,	1 mark
	and decreased emigration may also have an effect.	1 mark
13 b)	Increased predation on the mouse population will increase death rate,	1 mark
	so decreasing population growth and size.	1 mark
	Total:	70 marks

Introduction

This section looks at the effects that human activities are having on ecosystems as a result of agriculture to provide us with food, habitat destruction, pollution and conservation. Students will study a range of important ecosystems, and be expected to apply previous learning from several sections.

Where possible, introduce appropriate local examples of pollution and conservation, to help students fully appreciate the effect of human activity in their local area, and to help them consider what can be done to limit any damage.

Links to other topics

Topics	Essential background knowledge	Useful links
1 Characteristics and classification of living organisms	1.1 Characteristics of living organisms 1.2 Concept and uses of classification systems 1.3 Features of organisms	
6 Plant nutrition	6.1 Photosynthesis 6.2 Leaf structure	
7 Human nutrition	7.1 Diet	
8 Transport in plants	8.1 Xylem and phloem 8.2 Water uptake 8.3 Transpiration	
10 Diseases and immunity	10.1 Diseases and immunity	
12 Respiration	12.1 Respiration 12.2 Aerobic respiration 12.3 Anaerobic respiration	
13 Excretion in humans	13.1 Excretion in humans	
16 Reproduction	16.1 Asexual reproduction 16.2 Sexual reproduction 16.3 Sexual reproduction in plants	
18 Variation and selection	18.1 Variation 18.2 Adaptive features 18.3 Selection	
19 Organisms and their environment	19.1 Energy flow 19.2 Food chains and food webs 19.3 Nutrient cycles 19.4 Populations	
21 Biotechnology and genetic modification		21.2 Biotechnology 21.3 Genetic modification

Topic overview

B20.1	**Food supply**
	This learning episode looks at ways of increasing food production and possible impacts on the environment. A visit from a local farmer or farm advisor is suggested to give the learning episode local context. Examples from other places are also suggested.
B20.2	**Habitat destruction**
	Students study how habitat destruction, including deforestation, may be caused, and how it can affect and damage ecosystems and the environment.
B20.3	**Pollution**
	This learning episode introduces a range of types of pollution: untreated sewage, excess fertiliser, non-biodegradable plastics, and air pollution by greenhouse gases leading to climate change.
	Supplement Students learn about the causes and effects of eutrophication.
B20.4	**Conservation**
	In this learning episode, students learn about the conservation of endangered species and sustainable resources.
	Supplement Students look in more detail at specific conservation methods.
B20.5	**Consolidation and summary**
	This learning episode provides an opportunity for a quick recap on the ideas encountered in this section, as well as time for the students to answer the end of topic questions in the Student Book.

Careers links

These are some scientific careers that focus on this area of biology but careers in many other fields use the knowledge and skills gained studying science. **Environmental consultants** work to manage risks from floods, climate change and waste production. **Agroecologists** aim to ensure that crops and farm systems are sustainable and minimise their effect on the wider environment.

Starting points

The Student Book section opener puts the ideas in the section into context and sets the scene. It also allows students to acknowledge and value their prior learning, and provides a benchmark against which future learning can be compared.

- The questions provide a structure for introducing the section and can be used in a number of different ways:
- You could ask students to consider the questions as an introductory homework task.
- You could put students into groups to share their own ideas and understanding and then to report back to the whole class.
- Students could be given access to the Internet, preferably with a tight timescale, to find out the information required.

You could then use a spider chart or other form of wall chart to summarise everybody's ideas.

Recording these initial ideas allows you to retain them for reference as the individual topics are developed. In this way, your students' progress in learning can be readily acknowledged.

Learning episode B20.1 Food supply

Learning objectives

- Describe how humans have increased food production, limited to:
  (a) agricultural machinery to use larger areas of land and improve efficiency
  (b) chemical fertilisers to improve yields
  (c) insecticides to improve quality and yield
  (d) herbicides to reduce competition with weeds
  (e) selective breeding to improve production by crop plants and livestock
- Describe the advantages and disadvantages of large-scale monocultures of crop plant
- Describe the advantages and disadvantages of intensive livestock production

Common misconceptions

Some students have little idea of how the food that they eat is produced. They may be aware of farming, but have not made the connection to the food that is sold in supermarkets. If this is the case with your students, consider starting with an activity that allows students to follow the production of a crop plant or farm animal to a food product in a shop, such as wheat to produce bread or sheep to produce lamb chops. Be sensitive to students who are uncomfortable with the killing of animals and preparation of carcases for food production. This could be combined with task 1 below.

Resources

Student Book pages 421–424

Approach

1. Introduction

Ask students to work in pairs or small groups to write down examples of food that they eat, and where and how they are produced. Some students may have a better understanding of this than others. Take examples from around the class and encourage discussion. If students seem unsure about the sources of food, consider the suggested activity in the misconceptions section above.

Remind students of their work on population growth in Section 19.4: *Populations*, and explain that the human population is growing at an exponential rate. Give students a minute or so to think about the implications of this growth and how food production may need to change to feed everyone alive in e.g. 2050. Take examples from around the class, but do not expect detailed answers at this point. General ideas are sufficient. Make sure students explain any suggestion.

2. Increasing food production

If possible, ask a local farmer or farm advisory official to visit and talk to the students about intensive food production and other techniques used to increase food production in farming. Before the visit, students should make a list of questions that they need to ask. This could be done on a KWL grid, with three columns. The first column should identify what the students need to know, the middle column shows what they want to know, and the third shows what they have learned by asking their questions.

Make sure students include in the 'need to know' column aspects of farming that are required by the syllabus (i.e. use of agricultural machinery, chemical fertilisers, insecticides, herbicides, selective breeding). It may be helpful to check that students understand all the terms used in the learning objectives.

Note that unless the farmer has a mixed farm, it is probable that they will only be able to answer questions on either the production of plant crops or farm animals. Ask students to identify which questions their visitor is most likely to be able to answer. Remind students that their questions should be considerate. Intensive farming can be an emotive subject, so students may need reminding that this exercise is for gathering factual information, not for the airing of views. Be prepared to manage discussion during the visit if some students appear to forget this.

During the visit, students should record the answers to their questions, to complete their table later. After the visit, allow students time to research the unanswered questions, using the Student Book, other books or the internet.

3. Increasing food production project

If a visit is not possible, or if there is time, ask students to research a particular example of a way of increasing food production. This can be done individually or in small groups. Encourage different students or groups to investigate different examples, including the production of staple crops (e.g. rice, wheat, potatoes) or fruit and vegetables, the production of cattle for milk, the production of fish using fish farming methods, and the production of poultry for eggs and meat. During their research, they should record how production is intensified, including the use of chemicals, machinery and selective breeding, and how intensification is affecting the environment.

Students could use their findings to produce a poster or short presentation. Allow time for students to present their work to the class, and for others to ask questions that cover the syllabus requirements.

4. Advantages and disadvantages

Students should use their findings from tasks 2 or 3 to list the advantages and disadvantages of large-scale monocultures of crop plants, and of intensive livestock production.

5. Consolidation: essay notes

Gives students a few minutes to write headings for paragraphs in an essay on *Increasing food production*. Take examples from around the class and encourage discussion to cover all the main points of this learning episode.

Answers

Page 424

1. Any four from: agricultural machinery increases the area of land that can be farmed and improves efficiency; chemical fertilisers improve yield of crop plants; insecticides reduce damage to crops by pests and so increase crop quality and yield; herbicides kill competing weed plants, and so increase crop yields; selective breeding improves the plant and animals that we grow for food, so producing more food.

2. a) Advantage: more efficient in terms of ploughing and harvesting. Any one disadvantage from: increased need for pesticides and other chemicals; reduced biodiversity.
   b) Advantage: more efficient in terms of amount of food produced. Any one disadvantage from: soil erosion, habitat destruction, reduced biodiversity.

Learning episode B20.2 Habitat destruction

Learning objectives

- Describe biodiversity as the number of different species that live in an area
- Describe the reasons for habitat destruction, including:
  (a) increased area for housing, crop plant production and livestock production
  (b) extraction of natural resources
  (c) freshwater and marine pollution
- State that through altering food webs and food chains, humans can have a negative impact on habitats
- Explain the undesirable effects of deforestation as an example of habitat destruction, to include: reducing biodiversity, extinction, loss of soil, flooding and increase of carbon dioxide in the atmosphere

Common misconceptions

Students may recognise the immediate impacts of habitat destruction, such as the killing of a particular plant or animal species, but may not appreciate the wider implications as a result of interdependency between organisms and their environment. Encourage them to look beyond the obvious, to see what else might happen.

Resources

Student Book pages 424–427

Worksheet B20.2 Effects of deforestation

Approach

1. Introduction

Ask students to imagine that a small local wood is about to be cleared to make space for some new buildings. Give them a few minutes, working individually or in pairs, to write down as many of the effects of this on the environment as they can think of.

Take examples from around the class and discuss any suggestions that might appear to conflict in order to identify a resolution. Encourage them to use terms learned in Section 19: *Organisms and their environment*, including *habitat, food chain, food web, population* and (if appropriate) *community* and *ecosystem*. In addition, introduce and explain the term *biodiversity* as a measure of how many species live in an area.

2. Habitat destruction

Use the internet or books to provide students with a range of images of habitat destruction, such as:

- large-scale monocultures of crops or large-scale livestock production
- open-cast mining for rocks (e.g. limestone) or minerals
- freshwater or marine pollution (e.g. the results of eutrophication, oil spills or plastic waste).

(Avoid deforestation as an example, as this will be covered in greater detail later.) If possible, provide both local and global examples.

Ask students to choose one example, and to work in small groups to research details of why the destruction occurred, and what impact it had, or is having, on the environment and the organisms that lived there before destruction began. Encourage students to think beyond the immediate impact, to the long-term effects on food webs and how that might affect a larger area. Ask students to draw up a list of reasons for habitat destruction, and how it can affect organisms through changing food chains and food webs. Each group should present their findings to the rest of the class.

3. Reasons for large-scale deforestation

Give students access to the internet or to books to research an area of large-scale deforestation, such as the Amazon Basin, parts of Indonesia and Malaysia, or a local example if appropriate. They should try to find answers to the following questions:

- Why is the deforestation taking place? What will replace the forest?
- Who will benefit from what happens after the deforestation?
- Are there any people who will not benefit from the deforestation? Explain your answer.
- Is anything being done to minimise the impact of the deforestation on the environment? If so, what, and how will it help?
- Is there anything that people outside the area could do to help minimise the impact of deforestation within the area? (Note: the answer to this should highlight the role of global trade in creating markets for products from these areas, and indicate that it is not just a local issue.)

4. Effects of deforestation

Ask students to read about the effects of deforestation on pages 426–427 of the Student Book. Make appropriate links to previous learning, such as food chains and webs in Section 19: *Organisms and their environment*, when discussing extinction, and photosynthesis in Section 6: *Plant nutrition*, and the carbon cycle in Section 19, when discussing the increase of carbon dioxide in the atmosphere.

Worksheet B20.2 provides a set of cards with words relating to various aspects of deforestation and life processes. Students could use these in different ways, for example:

- to start a concept map on the topic, and then add words and annotations of their own
- to link aspects of deforestation to living processes, to help explain the impact of deforestation on the environment
- to organise ideas to use as the basis of a poster on the subject.

5. Consolidation

Ask students to write brief notes for a web article on habitat destruction and deforestation. They should organise their notes under main headings to cover the key points in this learning episode. Ask students to work in pairs to compare their headings and notes for content. They should decide on the best headings and notes. Then take examples around the class and encourage discussion to decide on which headings to use and which main points should be included in each section.

Answers

Page 427

1. a) Any two such as: using land for building; extracting natural resources; pollution by chemicals; using land for growing crops; livestock production

   b) any two such as: pollution by chemicals, including oil or discarded plastic waste; warming of oceans

2. a) The destruction/cutting down of large areas of forest and woodland.

   b) The washing away of soil by heavy rainfall.

   c) When there are no individuals of that species left alive.

3. a) There will be a reduction in biodiversity because of the loss of plant species as well as a loss of animal species because they use the plants for food or shelter.
   b) Soil is washed away and nutrients leached from soil by increased water flow through ground, so decreasing soil fertility.
   c) Increases atmospheric carbon dioxide concentration as less carbon dioxide taken from air through photosynthesis and stored as wood.

Worksheet B20.2 Effects of deforestation

The cards below include words related to deforestation and some of the effects that it can have.

Cut out the cards and use them as instructed by your teacher.

leaching	growth	nitrogen cycle
carbon cycle	respiration	atmospheric oxygen concentration
combustion	nutrients	soil erosion
rainfall	global climate change	atmospheric carbon dioxide concentration
photosynthesis	populations	enhanced greenhouse effect
local climate	transpiration	decrease in biodiversity
loss of habitats	extinction	flooding

Learning episode B20.3 Pollution

Learning objectives

- Describe the effects of untreated sewage and excess fertiliser on aquatic ecosystems
- Describe the effects of non-biodegradable plastics, in both aquatic and terrestrial ecosystems
- Describe the sources and effects of pollution of the air by methane and carbon dioxide, limited to: the enhanced greenhouse effect and climate change
- Supplement Explain the process of eutrophication of water, limited to:
    (a) increased availability of nitrate and other ions
    (b) increased growth of producers
    (c) increased decomposition after death of producers
    (d) increased aerobic respiration by decomposers
    (e) reduction in dissolved oxygen
    (f) death of organisms requiring dissolved oxygen in water

Common misconceptions

The term *eutrophication* is commonly misunderstood. This is easiest to resolve by looking at the source of the word: *eutrophic* means 'well-nourished'. Eutrophication is simply the addition of nutrients to an ecosystem, typically a water system. Whether this leads to pollution depends on the ecosystem, the organisms within it and the amount of nutrients added to it.

Eutrophication is a natural process in water systems such as ponds and lakes, as nutrients are brought in by streams and rivers. However, the most rapid eutrophication occurs as the result of human activity, such as run-off of fertilisers or the dumping of sewage. This rapid eutrophication may lead to a rapid change in the ecosystem, and therefore it may be defined as pollution.

Note that carbon dioxide at natural concentrations in the atmosphere is not considered a pollutant. It is only the enhanced levels resulting from human activities that are polluting.

If students carry out their own research for images showing air pollution, make sure they can distinguish between water vapour (as from air cooling towers of power stations), which is not an air pollutant, and emissions from chimneys that contain pollutant gases. Many people confuse the two.

The syllabus uses the term *enhanced greenhouse effect*, meaning an enhancement of the natural greenhouse effect due to the addition of greenhouse gases to the atmosphere. In science books and on the internet, this term is used interchangeably with *global warming*, which is the increase in atmospheric temperature currently being recorded. The terms *greenhouse effect* and *global warming/enhanced greenhouse effect* are frequently confused, leading to the misconception that the greenhouse effect is a 'bad thing'. Although the syllabus does not specifically mention the greenhouse effect, it is important for students to distinguish between the terms and to realise that the greenhouse effect is a result of natural processes that make life on Earth possible.

As with any contemporary scientific debate, there is not perfect agreement about the causes of current global warming. A few scientists believe that the enhanced greenhouse effect/global warming is due to natural causes. However, the consensus is that it is mainly the result of human activity. This is the approach used in these materials.

Resources

Student Book pages 428–435

Worksheet B20.3 Evaluating the evidence

B20.3 Technician's notes

Resources for demonstration (see Technician's notes)

Approach

1. Introduction

Give students several minutes to work in pairs or small groups to revise earlier work by thinking of as many human influences on the environment as they can. Ask them also to consider the impact of those influences, both for humans and for other organisms, in the short term and in the long term if appropriate. Take examples from around the class and ask students to justify their examples.

Introduce the word *pollution* and ask students to work in their pairs or groups to create a definition for it. Take a range of definitions from around the class and work together to produce a class definition for the word. Compare the class definition with the definition given on page 428 of the Student Book and discuss any differences.

If appropriate, ask students to suggest local examples of pollution and ask them to suggest what damage the pollution is causing to the environment and the organisms that live there.

Explain that students are going to study some different examples of pollution caused by human activities and their impact on ecosystems, but that there are many other examples too. Remind them of their work on food webs in Section 19: *Organisms and their environment*. Tell them to keep in mind that the effect of pollution on some organisms in a food web may have extended impacts throughout the ecosystem because organisms are interdependent.

2. Water pollution

Ask students to read the sections on water pollution and fertilisers on pages 428–429 in the Student Book. Provide students with recent examples of different types of water pollution, ideally both local and global examples, and ask students to identify the causes of the pollution and their effects on the environment, people and other organisms. Provide examples that include untreated sewage and fertiliser run-off. Alternatively give students time to research examples in books or on the internet.

Ask students to imagine that they work for a national water protection agency, which has responsibility for maintaining the quality of water in rivers, canals and lakes. They should produce a bullet point list for people who add substances to water systems, including industry, farmers and wastewater treatment plants, on how they can ensure that what they add to the water system has little impact on the environment. They could also research the legislation that is used to control water pollution.

Supplement 3. Eutrophication

Point out that fertilisers and sewage both add mineral ions (such as nitrates) to water, which leads to eutrophication. Ask students to read the section on fertilisers and eutrophication on pages 429-430 of the Student Book. They should use what they have learned to draw a flow chart or other diagram to answer the questions: What is eutrophication, how is it caused, what impact can it have on aquatic ecosystems, and why does it have this impact?

Supplement 4. Measuring dissolved oxygen in water: demonstration

Although biological oxygen demand (BOD) is beyond the requirements of the syllabus, you could introduce the idea to students – it is the amount of dissolved oxygen needed in water for organisms like bacteria to respire during the breakdown of organic material in water.

If you have an oxygen sensor that will measure dissolved oxygen concentration in water, then set up the following as a demonstration for students (see Technician's notes). You will also need several tall glass containers, pond/stream water from a source that is not naturally high in mineral ions, and soluble plant 'food'. (Safety note: always wash your hands thoroughly after touching pond/stream water and fully cover any cuts, etc. Also take care not to slip into the pond while collecting water.)

Gently add pond water to each container, so that it is nearly full. The water should contain microscopic pond organisms, including algae, and look a little green/murky and not completely clear. Label each of the containers to indicate the number of drops of plant 'food' you will add.

One container should have no addition, and the container with the most drops should be given about twice the amount recommended by the manufacturer.

Add the drops to each container and gently mix them into the water with a long stick. Allow the contents to settle. Then, using the oxygen sensor, measure the dissolved oxygen concentration at the bottom of each container.

Ask students to predict what will happen to the oxygen concentration over a week and to explain their predictions.

Leave the containers in a sunny place, without disturbance, and measure the oxygen concentration at the bottom of each container every day for a week.

Students should then analyse the results and check whether their predictions were correct. They should explain the importance of their results for the eutrophication of natural water systems.

5. Problems with plastics

Tell students that objects made of non-biodegradable plastics, which are disposed of in landfill tips, are estimated to take between 500 and 1000 years to decompose. (It may be worth pointing out that scientists do not really know how long it will take, because we have not had plastics for this long, and that this is very much an estimate.) Ask them to discuss and note down any problems that they think this will cause.

There is a range of problems, including the use of land for landfill (when it could be used for other purposes, as well as pollution of groundwater by decay products), the release of toxins as the plastic breaks down, and the large-scale pollution of oceans that harms aquatic organisms (including the 'North Atlantic garbage patch').

If there is time, ask students to research some of these problems. They could use their findings to produce a poster, or presentation, warning of the long-term damage to ecosystems caused by disposal of plastic.

6. Greenhouse gases

Write the term *global warming* in the middle of the board. Ask students to suggest related words and how to link them to start a concept map. It is likely that students will bring a wide range of ideas – not all of them accurate – from their previous studies and from what they have heard about in the media.

For this task, do not discuss any areas of misunderstanding, but note them for later in the learning episode. Introduce the term *enhanced greenhouse effect* as an alternative to *global warming* and explain that this is the term they should use in their learning.

Ask students to read about pollution by greenhouse gases and the enhanced greenhouse effect on pages 432–435 of the Student Book. They should take notes about the two key greenhouse gases, carbon dioxide and methane, how they are formed by natural processes, and how they are formed by human activities.

Then allow students the opportunity to research the effects of increasing amounts of these gases in the atmosphere. (Remind students they may need to search for *global warming* rather than the *enhanced greenhouse effect*.) This should lead to examples of climate change. Encourage students to work together to compile a range of examples of climate change. They should discuss the examples, and the evidence, in order to evaluate how reasonable are the claims for climate change and its effects.

7. Evaluating the evidence

Ask students to use the text on page 434 of the Student Book about the greenhouse effect to write definitions for the terms *greenhouse effect*, *enhanced greenhouse effect* and *global warming*. Their definitions should show the relationship between each term. They can then use the words in a sentence to explain how human activity could be causing climate change.

Worksheet B20.3 asks students to find graphs of global surface temperature and atmospheric carbon dioxide concentration, in order to discuss cause and correlation, and to evaluate the argument for emissions of carbon dioxide from human activity as a cause for the enhanced greenhouse effect and climate change. Alternatively, they could use the graphs on page 433 of the Student Book for the task.

This is a demanding worksheet and students may need explanation of some of the terms and techniques used in the sheet.

8. Consolidation

Ask students to write ten clues for a crossword on human influences on the environment, covering all of the activities so far looked at in this section. They should swap their clues with a neighbour to try them out, and comment on any clues that weren't clear. They should then try to improve any weak clues. Take examples from around the class, to cover all the key ideas of the last three activities (there are many crossword packages available on the internet, some of them free of charge).

Technician's notes

Be sure to check the latest safety notes on these resources before proceeding.

The following resources for the class demonstration on measuring dissolved oxygen in water:

tall glass containers and marker pen
pond or stream water to fill containers
long stick for mixing
oxygen sensor and recording equipment
soluble plant 'food' (as used for houseplants or garden crops)

Take the water sample from a pond or stream where mineral ion level is not high. Exclude any large animals, but microscopic animals such as water fleas can be included.

Water samples can be returned to their source after the investigation only if suitably diluted to reduce the concentration of added mineral ions to a minimum. This is better in flowing water, where natural dilution will occur more easily, than in a pond.

Answers

Page 431

1. Any two such as: untreated sewage, excess fertilisers from farmland.
2. The addition of nutrients to water.
3. Run-off of fertiliser into water as a result of heavy rainfall; leaching of soluble nutrients in fertiliser through soil into water systems.
4. Supplement Eutrophication leads to the rapid growth of algae which block light to plants deeper in the water which then die. These decay due to the action of decomposers which remove oxygen from the water for respiration. This does not leave enough oxygen in the water for the fish, so they die.

Page 432

1. They take a very long time to break down and they may leak poisonous chemicals into ground water, which can leak away into water systems.
2. The land cannot be used for many purposes for many years after the plastic has been dumped.
3. The plastics cause problems for wildlife in the oceans that eat them by accident, or entangle them.

Page 435 Science in context: Life on Mars or Venus?

Currently, Mars is too cold to support life as we know it. However, a thicker atmosphere in the past could have led to a much higher surface temperature because of the greenhouse effect. This would have made the conditions more favourable for life.

Page 435

1. a) Natural: respiration. Human: combustion of fossil fuels
   b) Natural: digestion of food in animal guts and decay of waterlogged vegetation. Human: increase in livestock and artificial waterlogged vegetation in rice paddy fields.

2. Any three from: increase in number and intensity of storms; more drought; more flooding; change to summer/winter temperatures, change to precipitation.
3. The greenhouse effect is a natural process that warms the Earth's surface when greenhouse gases in the atmosphere prevent longer wavelength radiation escaping into space. The enhanced greenhouse effect is the additional warming caused by the addition of greenhouse gases to the atmosphere as a result of human activity.

Worksheet B20.3

1. Make sure students have the standard graphs for this answer, showing a rapid increase in temperature and in carbon dioxide concentration over the past 50 years or so.
2. a) Temperature has been measured recently from direct measurements at fixed sites; for past temperatures, some measurements are direct measurements (such as in the UK and USA), others are calculated from proxies such as tree ring data and ice core data; carbon dioxide concentration since the late 1950s has been measured at Mauna Loa, and previously to that, from bubbles in ice cores.
3. b) Using proxies makes assumptions about the relationship between the proxy and the value being calculated, which can reduce the reliability of the calculated value.
4. The graphs should show a gradual rise since about 1750 and an increasingly rapid rise in the past 50 years or so.
5. The increasing carbon dioxide in the atmosphere is due to carbon dioxide released from human activity, which is enhancing the greenhouse effect and causing global temperatures to rise.
6. Much of the work has been centred on laboratory experiments, and, in particular, computer modelling.
7. The correlation is very close, and other natural factors have failed to produce good correlations with temperature, so it is becoming increasingly reasonable to accept this cause.
8. Many of the emissions come from industrial processes, or transport. Reducing emissions means persuading people to change what they do, which can be very expensive and difficult to achieve. While there is apparently doubt in the discussion, politicians and industry have used this as an excuse not to face difficult and costly decisions.

Worksheet B20.3 Evaluating the evidence

The term *global warming* is often used interchangeably with the term *enhanced greenhouse* effect. It was first used in the late 1970s, when scientists began to suggest a relationship between increased emissions of carbon dioxide from combustion of fossil fuels and an increase in the Earth's average surface temperature. Since then, there has been a major debate among scientists, politicians, industry leaders and the general public about whether human activities really do cause increased carbon dioxide emissions and higher temperatures on the Earth, and what we should do to avoid or minimise potential problems caused by global warming.

You are going to look at the evidence for some of this and make your own evaluations.

1. Use books or the internet to find graphs of the following:

    a. average global surface temperature over the past 250–400 years
    b. average atmospheric carbon dioxide concentration over the past 250–400 years.

    You will find graphs that extend beyond this time. However, these vary as the result of many other factors, including the relative positions of the Earth and Sun in the Solar System and global ice ages, and so do not help in the debate about global warming.

2. a) Annotate copies of the graphs to explain how those measurements were taken. (Note that different measurements on the same graph may have been gathered in different ways.)
    b) Were all the values measured directly, or have they been calculated from other measurements? If the latter, what assumptions were made in those calculations and how might those assumptions affect the reliability of the calculated values?

3. Compare the graph of surface temperature with that of atmospheric carbon dioxide concentration, and describe any similarities and/or differences.

4. When two factors vary in a similar way, we say they are *correlated*. Two factors may be correlated for various reasons:

    *   factor A is changing and causing factor B to change in a similar way
    *   factor B is changing and causing factor A to change in a similar way
    *   factor C is changing and causing factors A and B to change in a similar way.

    What causes are suggested for the increase in carbon dioxide concentration, and the correlation between carbon dioxide concentration and temperature?

5. It can be very difficult to prove that a correlation has a specific cause. Find out and describe what scientists have done in order to try to prove that the cause you gave in question 4 is real.

6. Given what you have found out, how reasonable do you think it is to say that the causes you gave in question 4 are the fault of human activity?

7. Suggest why some politicians and industry leaders have tried very hard to disagree with scientists about this.

Learning episode B20.4 Conservation

Learning objectives

- Describe a sustainable resource as one which is produced as rapidly as it is removed from the environment so that it does not run out
- State that some resources can be conserved and managed sustainably, limited to forests and fish stocks
- Explain why organisms become endangered or extinct, including: climate change, habitat destruction, hunting, overharvesting, pollution and introduced species
- Describe how endangered species can be conserved, limited to:
  (a) monitoring and protecting species and habitats
  (b) education
  (c) captive breeding programmes
  (d) seed banks
- Supplement Explain how forests can be conserved using: education, protected areas, quotas and replanting
- Supplement Explain how fish stocks can be conserved using: education, closed seasons, protected areas, controlled net types and mesh size, quotas and monitoring
- Supplement Describe the reasons for conservation programmes, limited to:
  (a) maintaining or increasing biodiversity
  (b) reducing extinction
  (c) protecting vulnerable ecosystems
  (d) maintaining ecosystem functions, limited to nutrient cycling and resource provision, including food, drugs, fuel and genes
- Supplement Describe the use of artificial insemination (AI) and in vitro fertilisation (IVF) in captive breeding programmes
- Supplement Explain the risks to a species if its population size decreases, reducing genetic variation (knowledge of genetic drift is **not** required)

Common misconceptions

Students may not appreciate that there are different levels of conservation, from international and national strategies for protecting organisms and ecosystems, down to the individual level, where there are things we can do in our daily lives to improve sustainability.

Resources

Student Book pages 435–442

Approach

1. Introduction

Show students an image of a species that is at risk of extinction as a result of human activity; for example, a tiger – tigers are now so limited in their habitats, and still hunted due to fear or for the trade in body parts for local medicine, that their population sizes are dangerously low. Identify the human-related activities that have put the tiger (or other species) at risk.

Give students a few minutes to suggest what we should do about it. Take examples from around the class and make a list of suggestions. Encourage discussion about how realistic each suggestion is and what more might be implied by each idea; for example, putting more animals into zoos for protection needs more room and more money to care for them.

Introduce the term *conservation* as referring to the methods used to protect species and the environment to prevent long-term damage. Explain that the term is also used to mean the sustainable use of resources from the environment. Introduce the term *sustainable resource* and its meaning using page 436 of the Student Book.

2. Sustainable use of forests and fish stocks

Identify forests and fish stocks as potentially sustainable resources. Give students a few minutes to consider the difference between managing each of these resources in an unsustainable and a sustainable manner, and to suggest how those differences might happen. Ask them to work in pairs or small groups to write a list of rules that could be used to help manage each resource sustainably. This may identify additional information that would be needed to make this possible, which can be developed in the supplement part of this task.

Supplement Assign each student to study the sustainable management of either forests or fish stocks. They should work in groups to research different examples, or different aspects (e.g. education, protected areas, quotas and replanting for forests; education, closed seasons, protected areas, controlled net types and mesh size, quotas and monitoring for fish stocks) of the same example. They could use information from pages 436–440 in the Student Book, other books or the internet. They should use what they find to explain the importance of the different conservation methods to the sustainability of the resource.

3. Extinction

Extend the task in the introduction by discussing reasons why organisms become endangered or extinct. This could be introduced by reading through the bulleted list on pages 441–442 of the Student Book.

Ask students to research other examples of organisms that are endangered or have become extinct as a result of human activity. For each example, they should identify the human activity responsible. Gather the examples and reasons on the board to compile a class list for students to share.

Supplement Students could include in their research, examples of species where the population size has dropped so low that lack of genetic variation is a conservation concern, for example cheetahs. They should link the lack of genetic variation to the increased risk of extinction, such as by infection. This can be related to the similar problem for organisms that breed by asexual reproduction (see Section 16: *Reproduction*).

4. Conservation of a species and habitat

Ask students to select an example of either an endangered species or habitat, and to research what has been done in terms of conservation, why it has been done and how successful it has been in its aims. Encourage students to consider local examples.

They should quickly realise that the protection of either the species or a habitat leads to protection of the other. For example, there is little reason to protect a species if you cannot also bring about protection of suitable habitat in which it can live permanently.

The Student Book provides an example of conservation of the Hawaiian goose on page 440. However, there are many examples that students could research, starting either from the perspective of a single species or from the perspective of a habitat (through protection of national parks, etc.).

Remind students of their earlier work in Section 18: *Variation and selection,* on how organisms are adapted to their environment if they seem unsure why this is important.

Students could present their findings in a poster or a presentation so that other students can learn about the example. Encourage other students to ask questions to clarify the role and purpose of conservation.

Supplement Students could include in their research, the use of artificial insemination (AI) and *in vitro* fertilisation (IVF) as part of captive breeding programmes.

5. Consolidation

Give students a few minutes to answer the question, 'What should we conserve and why should we conserve it?'

Supplement Students should write at least four different answers to the question.

Take examples from around the class and encourage discussion to select the four best answers.

Alternatively, ask students to prepare a presentation using presentation software or a poster that answers the question 'What should we conserve and why should we conserve it?'

Answers

Page 438

1. a) Protection so that something isn't damaged.
   b) A resource that is produced as rapidly as it is used so it can be used without running out and without long-term damage to the environment.
2. Supplement a) Replanting trees that have been cut down; only buying wood products from forests that are managed; having protected areas; having quotas
   Supplement b) monitoring to make sure that enough breeding fish remain after fishing to produce the next generation; controlling net types and mesh sizes; preventing overfishing by having fishing quotas; re-stocking rivers and lakes where fish numbers have decreased; having protected areas and closed seasons

Page 439 Science in context: The Hawaiian goose

The predators that caused its extinction on Hawaii are still present.

Page 442

1. a) At risk of extinction/dying out.
   b) Breeding animals in captivity, such as in zoos or wildlife parks.
   c) A collection of seeds of many plant species stored for use in the future.
2. Any three from: climate change, hunting, overharvesting, habitat destruction, pollution, introduction of foreign species.
3. Supplement Genetic variation will be reduced which means that the species will be less able to adapt to environmental change. It may also lead to inbreeding.
4. Supplement Artificial insemination: artificially placing sperm from a male into the uterus of a female *in vitro* fertilisation: when eggs are artificially fertilised by sperm, in a laboratory dish.

Learning episode B20.5 Consolidation and summary

Learning aims

- Review the learning points of the topic summarised in the end of topic checklist
- Test understanding of the topic content by answering the end of topic questions

Resources

Student Book pages 443–447

Approach

Introduce the learning episode

Ask students to work with a partner to make a list of key words from this topic. They could then work together to produce a spider diagram showing how the different concepts are linked. They could compare their list with the list of key terms given on page 443 of the Student Book. Discuss the checklist on pages 443–444 and use questioning to see how much of the content they are comfortable with.

Students could make flashcards of the key content and then use the flashcards to quiz each other on the information.

Develop the learning episode

Ask students to work individually through the end of topic questions on page 445–447 of the Student Book without looking at the text. As they work, walk around the classroom observing their answers and asking questions as necessary to find out which questions are causing difficulties.

Finish the learning episode

After a set period, ask the students to stop working and discuss any areas of difficulty you observed as you walked round the class. Students should complete any unanswered questions for homework, but you should stress that they should try to answer the questions without looking at the text, so that they can see how much they have remembered.

Answers

End of topic questions mark scheme

The marks available for a question can indicate the level of detail you need to provide in your answer.

Question	Correct answer	Marks
1	C	1 mark
2	Food production per person per year has increased continuously in developing countries since 1960.	1 mark
	This is probably due to an increase in the use of intensive farming techniques in these areas.	1 mark
3 a)	They can earn more money from selling the livestock than they can get for products from the rainforest.	1 mark
3 b)	Suitable advantage, e.g. more money in the economy.	1 mark
	Any suitable explanation such as so raised standards of living or better infrastructure.	1 mark
3 c)	Any two from: habitat destruction; soil erosion or flooding where trees removed; reduced biodiversity, even extinction of some species, as mostly grassland; increased atmospheric carbon dioxide leading to climate change; reduced availability of local resources such as firewood and bushmeat for local people	1 mark per example to max. of 2 marks

4	Soil fertility decreases after the trees are removed as more rainwater seeps through the soil, taking mineral ions with it.	1 mark
	With fewer trees photosynthesising, atmospheric carbon dioxide levels increase leading to climate change.	1 mark
	Less water is returned to the air through transpiration, increasing the risk of flooding and soil erosion.	1 mark
	Animals and other plants that depended on the trees may become locally extinct, so there is a loss of biodiversity. So there are not the same plants and animals in the area to recolonise or the same conditions for the trees to grow, and other tree species that are better adapted to those conditions will move in.	1 mark
Supplement 5 a)	The oxygen is used by the microorganisms in the water, for (aerobic) respiration.	1 mark 1 mark
Supplement 5 b)	Polluted water would use more oxygen, because it contains a greater number of microorganisms.	1 mark 1 mark
Supplement 5 c)	Worms, because they can survive better in the low oxygen conditions found in polluted water	1 mark 1 mark
Supplement 5 d)	Organism has characteristics that enable it to survive and reproduce well in that habitat	1 mark
Supplement 5 e)	The organisms have to live in the water all the time, so they will show the condition of the water over a long period.	1 mark
	Sampling for oxygen demand only gives a result for the time that the water was sampled, and it may change from one period to another.	1 mark
6 a)	to increase crop growth	1 mark
6 b)	Some of the fertiliser will not be used by the plants, and so it will run off into the water system/nearby streams or rivers.	1 mark
Supplement 6 c)	Fertiliser in the water will increase the mineral ions, so plants and algae will grow faster.	1 mark
	This will cover the surface of the water, blocking light from reaching plants deeper in the water.	1 mark
	Plants lower down will die, which will increase the number of microorganisms in the water as they decay the plants.	1 mark
	Respiration of the microorganisms removes oxygen from the water, decreasing the oxygen available for other organisms, which might die.	1 mark
6 d)	The farmer will be able to see if there are any areas where the crops are not growing as well as expected.	1 mark 1 mark
	These areas might need more fertiliser.	
Supplement 7 a)	Over the period 1966 to 2009, numbers of farmland birds decreased by about 50%.	1 mark
	The major cause of this decrease was in the specialist species, as generalist species varied a little in number but remained at about the same level.	1 mark
Supplement 7 b)	If a food or another resource is lacking for a generalist species, it can use a different type, or go to another habitat to find it,	1 mark
	but specialist species cannot.	1 mark
Supplement 7 c)	Any suitable answer such as: less available food because of use of pesticides and herbicides; other species have not made the change to feeding in gardens over winter, or garden feeding does not provide the right sort of food for them.	1 mark
Supplement 7 d)	This will help maintain biodiversity, and help prevent individual species populations from decreasing/dying out.	1 mark 1 mark
	Total:	36 marks

Introduction

This section begins with a discussion of why bacteria are useful in biotechnology and genetic modification. It continues with examples of the use of bacteria and other microorganisms to make products, and concludes with how genetic modification is carried out and what it is used for.

Links to other topics

Topics	Essential background knowledge	Useful links
1 Characteristics and classification of living organisms	1.1 Characteristics of living organisms 1.3 Features of organisms	
2 Organisation of the organism	2.1 Cell structure and size of specimens	
4 Biological molecules	4.1 Biological molecules	
5 Enzymes	5.1 Enzymes	
12 Respiration	12.1 Respiration 12.2 Aerobic respiration 12.3 Anaerobic respiration	
14 Coordination and response	14.3 Hormones 14.4 Homeostasis	
15 Drugs	15.1 Drugs	
16 Reproduction	16.1 Asexual reproduction 16.2 Sexual reproduction	
17 Inheritance	17.1 Chromosomes, genes and proteins 17.2 Mitosis	
18 Variation and selection	18.1 Variation 18.3 Selection	
19 Organisms and their environment	19.4 Populations	
20 Human influences on ecosystems	20.1 Food supply	

Topic overview

B21.1	Biotechnology
	This learning episode looks at why bacteria are used in biotechnology and genetic modification. It then covers how yeast is used in producing ethanol for biofuels, and in bread-making, and how enzymes are used in fruit juice production and in biological washing powders. Supplement Students will look in more detail at why bacteria are useful in biotechnology and genetic modification. They will look at the production of lactose-free milk, and learn about the use of fermenters for the large-scale production of a range of substances.

B21.2	**Genetic modification**
	In this short learning episode , students learn about the basic process of genetic modification and some examples of how it can been used.
	Supplement Students learn in more detail about genetically modifying bacteria to produce human proteins, and the advantages and disadvantages of genetically modifying crops.
B21.3	**Consolidation and summary**
	This learning episode provides an opportunity for a quick recap on the ideas encountered in this section, as well as time for the students to answer the end of topic questions in the Student Book.

Careers links

These are some scientific careers that focus on this area of biology but careers in many other fields use the knowledge and skills gained studying science. **Synthetic biologists** are responsible for engineering the genetic code of organisms to form useful products. **Biotechnology technicians** work in laboratories and may collect samples, prepare experiments and record data.

Starting points

The Student Book section opener puts the ideas in the section into context and sets the scene. It also allows students to acknowledge and value their prior learning, and provides a benchmark against which future learning can be compared.

- The questions provide a structure for introducing the section and can be used in a number of different ways:
- You could ask students to consider the questions as an introductory homework task.
- You could put students into groups to share their own ideas and understanding and then to report back to the whole class.
- Students could be given access to the Internet, preferably with a tight timescale, to find out the information required.

You could then use a spider chart or other form of wall chart to summarise everybody's ideas.

Recording these initial ideas allows you to retain them for reference as the individual topics are developed. In this way, your students' progress in learning can be readily acknowledged.

Learning episode B21.1 Biotechnology

Learning objectives

- State that bacteria are useful in biotechnology and genetic modification due to their rapid reproduction rate and their ability to make complex molecules
- Describe the role of anaerobic respiration in yeast during the production of ethanol for biofuels
- Describe the role of anaerobic respiration in yeast during bread-making
- Describe the use of pectinase in fruit juice production
- Investigate and describe the use of biological washing powders that contain enzymes
- **Supplement** Discuss why bacteria are useful in biotechnology and genetic modification, limited to:
  1. few ethical concerns over their manipulation and growth
  2. the presence of plasmids
- **Supplement** Explain the use of lactase to produce lactose-free milk
- **Supplement** Describe how fermenters can be used for the large-scale production of useful products by bacteria and fungi, including insulin, penicillin and mycoprotein
- **Supplement** Describe and explain the conditions that need to be controlled in a fermenter, including: temperature, pH, oxygen, nutrient supply and waste products

Common misconceptions

Note that the role of anaerobic respiration of yeast in bread-making is limited. Aerobic respiration of yeast produces much of the carbon dioxide that causes bread dough to rise. However, as oxygen concentration inside the dough falls, yeast will replace aerobic respiration with anaerobic respiration, and this continues to release carbon dioxide. The alcohol produced by anaerobic respiration evaporates from the dough during baking.

Resources

Student Book pages 451–459

Worksheet B21.1a Extracting juice from fruit

B21.1a Technician's notes

Worksheet B21.1b Biological detergents

B21.1b Technician's notes

Worksheet B21.1c Immobilising enzymes

B21.1c Technician's notes

Supplement Worksheet B21.1d Making lactose-free milk

B21.1d Technician's notes

Supplement Worksheet B21.1e Constructing a fermenter

Resources for class practicals (see Technician's notes)

Approach

1. Introduction

Get students to revise what they have learned earlier about bacteria by asking them to draw up a concept map around the word *bacteria*. This could be done individually, in small groups or as a class, with students suggesting related words and how they should be arranged to prepare the map. Remind them to consider the size of bacterial cells and the rate at which they can divide.

Supplement Students should also focus on the key features of a prokaryotic cell. If needed, prompt with words such as *circular DNA*, and *plasmids*.

Keep the maps for the consolidation task at the end of this learning episode .

2. Using bacteria

Ask students to read pages 451–452 of the Student Book on the use of bacteria in biotechnology and genetic modification. They should use what they have read to write a definition for the term *biotechnology*. They should also note down the advantages of using bacteria to make products.

Ask students to write a three mark question about the advantages of using bacteria in biotechnology and genetic modification, and the marking scheme to go with it. They should exchange questions with another student, and identify one good point and one weak point about the question and answer they received. They should return the question and answer and decide how to improve their own question based on the comments they received.

`Supplement` Students should include references to plasmids and/or there being few ethical concerns over the use of bacteria.

3. Ask the expert: using yeast

Split the class into groups of three, and assign each group to research the use of yeast either in making biofuel or in making bread. The students within each group should then share out the following research questions:

- What are the stages in production of bioethanol/bread?
- What is the role of anaerobic respiration in this process?
- Compare the yeast-based product with a similar product (e.g. fossil fuel/unleavened bread) to identify any advantages and any disadvantages with using yeast.

Each student should then have time to research the answer(s) to their question from the Student Book, other books or the internet. The students should return to their groups to share what they have found out, and use it to prepare notes for a presentation.

Ask groups that have worked on the same use of yeast to compare and combine notes to select the best answer for each question. Allow time for students from the other yeast use group to ask questions from the 'experts' to cover all the key learning points.

4. Extracting juice from fruit

Worksheet B21.1a asks students to investigate the effect of pectinase enzyme on the release of juice from fruit (see Technician's notes). At the end of the investigation, explain that pectin is one of the substances in a plant cell wall that help to give the wall its strength. By digesting the pectin with the enzyme, it is easier for the plant cell membranes to break open and release the juice inside.

The investigation could be extended by using different types of apple, tinned fruit juice or apple puree at different temperatures.

SAFETY INFORMATION: Wear disposable gloves and eye protection. Do not allow students to eat the fruit or drink the juice produced in this practical. Extra care should be taken with the high concentrations of pectinase and when chopping apples. Use proper knives, after safe practice has been demonstrated (see Technician's notes).

5. Biological detergents

Worksheet B21.1b asks students to plan an investigation into the effectiveness of biological and non-biological detergents at different temperatures. You should check their suggested method and ask students to improve any weaknesses before they are allowed to carry it out.

You will need to provide identically stained cloths for students to use in their tests. These should be prepared the previous day, to allow the 'stains' to dry. If the number of cloths is limited, then ask each student or group to test at one particular temperature, and then pool results. Water baths at each temperature will be needed.

SAFETY INFORMATION: Students should wear washing up gloves and eye protection. Water temperature should be no more than 50 °C, to avoid scalding, unless tools are provided for stirring the cloths in hotter water. In that case, warn students of the hazards of hot water, and limit the temperature to no more than 80 °C. The teacher needs to risk assess individual student-proposed methods.

Note that the results of this investigation may vary from what is expected. This may be due to experimental error, but may also be due to the quality of detergent used. If possible, use a non-biological and a biological version of the same type and make of detergent to help limit variation. The enthusiasm of students in washing the cloths will also have an impact on results.

If results vary too much from expected, make sure students understand what theory says should have happened and why.

Supplement 6. Investigating lactase

This practical may be carried out by students, but could also be done as a demonstration if equipment is limited (see Technician's notes). Worksheet B21.1c provides instructions for making alginate beads containing lactase. If preferred, the beads could be prepared ahead of the lesson so that students only carry out the production of lactose-free milk. Prepared beads should be kept refrigerated and used within a few days.

Worksheet B21.1d supports the activity to produce lactose-free milk. The method used in both sheets is based on that provided by the NCBE (National Centre for Biotechnology Education), University of Reading.

SAFETY INFORMATION: Extra care should be taken with the use of concentrated enzyme solution. Students should not touch the coloured test region of the glucose test strips. If cutting strips up before use, avoid raising dust, or carry out the cutting inside a large plastic bag that will retain any dust particles from the strip. Wash hands after use.

Supplement 7. Constructing a fermenter

Worksheet B21.1e contains an unlabelled diagram of a fermenter for students to label and annotate. This can be used with page 458 of the Student Book to help students remember the conditions required for producing microorganisms on a large scale.

Supplement 8. Use of fermenters

Ask students to use the internet, books or the Student Book, to research the various products that can be made in fermenters (such as insulin, penicillin and mycoprotein) and the different microorganisms used (bacteria and fungi).

Students could present their findings in a poster, slide presentation or short report. Ask them to consider the benefits of producing these products in this way (compared with alternative methods).

9. Consolidation

Ask students to add to the map that they produced in the introduction task any new words that they have learned and decide how best to arrange them to provide a useful revision tool for this section.

If there is time, students could compare maps and discuss any differences between them. They could use this discussion to update and improve their own map.

Technician's notes

Be sure to check the latest safety notes on these resources before proceeding.

The following resources are needed for the class practical B21.1a, per group:

apples or soft juicy fruit – about 100 g will be needed per group (some varieties of apples, e.g. Golden Delicious, do not work well in this practical so the apple material needs to be trialled)
knife and chopping board, or food blender or processor (if necessary, the fruit could be prepared before the lesson)
mass balance
2 filter papers (coffee filter papers work well for this) and 2 funnels
2 100 cm³ beakers and stirring rods
marker pen
4 cm³ pectinase enzyme solution (prepare according to manufacturer's instructions)
4 cm³ distilled water
water bath at 40 °C
2 measuring cylinders
stopwatch or clock
eye protection and disposable gloves

In this practical, students extract juice from fruit with and without the use of pectinase enzyme. They should find that more juice is extracted when the enzyme is used. They should be able to relate this to the temperature of the water bath and suggest that 40 °C is near to the optimum temperature of the enzyme, and so allow the best breakdown of cell walls during the 30 minutes in the water bath.

Apples work very well for the experiment, as they contain lots of calcium pectate. Do not use *Golden Delicious* apples as they do not work well with this practical. If liquid enzyme is obtained from the UK National Centre for Biotechnology Education, NCBE (Pectinex), then a 50%(v/v) dilution (immediately before use) is recommended. The NCBE method suggests the use of 2 cm³ rather than 4 cm³, so a 25% solution may be more appropriate.

Some people are allergic to enzymes. If you suspect you are allergic, take great care when preparing the solution for students and wear disposable gloves and eye protection. Use a fume cupboard if preparing solutions from powdered enzymes. At the low concentration used in this experiment, students with an allergy to the enzyme are less likely to have a severe reaction, but gloves and eye protection are still recommended.

The following resources are needed for the class practical B21.1b, per group

identically stained cloths, two cloths per temperature to be tested (one for biological and one for non-biological test – see note below about preparing cloths)
biological clothes detergent, e.g. washing powder, liquid, gel or capsules
non-biological clothes detergent, e.g. washing powder, liquid, gel or capsules
water baths at different temperatures (for washing clothes in), e.g. 10 °C, 20 °C, 30 °C, 40 °C – only provide higher temperatures (maximum 80 °C) if suitable tools are available (e.g. washing tongs) for moving the cloths without putting hands in water
stopwatch
eye protection and rubber gloves

Preparing 'stained' cloths: Cut up an old cotton sheet (or similar) into pieces about 20 cm square. Treat each piece identically, placing a small amount of staining substance in the same position on each square. Suitable staining substances include: chocolate (best melted), coffee, tomato ketchup, orange squash, cold tea, grass juice (prepare by crushing grass or other leaves with ethanol on a mortar and pestle), diluted food colouring. **Do not use blood.**

Leave the stains at least overnight to dry, as this will make it more difficult for them to be washed out.

Ideally, use a non-biological and a biological version of the same type and make of detergent.

The following resources are needed for class practicals B21.1c and B21.1d, per group or demonstration

2% sodium alginate solution
lactase enzyme solution, e.g. Novozymes Lactozym®
1.5% calcium chloride solution
2 plastic syringes, clamp and stand
small beakers, stirring rod, tubing and clip
small sieve or strainer
nylon gauze (small piece to fit inside syringe to prevent beads getting stuck in nozzle)
milk (not UHT)
glucose test strips available from science equipment suppliers
eye protection and disposable gloves

SAFETY INFORMATION: Some people are allergic to enzymes. If you suspect you are allergic, take great care when preparing the solution for students and wear disposable gloves and eye protection. Use a fume cupboard if preparing solutions from powdered enzymes. At the low concentration used in this experiment, students with an allergy to the enzyme are less likely to have a severe reaction, but gloves and eye protection are still recommended. Extra care should be taken with the use of concentrated enzyme solution. All solutions should be made up using distilled or deionised water. Sodium alginate is not readily soluble and needs warm water and stirring to dissolve it.

Answers

Page 452

1. The use of organisms to make products, such as making bread or cheese or in genetic modification.
2. They produce many complex chemicals and they grow and reproduce rapidly.
3. Bacteria do not have a nervous system like most animals do, so it is not believed that they can feel pain or suffer.

Page 455 Developing Practical Skills

1. Any suitable plan that includes the following:

   - sample tested with pectinase and control sample without pectinase
   - apple crushed and same mass carefully measured into each sample container
   - test and control samples placed in water bath for same amount of time, to allow enzyme time to break down plant cell walls
   - samples filtered separately and volume of juice produced then measured.

   Other variations could be considered, such as comparing volume of juice produced at room temperature and at 30 °C for samples with and without pectinase, or comparing the volume of juice produced using a range of enzyme concentrations.

2. Prediction along the lines of: more juice from sample with pectinase because the enzyme breaks down cell walls, allowing more cell sap/juice to be released from uncrushed cells.

3. The results show that more juice was extracted, at both temperatures, using enzyme compared with using no enzyme. More juice was extracted, both with and without using enzymes, at 30 °C than at 20 °C.
4. More juice was released from the samples with enzyme, because the enzyme broke down the cell walls of uncrushed cells and released more juice. More juice was released by the enzyme at 30 °C than at 20 °C because the enzyme works faster at 30 °C. More juice was also released at 30 °C than at 20 °C when no enzyme was used, because even without an enzyme, chemical processes work faster at higher temperatures.
5. Using pectinase near its optimum temperature allows more juice to be extracted from the same mass of apple pulp more quickly than at cooler temperatures or without the enzyme.
6. At a higher temperature, the enzyme might be denatured, and so would not break down cell walls and release juice so well.
7. This produces a fair comparison between the results from the different samples – the mass of apple is a control variable that needs to be kept constant, because if it were not, that would affect the results.

Page 457

1. a) carbon dioxide
   b) ethanol (alcohol)
2. It causes more juice to be released from a mass of fruit pulp because it breaks down the cell walls of uncrushed cells.
3. They contain enzymes that break down organic chemicals in stains more quickly and at lower temperatures than soap and hot water.
4. Supplement Lactase breaks down the lactose in milk, making it more suitable for drinking by people who are lactose-intolerant.

Supplement Page 459

1. A large vessel in which microorganisms are grown in large numbers under controlled conditions.
2. a) Any four from: temperature, pH, oxygenation, nutrient concentration, waste products
   b) Any four from the following:
   - Temperature will increase due to the reactions of respiration and other reactions of the microorganisms. Changes in temperature will affect the activity of enzymes in the microorganisms; If temperature rises too high, it may reduce rate of growth or kill the microorganisms.
   - pH may change because of substances released by the microorganisms into the solution. Changes in pH will affect the activity of enzymes in the microorganisms; this may reduce rate of growth.
   - Oxygen concentration might fall as oxygen is used for respiration. Microorganisms are aerobic, so rate of growth will reduce if oxygen concentration falls as this will reduce the rate of growth.
   - Nutrient concentration will fall as microorganisms use nutrients to make new cells. Rate of growth will fall if nutrients are not added to replace what is used.
   - Waste products will build up as they are produced by the microorganisms. Many waste products are toxic and may reduce the rate of growth or kill the microorganisms if they are not removed.
3. a) Keeping things sterile.
   b) It prevents other microorganisms growing rapidly in the fermenter and competing with the added microorganisms. It also prevents the growth of any potentially harmful microorganisms.

Worksheet B21.1a Extracting juice from fruit

Many fruits are pressed (squeezed) to extract their juice, which is used as a drink or to make other products such as jams or jellies. The fruit is first mashed and then put through a press to squeeze out the juice. However, a lot of juice is trapped within the plant cells. If the plant cell walls are first broken down chemically, more juice can be extracted from the same amount of fruit. In this practical you will compare the volume of juice extracted from the same amount of fruit with and without the use of an enzyme.

Apparatus

apples or other fruit

knife and chopping board, or food blender or processer

mass balance

2 filter papers

2 funnels

beakers

marker pen

pectinase enzyme solution

distilled water

stirrers

water bath at 40 °C

2 measuring cylinders

stopwatch or clock

eye protection and disposable gloves

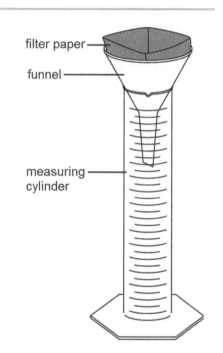

filter paper

funnel

measuring cylinder

SAFETY INFORMATION
Wear disposable gloves and eye protection.
Do not eat the fruit or drink the juice produced in this practical.

Method

1. Put on eye protection and gloves.
2. Peel one or two apples and discard the peel. (You will need 100 g of apple flesh.)
3. Chop the apple into as small pieces as you can manage with the knife. Alternatively, liquidise it in a food blender.
4. Use the balance to measure 50 g of apple into each of two beakers. Label one beaker 'with enzyme' and the other 'without enzyme'.
5. To the 'with enzyme' beaker add 4 cm³ of enzyme solution. Use a stirrer to mix the apple and enzyme thoroughly.
6. To the 'without enzyme' beaker add 4 cm³ of water. Use a clean stirrer to mix the apple and water thoroughly.
7. Place both beakers in the water bath for 25–30 minutes. Stir each beaker with its own stirrer from time to time.

Page 1 of 2

8. Set up two measuring cylinders as shown in the diagram. Place a funnel in the neck of each cylinder and place filter paper into each funnel.
9. After the beakers have finished in the water bath, pour the contents of each beaker into a separate funnel and start the clock.
10. Measure the volume of juice in each measuring cylinder every 5 minutes for up to 30 minutes.
11. Draw up a suitable table to display your results.

Handling experimental observations and data

12. Use your table to produce a graph. Show both sets of results on the same graph.
13. Describe the results shown on your graph.
14. Suggest what effect the enzyme has on plant cells. Explain your answer.
15. Explain why enzymes are used to produce fruit juices in commercial production.

Planning and evaluating investigations

16. Why was it important to chop the fruit into as small pieces as possible?
17. Explain why the mixtures were left in the water bath for about 30 minutes before they were filtered.
18. Why was a temperature of 40 °C used in the water bath?
19. Describe any problems you had with this investigation. How do you think they affected your results?
20. Suggest how the method could be adjusted to avoid these problems.

Page 2 of 2

Worksheet B21.1b Biological detergents

Many detergents that we use for cleaning clothes contain enzymes. The enzymes are similar to the ones that break down food in your digestive system. However, detergent enzymes mainly come from bacteria, and may have a different optimum temperature to human enzymes.

You will be given two detergents: one 'biological' (containing enzymes) and one 'non-biological' (contains no enzymes). You will also be given some pieces of cloth stained with a range of substances that are often the cause of stains on clothing.

SAFETY INFORMATION
Wear rubber gloves and eye protection when working with enzymes.

Plan an investigation into the effectiveness of biological and non-biological detergents at different temperatures.

You will need to consider:

- what equipment you will need
- how to manage risks so that the investigation can be carried out as safely as possible
- which temperatures you will use
- how to control all other variables, such as amount of detergent, length of time cloths are in water, how 'washing' will be carried out
- how to record results so that they can easily be compared (e.g. a digital camera, or a colour comparison chart).

Write out the method for your investigation and show it to your teacher. When your method has been approved, you should carry it out. Use your results to answer the following questions.

Handling experimental observations and data

1. Which was the best temperature for removing stains using non-biological detergent?
2. Non-biological detergent is mainly soap. Suggest a reason why the non-biological detergent worked best at that particular temperature.
3. Which was the best temperature for removing stains using the biological detergent?
4. Using what you know about enzymes, suggest a reason why the biological detergent worked best at that temperature.
5. Comparing the two detergents, which gave the cleanest results overall?
6. Suggest a reason for your result in question 5.

Planning and evaluating investigations

7. Describe any problems you had with this investigation, and suggest why they happened.
8. Describe how you would change the investigation in order to get better results.

Worksheet B21.1c Immobilising enzymes

Immobilised enzymes are enzymes that are trapped in some way, so that they are not washed away when liquids flow over them. This is particularly useful when using enzymes in a process such as to produce lactose-free milk.

Use the following method to prepare beads containing lactase enzyme for use in Practical B21.1d.

Apparatus

8 ml sodium alginate solution

2 ml lactase enzyme solution

100 ml calcium chloride solution

10 ml plastic syringe barrel

retort stand

small beaker

small sieve or strainer

distilled water

stirring rod

SAFETY INFORMATION
Wear eye protection when working with enzymes.

Method

1. Pour the enzyme solution into the alginate solution and stir thoroughly using the stirring rod.
2. Set up a 10ml syringe barrel vertically on a retort stand about 10cm above the beaker of calcium chloride solution.
3. Slowly pour the alginate solution into the syringe barrel so that drops of the alginate solution fall into the calcium chloride solution.
4. Allow one drop of mixture at a time to fall into the calcium chloride solution. Do not allow the tip of the syringe to touch the calcium chloride solution, as the alginate will harden and block the syringe nozzle.
5. Leave the beads in the solution for a few minutes, so that they harden.
6. When ready to start Practical B21.1d, separate the beads from the solution by pouring the mixture through the sieve over a sink or other container.

Supplement Worksheet B21.1d Making lactose-free milk

This investigation uses lactase enzyme trapped in alginate beads to break down lactose in milk to the simpler sugars, glucose and galactose. A similar procedure is used on a manufacturing scale to produce lactose-free milk for people who are intolerant to lactose.

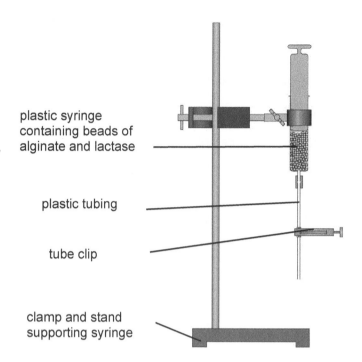

plastic syringe containing beads of alginate and lactase

plastic tubing

tube clip

clamp and stand supporting syringe

Apparatus

alginate beads containing lactase enzyme

plastic syringe

beakers

clamp and stand

nylon gauze

stirring rod

plastic tubing + clip

milk

glucose test strips

SAFETY INFORMATION

Wear disposable gloves and eye protection when working with enzymes.

Wash spills immediately with clean water.

Do not drink the milk.

Method

1. Attach the tube to the nozzle of the syringe, and close the tubing with the clip, as shown in the diagram.
2. Push the nylon gauze into the wide end of the syringe and down to the nozzle end. This will stop beads blocking the nozzle.
3. Support the syringe using a clamp and stand, and place a small beaker under the end of the tubing.
4. Pour the beads into the open end of the syringe.
5. Use a test strip to test the milk and record the result. Fill the syringe with milk.
6. Leave the milk in the syringe for 2 minutes.
7. Open the tube clip so that the milk runs out into the beaker.
8. Test the milk with another test strip and record the result.
9. Close the tube clip, then pour the milk from the beaker back into the syringe.
10. Repeat steps 6–9 until the test strip records maximum glucose.

Handling experimental observations and data

11. Display your results in a suitable way.
12. Describe your results, and explain them using your understanding of enzymes.
13. Explain the value of immobilising the enzymes for this process.

Supplement Worksheet B21.1e Constructing a fermenter

The diagram below shows the structure of a large fermenter, used for growing microorganisms to produce products made by the microorganisms, such as the antibiotic penicillin.

Label the parts of the fermenter, then annotate the parts to explain why they are needed to make sure as much product is formed as quickly as possible.

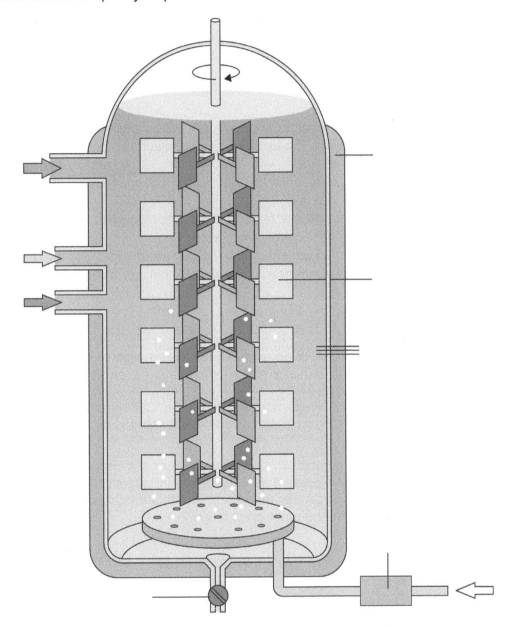

Learning episode B21.2 Genetic modification

Learning objectives

- Describe genetic modification as changing the genetic material of an organism by removing, changing or inserting individual genes
- Outline examples of genetic modification:
  - (a) the insertion of human genes into bacteria to produce human proteins
  - (b) the insertion of genes into crop plants to confer resistance to herbicides
  - (c) the insertion of genes into crop plants to confer resistance to insect pests
  - (d) the insertion of genes into crop plants to improve nutritional qualities
- Supplement Outline the process of genetic modification using bacterial production of a human protein as an example, limited to:
  - (a) isolation of the DNA making up a human gene using restriction enzymes, forming sticky ends
  - (b) cutting of bacterial plasmid DNA with the same restriction enzymes, forming complementary sticky ends
  - (c) insertion of human DNA into bacterial plasmid DNA using DNA ligase to form a recombinant plasmid
  - (d) insertion of recombinant plasmids into bacteria (specific details are **not** required)
  - (e) multiplication of bacteria containing recombinant plasmids
  - (f) expression in bacteria of the human gene to make the human protein
- Supplement Discuss the advantages and disadvantages of genetically modifying crops, including soya, maize and rice

Common misconceptions

The discussion of genetically modified (GM) or genetically engineered organisms can produce some very polarised debates among politicians, industry spokespeople and the general public. GM products are sometimes presented as dangerous by the media, and by some environmental activists, with descriptions used, such as 'Frankenstein foods', which aim to scare. Conversely, companies who produce genetically modified organisms for profit may state the advantages of their products but mention little of any problems.

The advantages and disadvantages of genetically modifying organisms is beyond the core syllabus. However, it is important for students' development as scientifically literate citizens that they develop a balanced understanding of the process of genetic modification so that they can see both sides of the argument and offer scientific evidence for their own point of view.

Resources

Student Book pages 459–464

Supplement Worksheet B21.2 Genetic modification

Approach

1. Introduction

Remind students of their work on DNA and its role in organisms. Explain that all organisms and bacteria have DNA in their cells. Ask students to suggest what this potentially means for transferring DNA from one organism to another. They should be able to suggest that a gene, made of DNA, from one organism, could produce its protein/characteristic in a different organism – that includes transferring genes between species as different as plants, animals and bacteria.

Give students a few minutes to work in pairs or small groups to suggest useful genes to transfer between species. Take examples from around the class and discuss what would be useful about the transfer.

2. Examples of genetic modification

Ask students to draw up a table with these headings:

Source of genes   *Organism genes inserted into*   *Effect of genes*   *Notes*

Then ask students to read pages 459–464 of the Student Book for examples of genetic modification. They should use the text to complete their table with four examples. If there is time, and students have access to the internet, students could look for other examples (although they do not need to know more than is in the Student Book as it is beyond the core syllabus).

Supplement 3. The process of genetic modification

Worksheet B21.2 offers a cut-and-paste activity that builds a diagram of the process of genetic modification in bacteria. Students cut out diagrams, arrange them in the correct order, paste them into their books and annotate them to explain the process. They should add details to show how a human protein (e.g. insulin) is produced from these genetically modified bacteria, and explain why this process was developed.

If students are having difficulty with the order, ask them to read page 461 of the Student Book to help them, and to help with annotation of their diagram.

Make sure students understand why the genetically modified bacterial cell produces the human protein. (Because the inserted human gene is within the genetic material of the genetically modified bacterial cell and will be copied to produce its protein in the same way as any other bacterial genes present.)

Supplement 4. Advantages and disadvantages

Split the class in two and ask one half to research the advantages of growing GM crops and the other half to research the disadvantages. Encourage each half to work cooperatively, to collect as many examples to support their case as they can, and to construct one or two sentences to describe each argument as clearly as possible.

Then allow time for a debate. During the debate, one group should present one example. The other group should then ask questions about the example. This should be followed by an evaluation (e.g. by scoring between 1 and 5 stars) of how important that example is. After all well-argued examples have been presented, use the stars to make a final evaluation about how the class feels about GM crops.

Remind students that this is a subjective evaluation, built on the evidence available but affected by the views of those involved in the debate. If there is time, encourage them to consider how the debate might produce a different result in a different context, e.g. among a group of farmers who are considering which seed to buy, or within a country's government that needs to consider everyone's views, including those of farmers, food manufacturers, possible sales of surplus food to other countries, and of people buying food in shops.

5. Consolidation

Ask students to write an entry for an online encyclopaedia for the term *genetic modification*. Give them a few minutes to do this and then take examples from around the class for discussion.

Answers

Page 463 Science in context: The advantages of GM insulin

There would be no difference. In both cases, the insulin has been produced using identical human insulin genes.

Page 464

1. Changing the genetic material of an organism by removing, changing or inserting individual genes. This will change the proteins produced by the organism.
2. Any two suitable examples, such as: bacteria that produce human insulin, crop plants that are resistant to herbicides, crop plants that are resistant to pest damage, crop plants that produce additional vitamins.
3. Supplement The human insulin gene is isolated using a restriction enzyme > a bacterial plasmid is cut open using the same restriction enzyme > the gene is inserted into the plasmid using DNA ligase > the plasmid is inserted back into a bacterium > the insulin gene causes the bacterium to make human insulin > when the bacterium divides, all the cells it produces contain the insulin gene and produce human insulin.
4. Supplement Any suitable advantage, such as: less use of chemicals so reduce effort and cost, less damage to environment, increased yield of crop.
   Any suitable disadvantage, such as: reduced biodiversity, increased seed cost, increased dependency on particular chemicals, risk of gene transfer to wild plants by pollination, possible health effects to people eating the genetically modified products (although not proven).

Supplement Worksheet B21.2 Genetic modification

The diagrams below, when arranged together in the right order, show the process used to genetically modify bacteria.

Cut out the boxes and arrange them in the correct order to show the stages in the process. When you are happy with the order, paste them into your workbook.

Add any labels and annotations, such as which enzymes are used in which stages, needed to explain clearly how the process is carried out.

Then explain why the genetically modified bacterium will produce the protein coded for by the inserted gene.

required gene isolated	genetically modified bacterium with plasmid containing required gene
required gene inserted into plasmid	required gene identified in original cell (e.g. human insulin gene)
bacterium with plasmid	bacterial plasmid cut open

Learning episode B21.3 Consolidation and summary

Learning aims

- Review the learning points of the topic summarised in the end of topic checklist
- Test understanding of the topic content by answering the end of topic questions

Resources

Student Book pages 465–469.

Approach

Introduce the learning episode

Ask students to work with a partner to make a list of key words from this topic. They could then work together to produce a spider diagram showing how the different concepts are linked. They could compare their list with the list of key terms given on page 465 of the Student Book. Discuss the checklist on pages 465–466 and use questioning to see how much of the content they are comfortable with.

Students could make flashcards of the key content and then use the flashcards to quiz each other on the information.

Develop the learning episode

Ask students to work individually through the end of topic questions on pages 467–469 of the Student Book without looking at the text. As they work, walk around the classroom observing their answers and asking questions as necessary to find out which questions are causing difficulties.

Finish the learning episode

After a set period, ask the students to stop working and discuss any areas of difficulty you observed as you walked round the class. Students should complete any unanswered questions for homework, but you should stress that they should try to answer the questions without looking at the text, so that they can see how much they have remembered.

Answers

End of topic questions mark scheme

The marks available for a question can indicate the level of detail you need to provide in your answer.

Question	Correct answer	Marks
1	C	1 mark
2 a)	fuel made from dead plant or animal material	1 mark
2 b)	glucose/sugar $\rightarrow$ ethanol/alcohol + carbon dioxide (+ energy)	1 mark
	Ethanol/alcohol is the biofuel.	1 mark
2 c)	Any one suitable material, such as: sugarcane waste, wood chips, straw.	1 mark
2 d)	Reduces the rate at which we are using up a non-renewable fuel,	1 mark
	balances the amount of carbon dioxide in the air (as only the carbon dioxide taken in by the plants is returned to the air when the fuel is burned).	1 mark
3 a)	Washing powder that contains enzymes.	1 mark
3 b)	Either of: grow rapidly, can produce complex molecules such as enzymes.	1 mark

3 c)	Enzymes digest the organic molecules in the stains quickly and at a low temperature,	1 mark
	and the molecules then dissolve in the water and can be washed away.	1 mark
3 d)	High temperatures denature the enzymes,	1 mark
	so they do not work as well at temperatures above the optimum for the enzyme.	1 mark
4 a) **Supplement**	antibiotic/drug used to kill bacteria that have infected the body	1 mark
4 b) **Supplement**	fungus/*Penicillium*	1 mark
4 c) **Supplement**	conditions are well controlled, so fungus grows as quickly as possible	1 mark
4 d) **Supplement**	Heat is released by the fungus as it respires.	1 mark
	If the mixture gets too hot, enzymes in the fungal cells will not work so well/become denatured and the production of penicillin will slow down.	1 mark
5 a)	A gene from a different species is inserted into an organism so that it produces the protein/characteristic it codes for.	1 mark
5 b)	Any one suitable example, such as: gene for herbicide resistance, gene that gives the plant resistance to an insect pest, gene that improves nutritional qualities.	1 mark
5 c)	Any one suitable example, such as: insertion of human insulin gene into bacteria so they make human insulin.	1 mark
6 a)	To help people who eat diets that are naturally low in vitamin A,	1 mark
	and prevent deficiency diseases of vitamin A.	1 mark
6 b)	Genes from other organisms were inserted into the rice DNA,	1 mark
	so that the rice grains produced a chemical that is turned into vitamin A in the human body.	1 mark
6 c) **Supplement**	Any one suitable example with explanation, such as: more costly than normal rice,	1 mark
	so the people who need it may not be able to afford it.	1 mark
6 d) **Supplement**	They are concerned that the GM crops may cause health problems that we do not know about yet.	1 mark
7 a) **Supplement**	Bacteria do not normally produce human insulin/they do not normally have the gene for human insulin.	1 mark
	Inserting the human insulin gene into the plasmid DNA in bacteria means they are able to produce human insulin.	1 mark
7 b) **Supplement**	The process for using information coded in DNA to make proteins is the same in humans and bacteria (and all other organisms),	1 mark
	so a gene for a protein in one species will produce the same protein in a different species.	1 mark

7 c) Supplement	i) Used to cut the gene out of the human DNA and to cut open the bacterial plasmids.	1 mark
	ii) Joins the human gene to the plasmid DNA	1 mark
	iii) Short length of unpaired bases, left at the ends of DNA after it's been cut by some restriction enzymes, which can join to other 'sticky ends'.	1 mark
	iv) Small loops of DNA in a bacterium into which the new gene is inserted before being returned to the bacterium.	1 mark
	Total:	36 marks

Exam-style questions and sample answers have been written by the authors. In examinations, the way marks are awarded may be different. References to assessment and/or assessment preparation are the publisher's interpretation of the syllabus requirements and may not fully reflect the approach of Cambridge Assessment International Education.

The marks available for a question can often indicate the level of detail you need to provide in your answer.

Pages 470–481

Question	Correct answer	Marks
1	A	1 mark
2	D	1 mark
Supplement 3	D	1 mark
4 a) i)	microscopic algae	1 mark
	brown seaweed	1 mark
ii)	topshell	1 mark
	limpet	1 mark
	periwinkle	1 mark
iii)	crab	1 mark
	dog whelk	1 mark
	gull	1 mark
4 b) i)	Any one from: introducing foreign species; overharvesting food crops; habitat destruction; disease; hunting	1 mark
4 b) ii)	Dog whelks: increase in number as less predation from crabs	1 mark
	Limpets: decrease in number as being eaten by more dog whelks	1 mark
	Gulls: decrease in number as fewer crabs to feed on	1 mark
4 c)	periwinkles feed on brown seaweed	1 mark
	periwinkles gain protection/shelter from brown seaweed (or any other sensible suggestion)	1 mark
5 a)	**Carbon dioxide** from the atmosphere is converted to complex carbon compounds in **plants** by the process of **photosynthesis**. This is often called carbon **fixation**. Plants are then often eaten by **animals**, which build up their own complex carbon compounds. The process of **respiration**, in both plants and animals, returns some of this carbon back to the atmosphere as **carbon dioxide**. When organisms die, their bodies decay as they are broken down by **decomposers**. Some of the complex carbon compounds are taken into the bodies of these organisms, where some may be converted to carbon dioxide during their **respiration**.	9 marks – 1 mark for each correct word in **bold**
5 b)	Combustion	1 mark

6 a)	Developing countries:	
	gradual increase in population from 1750 to early 1900s,	1 mark
	followed by rapid increase from around 1940,	1 mark
	followed by slight slowing of the increase from around 2030	1 mark
	Developed countries:	
	slow increase in population from 1750 to around 1940,	1 mark
	followed by a slight decrease from around 1940	1 mark
	followed by slight increase from around 2020	1 mark
6 b) i)	The birth rate/survival rate is only slightly greater than the death rate,	1 mark
	because of shortage of food supply/famine/disease/high infant mortality.	1 mark
6 b) ii)	The birth date/survival rate is much greater than the death rate,	1 mark
	because of plentiful food/low infant mortality/medical care.	1 mark
6 c) i)	They follow a similar pattern,	1 mark
	e.g. when there is a rapid increase both from around 1940	1 mark
6 c) ii)	Accurate records have not always been kept, particularly historically.	1 mark
	Not all species of organism on the planet have been identified, so extinctions might go unnoticed.	1 mark
6 c) iii)	Any five from: climate change; habitat destruction; hunting; overharvesting; pollution; introduced species.	5 marks – 1 mark for each
7 a)	oak trees → aphids → ladybirds	1 mark for order, 1 mark for arrows
7 b)	pyramid of numbers: pyramid of biomass: ladybirds aphids oak trees 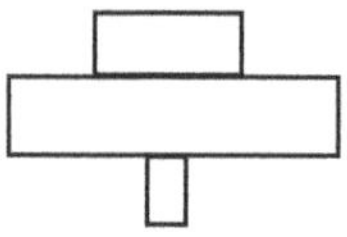 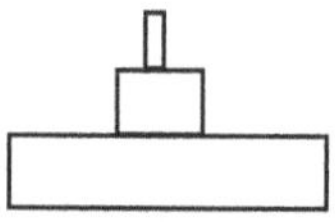	2 marks – 1 mark for each pyramid
7 c)	Advantage: overcomes problem (seen above with oak tree) where the pyramid is not a typical pyramid shape because a very large producer is involved	1 mark
	Disadvantage: the biomass of the organisms in an ecosystem/feeding relationship is more difficult to measure than their number	1 mark
7 d) i)	oak trees → aphids → ladybirds → spiders → blackbirds	1 mark
Supplement 7 d) ii)	From the Sun's energy falling on a leaf, plants only convert a maximum of 1–2% of this light energy into chemical energy in biomass.	1 mark
	Energy is lost as it is transferred from one trophic level to the next,	1 mark
	through excretion/ egestion (undigested food)/ heat from respiration.	1 mark
	The efficiency of energy transfer from one level to another is only 5–20%	1 mark
	(depending on the organisms involved).	1 mark
	There is not enough energy in the highest trophic level to support a further level.	1 mark

8	A: nitrogen-fixing (bacteria)	1 mark
	B: decomposers	1 mark
	C: denitrifying (bacteria)	1 mark
	D: nitrifying (bacteria)	1 mark
9 a)	Brazil: 43.9	1 mark
	Colombia: 46.7	1 mark
	French Guiana: 10.0	1 mark
	Peru: 45.3	1 mark
	Suriname: 9.2	1 mark
	Venezuela: 47.7	1 mark
9 b) i)	Venezuela	1 mark
9 b) ii)	Suriname	1 mark
9 c)	Any two from: timber/ land for crops/ land to graze cattle/ mining	2 marks
9 d)	(Deforestation causes habitat destruction, which can lead to	
	reduced biodiversity,	1 mark
	extinctions,	1 mark
	loss of soil/soil erosion,	1 mark
	flooding,	1 mark
	increased atmospheric carbon dioxide.	1 mark
10	Answer given in Student Book	
11 a)	Increased availability of mineral ions in the water,	1 mark
	causes increased growth of algae,	1 mark
	which can lead to the death of fish and other aquatic animals.	1 mark
11 b) i)	Insecticides improve crop yields,	1 mark
	by killing insect pests/insects that feed on crop plants,	1 mark
	so damaging them/ reducing their growth.	1 mark
11 b) ii)	Any two from:	
	Harmless insects may be killed, reducing the amount of food available for any animals that specialise on eating insects, such as insectivorous birds. This will affect other organisms in the food web, because of interdependence.	2 marks – 1 mark each
	If the other species killed are predators of the pest then, once the insecticide has been washed away by rain, it is possible for the pest species to return and increase in number even more rapidly, causing even more damage to the crop.	
	Insecticides may build up in food chains and be toxic to other organisms at higher levels.	
12 a) i)	Sewage concentration shows an ever-increasing concentration.	1 mark
	Oxygen concentration was stable from 2006 to 2007, showed a sharp decrease through to 2009, then decreased more slowly to 2010.	1 mark

Supplement 12 a) ii)	Because of its organic content	1 mark
	sewage is fed upon by bacteria/microorganisms in the water.	1 mark
	The bacteria/microorganisms use oxygen for respiration, so this is removed from the water.	1 mark
Supplement 12 b)	Trout are most sensitive to reduced oxygen concentrations so their population in the river disappears first.	1 mark
	Perch are only slightly affected by the initial decrease in oxygen concentration,	1 mark
	but when oxygen concentration decreases beyond a certain level, few perch can survive.	1 mark
	Some perch are tolerant of low oxygen concentrations, so their population does not completely die out.	1 mark
	Carp are most tolerant of low oxygen concentrations and their numbers increase as oxygen concentrations fall, possibly due to less competition from other fish.	1 mark
	As oxygen concentrations eventually fall to a certain level, carp numbers start to level off.	1 mark
Supplement 13 a)	Many drugs are complex molecules that are the natural products of microorganisms but are difficult to produce in the laboratory/ synthetically.	1 mark
	Microorganisms also reproduce very rapidly, meaning that the drugs can be produced at a rapid rate.	1 mark
Supplement 13 b) i)	nutrients	1 mark
Supplement 13 b) ii)	Two from: temperature; pH; oxygen level	2 marks
Supplement 13 c) i)	To mix the contents of the fermenter,	1 mark
	and ensure that the microorganisms/nutrients/oxygen are distributed evenly throughout the fermenter.	1 mark
Supplement 13 c) ii)	The respiration of the microorganism releases heat.	1 mark
	It is important that the temperature is kept constant, so that the microorganism's enzymes work to their optimum.	1 mark
	Circulating water is used to distribute heat/ regulate the temperature when it becomes raised.	1 mark
Supplement 13 d)	air filter	1 mark
Supplement 13 e)	aerobic (respiration),	1 mark
	using the air pumped into the fermenter	1 mark

14	Apply a stain (e.g. tomato sauce) to a piece of material.	1 mark
	Cut the material into many same-sized pieces.	1 mark
	Keep one piece unwashed.	1 mark
	Handwash a piece in water with biological washing powder, and one piece with non-biological washing powder.	1 mark
	Repeat this at a range of temperatures.	1 mark
	Control other variables: duration of wash/ how vigorous the wash/ amount of washing powder used/ use washing powders of the same brand the only difference being whether they are biological or not	1 mark
	Leave washed pieces to dry.	1 mark
	Compare them with the unwashed piece to decide how effectively the stain was removed from each piece.	1 mark
15 a)	Changing the genetic material of an organism,	1 mark
	by removing genes,	1 mark
	changing genes,	1 mark
	or inserting genes.	1 mark
15 b)	It is quicker,	1 mark
	the development of new varieties of crop or breeds of animal takes place by breeding over many/several generations.	1 mark
	(Allow other valid answer, e.g. can introduce characteristics not previously present in the species.)	
15 c)	Any one from: producing human insulin from bacteria; producing antibiotics	1 mark
Supplement 16 a)	The gene for the production of insulin is identified in human DNA,	1 mark
	and cut out using a restriction enzyme.	1 mark
	Bacterial plasmid are cut using the same restriction enzyme.	1 mark
	The human gene is inserted into the bacterial plasmids using ligase enzyme	1 mark
	The recombinant plasmids are inserted into bacteria.	1 mark
	The bacteria are cultured/multiplied to produce human insulin.	1 mark
Supplement 16 b)	Insulin produced by genetically modified bacteria is identical to human insulin,	1 mark
	because it is made using the human gene.	1 mark
	Production by bacteria is quicker/ can be made in large amounts,	1 mark
	because bacteria can be grown quickly/in large numbers in fermenters.	1 mark
	(Allow other valid answers, e.g. animals do not need to be killed.)	

wonderful wire

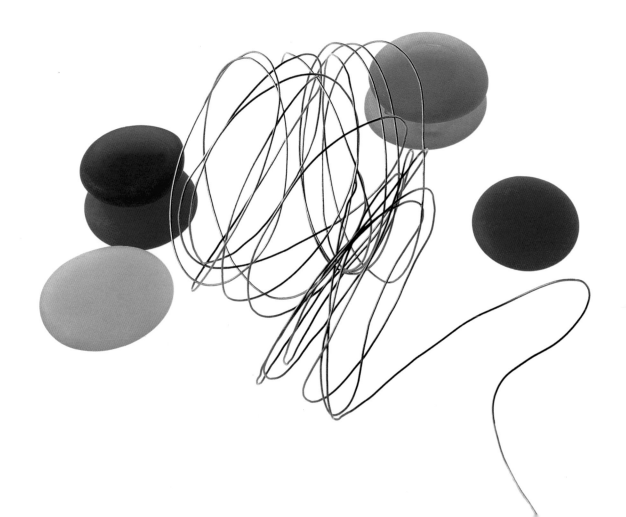

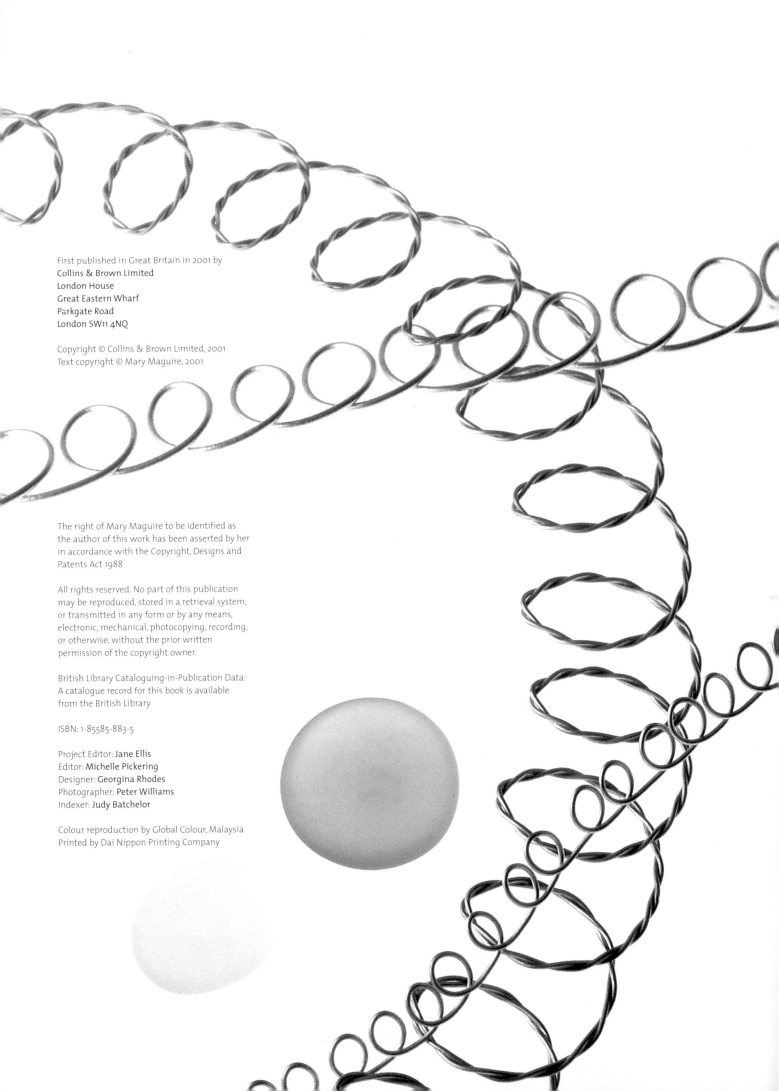

First published in Great Britain in 2001 by
Collins & Brown Limited
London House
Great Eastern Wharf
Parkgate Road
London SW11 4NQ

British Library Cataloguing-in-Publication Data:
A catalogue record for this book is available
from the British Library

ISBN: 1-85585-883-5

Project Editor: **Jane Ellis**
Editor: **Michelle Pickering**
Designer: **Georgina Rhodes**
Photographer: **Peter Williams**
Indexer: **Judy Batchelor**

Colour reproduction by Global Colour, Malaysia
Printed by Dai Nippon Printing Company

wonderful wire

Mary Maguire

Photographs Peter Williams

COLLINS & BROWN

CONTENTS

INTRODUCTION

The ancient but long-neglected craft of wirework has recently re-emerged with many new applications, fresher and brighter than ever before. It has become an extremely fashionable choice for both functional and decorative objects around the home. In this book we combine the hard linear qualities of wire with the translucent beauty of glass to create a surprizing and delightfully inspiring range of projects – giving the craft of wirework a more contemporary edge. The 22 projects are designed to suit a range of abilities: for the beginner, the butterflies, wall hooks and napkin rings are a good starting place. The more experienced maker may enjoy the challenging lampshade or glass carrier projects. The Basic Techniques section (page 12) shows you how to manipulate wire for both structural and decorative purposes. By bending, joining, binding and coiling, you can add both strength and beauty to a design.

The projects incorporate a range of ready made glass objects, such as marbles and beads, etc., all smoothed edged and user friendly. A list of useful tools is provided on page 8, but for most projects a pair of wire cutters, ordinary and round-nose pliers will suffice. Use this book not only as a pattern book to copy from, but as an inspiration. When you have started bending wire you will see its possibilities: it is an inexpensive, easily available material, so feel free to experiment. There are all sorts of items to make for your home: soap dishes, toothbrush holders, plate racks, baskets, string dispensers, fruit bowls and picture frames. For the garden, bird feeders, topiary framers and plant supports. Once you have got the knack you can go ahead and create your own wonderful wire works.

TOOLS

Here is a selection of useful tools for

making wire objects. Although

several different types of pliers are

specified in the project text because

they are the easiest to use for the

task in hand, in many cases a pair of

ordinary household pliers with

cutting edges will be sufficient.

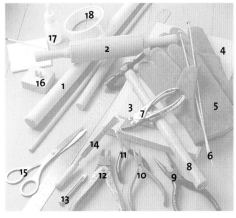

KEY

1 Assorted dowel rods	10 Round-nose pliers
2 Rolling pin	11 Miricle pliers
3 Metal plate	12 Wire cutters
4 Wooden block	13 Glass drill bit
5 Protective gloves	14 Metal file
6 Knitting needles	15 Scissors
7 Parallel pliers	16 A simple gig
8 Hammer	17 Strong glue
9 Chain-nose pliers	18 Masking tape

MATERIALS

Wire is available in many colors and thicknesses. Some are fine and soft enough to knit with like the enameled copper wire shown here. Others, such as galvanized wire, are hard and have to be bent with pliers. The glass used in the projects ranges from sea-worn fragments to Christmas tree baubles.

KEY

1 Enameled copper wire
2 Frosted glass droplets
3 Sea-worn glass shapes
4 Mirror
5 Assorted beads
6 Glass stars
7 Glass disk beads
8 Marbles
9 Chicken wire
10 Christmas tree baubles
11 Recycled glass beads
12 Galvanized wire
13 Florist's wire

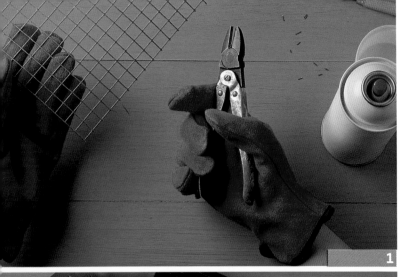

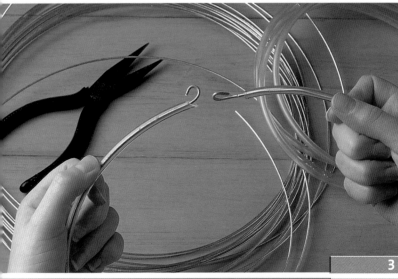

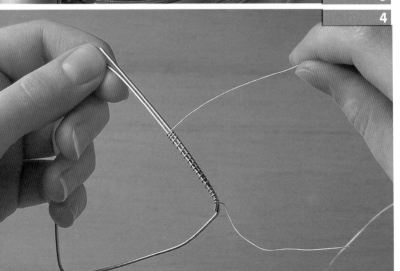

BASIC TECHNIQUES

1

WORKING WITH WIRE

Caution must be taken when working with wire. Use gloves to prevent scratches from sharp ends, and wear a pair of goggles to protect your eyes in case pieces of wire fly into the air when it is cut.

2

DRILLING GLASS

Wear goggles to protect your eyes. Stick a piece of masking tape over the area to be drilled to keep the drill from slipping and hold it securely. If appropriate, clamp the glass with protective cushioning. When drilling a difficult object like this saucer, use removable adhesive mounting pads (Blu-Tack) to hold it in place while drilling. Follow the instructions on the drill bit packaging.

3

JOINING WIRES

The simplest way to join wires is to use a pair of chain-nose (snipe-nose) or round-nose pliers to form a small loop at the each end of the wire. Link the loops together and squeeze them closed with the pliers.

4

BINDING WIRES

Another way to join wires is to bind them together. Place the wires side-by-side, then bind them together by wrapping a length of finer wire tightly around them.

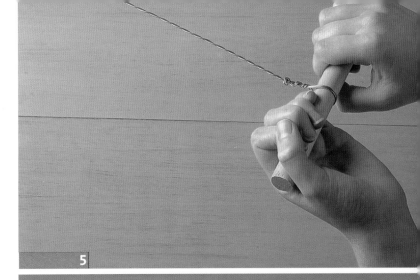

5
TWISTING WIRE

Take a length of wire and loop it around a door handle. Wrap the loose ends securely around a piece of dowel, a coat hanger, or something similar. Keeping the wire taut, turn the dowel around until you achieve the required tension in the wire. Do not overwind or the wire may snap. Take care when releasing the wire because the excess tension will cause it to spin.

6
FORMING SCROLLS

Use round-nose pliers to form a small circle at both ends of a piece of wire. Holding one of the circles with the pliers, wind the length of wire around to form an evenly spaced spiral. Do the same at the other end, working in the opposite direction.

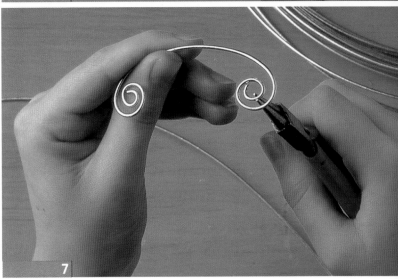

7
FORMING DOUBLE-ENDED SPIRALS

Make double-ended spirals in the same way as scrolls, but bend both ends of the wire in the same direction so that the spirals face each other. If you need to produce spirals of a precise size and shape, draw a template that you can check the wire against.

8
FORMING HEARTS

Make a double-ended spiral with sufficient straight wire between the two spirals to form the body of the heart. Place a finger in the center of the straight length of wire and push the coiled ends together with your thumbs. For a more pointed tip, bend the wire around round-nose pliers instead of your fingers.

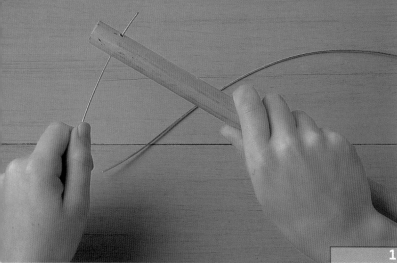

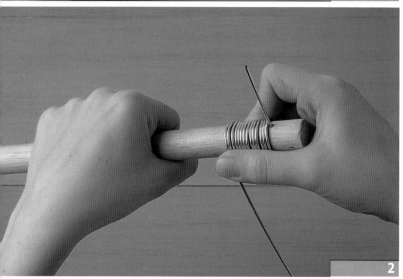

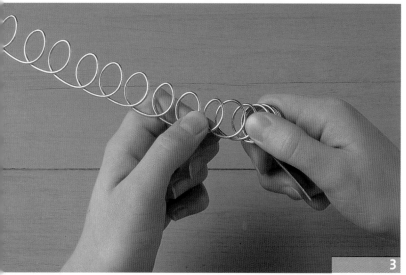

1

MAKING A FLATTENED COIL

Drill a hole through the end of a piece of dowel, then rub a piece of candle over the dowel so that the wire will slide off it more easily after coiling. Thread the end of the wire through the hole in the dowel to keep it in place while coiling.

2

Using your thumb to brace the wire against the dowel, twist the dowel around until you achieve a tight, even coil of wire. Gradually slide your thumb along the dowel as the coil lengthens. To remove the finished coil, cut off the end of the wire that is threaded through the hole in the dowel and twist the coil off.

3

Extend the coil out sidewise by gripping it firmly between both thumbs and pulling it out to the side so that the loops lie side by side.

4

Place the extended coil of wire on a cloth (to protect your work surface) and flatten it with a rolling pin, as the wire may emboss its pattern onto the wood.

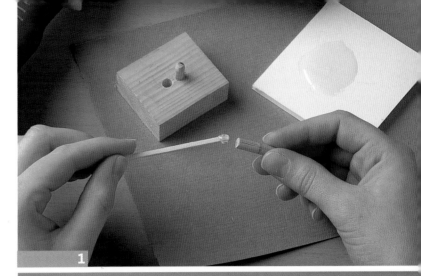

1

MAKING WIGGLY WIRE

Decide the depth of curves you require, then mark two points on a piece of wood this distance apart. Make two small pegs from dowel and drill two holes in the piece of wood at the marked points large enough to accommodate the pegs. Glue the pegs in position.

2

The finished wooden structure is called a gig. For small loops, use pegs made from thin dowel; for larger loops, use thicker dowel. Try using different thicknesses of wire for different effects.

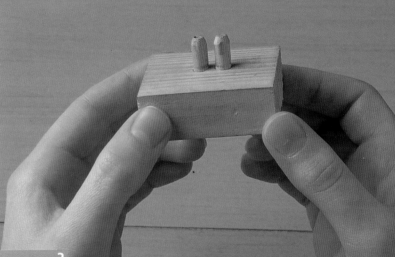

3

Hold the gig firmly in one hand, with the two pegs running in a perpendicular line to your body. Using your thumb to hold the end of the wire firmly against the block of wood, bring the remaining length of wire down toward the bottom peg at a right angle and bend it around the base of the peg. Then bend the wire up between the two pegs and around the top peg. Lift the curled wire off the pegs and move it to the side so that you can repeat the process to create two more curls. Continue until you have achieved the required length of wiggly wire.

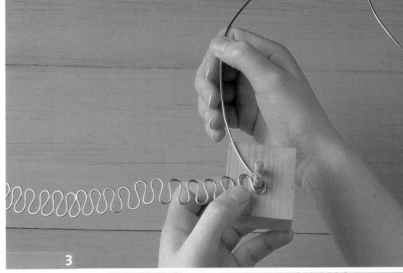

4

Instead of using a gig to make wiggly wire, you can do it freehand, using round-nose pliers to bend the wire. To make sure that the depth of the curls is consistent throughout, mark the required depth at regular intervals along the straight wire before bending it. Use your fingers to gauge the width of the loops.

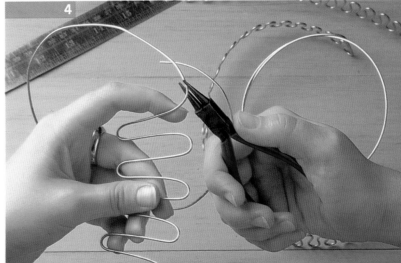

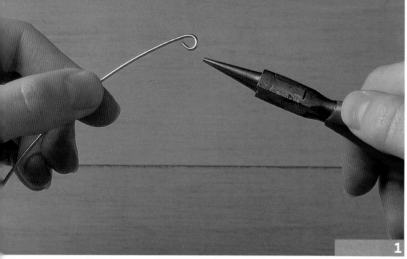

1

MAKING SPIRALS

Hold the end of the wire with a pair of round-nose pliers and twist the length of the wire around the tip of the pliers to form a small loop.

2

For an open spiral, hold the central loop with the pliers and continue to twist the wire into a spiral. Brace the length of the wire between your thumb and forefinger to control it and keep the spiral smooth and even.

3

For a closed spiral, hold the central loop flat between a pair of parallel pliers (ordinary household pliers will do if you do not have a pair of parallel pliers). Wind the wire into a spiral, repositioning the wire between the jaws of the pliers as necessary to keep it flat.

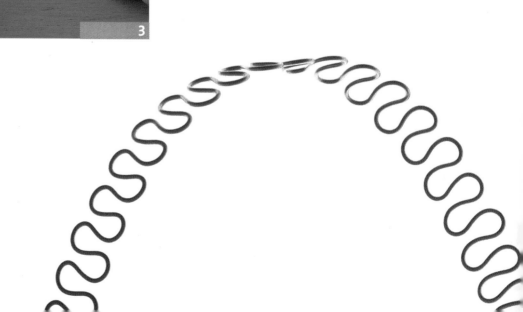

 1

MAKING A COILED SPIRAL

Drill a hole through the end of a piece of dowel, rub the dowel with some candle wax, and thread the end of the wire through the hole (see page 14, step 1). Twist the wire around the dowel to form a tight coil, then remove it from the dowel (see page 14, step 2).

2

Holding the coil firmly in one hand, use the thumb and index finger of your other hand to bend each loop off the tight coil of wire. Keep the space between each loop even by gauging it with your fingertip.

3

As you bend off each loop of the coil, the wire will naturally work itself into a spiral. Continue until you have finished the whole coil.

BEADED BUTTERFLIES

These beautiful beaded

butterflies are made from

enameled copper wire,

which is soft enough to

manipulate by hand and

therefore ideal for

beginners. They can be worn

as brooches or hair slides

with the appropriate jewelry

attachment fixed to the

back, or you could make

them into decorative

houseplant supports by

attaching them to canes.

They also make ideal gift-

wrapping decoration for

special presents.

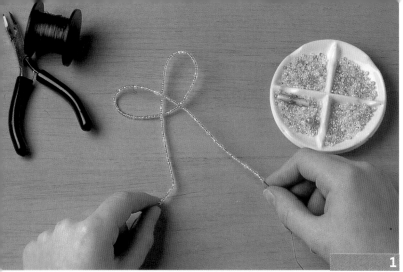

MATERIALS

(for each butterfly)

2ft (60cm) enameled copper wire

Assortment of seed beads

Long bead for the body

Round bead for the head

Strong glue

TOOLS

Household pliers

1

Cut a 2ft (60cm) length of wire with pliers. Thread seed beads onto the wire until approximately 16in (40cm) is covered with beads. Push the beads into the central section of the wire, then shape the wire into two loops to form the outline of the top section of the wings, as shown.

2

Loop the wire around again to form inner wings, then wrap each end of the wire tightly around the base of the wings to hold them in position. Form two more loops for the lower wings, then wrap the wires around the center point to secure.

3

Twist the remaining lengths of unbeaded wire together to form the body of the butterfly. Slide the long bead for the body and then the round bead for the head onto this twisted length of wire.

4

Bend the body section upward so that it lies flat on top of the wings. Wrap the ends of the wire around the wings on either side to secure the body in place.

5

Spread out the two ends of wire to form antennae and thread a seed bead onto the very end of each one. Secure the beads in place with strong glue.

6

Coil the end wires around these beads to finish off the antennae.

NAPKIN RINGS

These elegant napkin rings are made simply by threading beads onto wire and coiling it – nothing could be easier. Choose the beads to complement your tableware, or make the napkin rings as a gift for someone else, perhaps a Christmas or wedding present. The beads do not have to be all of the same color, but it is best to use the same type of bead throughout and the same tonal range.

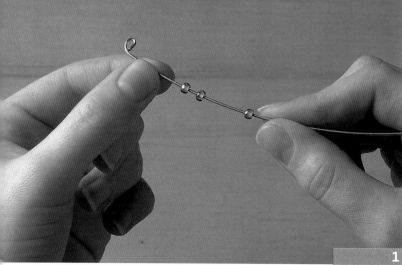

MATERIALS
(for each ring)
0.8mm galvanized wire
Assortment of small beads
2 larger end beads
Strong glue

TOOLS
Wire cutters
Round-nose pliers
Thick dowel

 1

Cut a 24in (60cm) length of wire. Holding one end of the wire with pliers, twist the wire around to form a small loop. This will keep the beads from falling off. Start threading the beads onto the wire.

 2

Continue to thread beads onto the wire, making sure there are no kinks in the wire to keep the beads from sliding on easily. Leave around 6in (15cm) of wire unbeaded at the end to allow for movement of the beads when the wire is being molded in the next step.

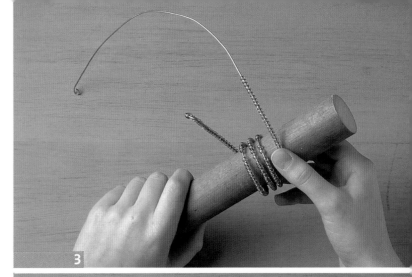

3

Mold the beaded wire into shape by wrapping the middle section three times around a thick piece of dowel. Gently ease the coil of beaded wire off the dowel.

4

Cut off most of the excess unbeaded wire, then use pliers to form a small loop at the end. Holding the loop firmly with the pliers, form the end of the beaded wire into a scroll. Repeat at the other end of the wire.

5

Snip off the end loops and straighten sufficient wire at each end of the napkin ring to hold the end beads.

6

Fix the end beads in position using strong glue.

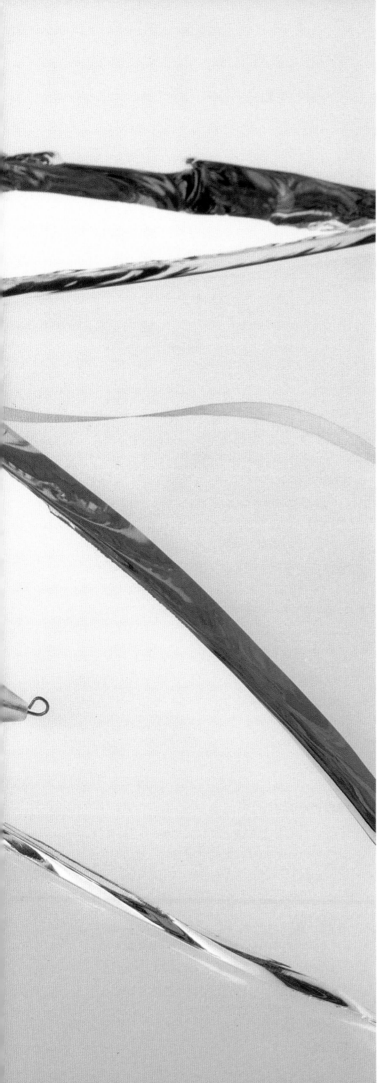

CHRISTMAS STAR

Making Christmas decorations can be very satisfying, whether they are for yourself or a gift. Make sure that the central bead you choose has a hole large enough to accommodate the necessary wires. You can make the star simpler by using fewer wires, or more complex by adding even more wires. A collection of different colored stars looks wonderful hanging on a tree, or catching the light at a window.

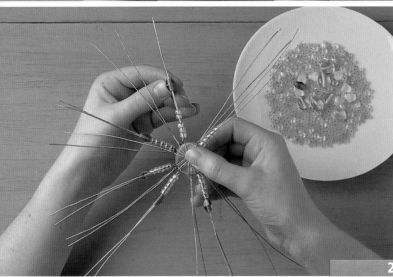

MATERIALS
0.8mm galvanized wire
Central bead with hole big enough to take all the wires
Assortment of small beads and seed beads
12 large beads (half in one color and design,
half in another)
6 cone-shaped beads
Crystal droplet

TOOLS
Wire cutters
Round-nose pliers

1

Cut twelve 12in (30cm) lengths of wire. Thread one of the wires through the central bead, bending the wire in half so that the bead is in the middle of the wire. Twist the two ends of the wire together close to the bead to secure it in position. Repeat with the remaining lengths of wire. Spread the wires evenly around the bead.

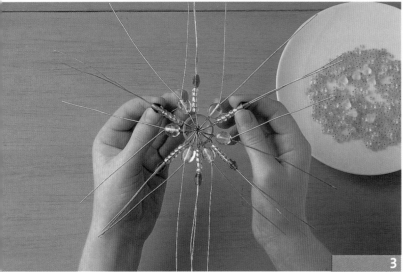

2

Thread a selection of small beads onto the two ends of every alternate wire, then add a large bead (use the same type of bead for all six wires). If necessary, spread out the two ends of the wires slightly to keep the beads from falling off.

3

Thread the two ends of the remaining empty wires through the remaining large beads, then splay out the two ends to form a V shape.

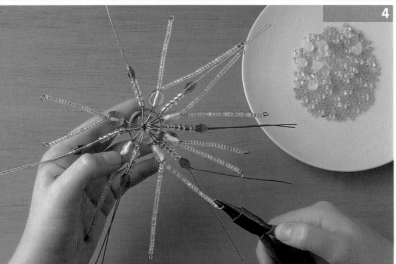

4

Thread seed beads onto each arm of the six V shapes. Once beaded, use pliers to form a small loop at the end of each wire and bend to a 45-degree angle.

5

Arrange the arms of each V shape so that they meet the adjacent arm of the next V shape to form the points of the star. Fill the remaining doubled wires with seed beads to fit inside the points of the star. If the doubled wire is too thick to go through the beads, cut off one of the wires, wrapping the end tightly around the remaining wire to secure it. When you have finished adding beads, thread the ends of these wires through the overlapping loops of the V shapes, as shown.

6

Thread a cone-shaped bead onto the wire projecting through each point of the star and fix in place by forming a loop at the end of each wire using pliers. Hang a crystal droplet from the loop at base point of the star, and attach a ribbon or something similar to the top point for suspending the star.

WHIMSICAL FAIRY

This pretty fairy will bring

a touch of magic to any

occasion. At Christmas she

can grace your tree, or

suspend her above a

birthday cake to preside

over the celebrations. Or

why not hang her in front of

a window where she will be

brought to life by passing

sunbeams. Her body is made

from a wire frame, which

holds her delicate glass head

in place. Her dress is knitted

wire, her glass slippers are

beads, and her wings are

made from a plastic bottle.

MATERIALS

Seed beads – silver, green and beads for shoes/heart/eyes
Enameled copper wire in pink and silver
Fabric and thin metal foil for dress
2 frosted glass droplets for head
0.7mm, 1.1mm, and 1.5mm galvanized wire
Needle and silver thread
Plastic bottle and glass star

TOOLS

3mm and 4mm knitting needles, scissors

1

For the dress, thread 220 seed beads onto the pink wire and
1,964 beads onto the silver wire (add a few extra beads to
allow for errors). Cast on 44 stitches using 3mm needles
and pink wire. Knit one row with no beads, then knit four
more rows, adding a bead in each stitch on each even-
numbered row; knit another ten rows using silver wire,
again adding a bead in each stitch on each even-numbered
row. Change to 4mm needles and follow the instructions
for the evening purse on pages 80–81, starting at the point
where 4mm are first used in the purse pattern and using
silver wire throughout. Push the finished triangle into
shape with your fingers. Cut a triangle of backing fabric
and foil the same size as the knitted triangle.

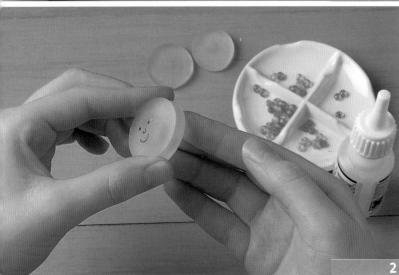

2

To make the face, stick two frosted glass droplets together
with strong glue. The groove between the pieces will hold
the body in place later. Draw the face using a metallic pen.

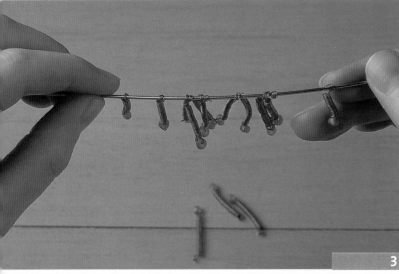

3

Wrap lengths of pink wire around a knitting needle to
make a long coil. Cut the coil into 1in (2.5cm) lengths.
Thread a bead onto one end of each wire coil at a 45-
degree angle and bend the end of the wire to secure it in
place. Thread the unbeaded ends of each coil onto a length
of 1.1mm wire, as shown.

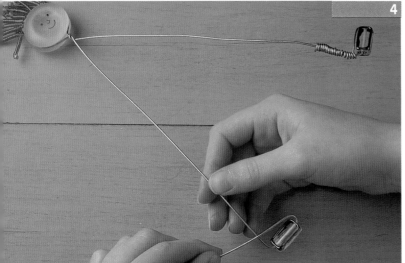

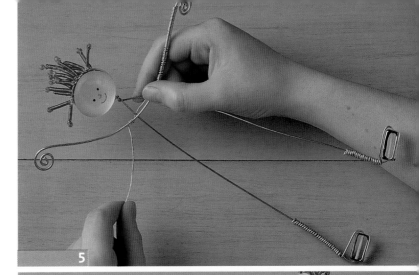

4

Group the coils into a cluster and position them on the center of the wire. Wrap this around the glass head; the wire will fit neatly into the groove between the droplets. Twist the wire together once at the neck position to secure. Spread the wire ends into a V shape, allowing sufficient length for the dress and legs. Thread a shoe bead onto each wire, bending the beads to a 45-degree angle. Wrap the remaining wire tightly around each leg.

5

Cut a length of 1.5mm wire and form a double-ended spiral (see Basic Techniques, page 13). This will become the arms and hands. Attach it near the top of the body by wrapping 0.7mm wire tightly around the arms from wrist to wrist, incorporating the body as shown.

6

Cover the body with the triangle of foil, then sandwich it between the knitted dress and the backing fabric. Stitch these together around the edge with silver thread.

7

Wrap a piece of 0.7mm wire around a glass star and thread seed beads onto the remaining length of the wire to make the handle of a magic wand. For the fairy wings, fold a piece of paper in half and draw one wing from the fold. Cut out and unfold the paper to produce two wings. Draw around the pattern on a plastic bottle and cut out. Make four holes in the center of the wings. Spray paint the wings a different color or give them a frosted edge with glitter paint.

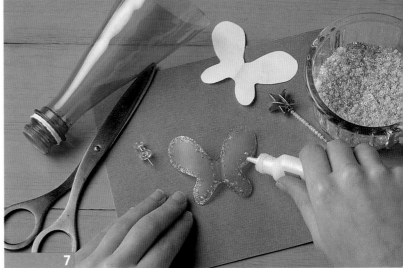

8

Attach the wings to the fairy's back by stitching a big cross-stitch through the holes with thin wire or thread. Attach the wand to the fairy's hand by binding the end of the wire wand around the hand. Suspend a heart bead from the other hand with thread. Gently stretch out the hair coils.

NAME PLAQUE

All you need to make your own name plaque is wire and jewelry pliers. Start by deciding which type of lettering you prefer (bearing in mind that all the letters must be made from a continuous piece of wire). Next practice drawing it out on paper (or computer) until you have a design that you are pleased with. Practice manipulating the wire to get a smooth, even result. The plaque can be used for a door or on a box as shown here, or you could make smaller versions to put on a bag, book or gift. Try incorporating motifs such as flowers, musical instruments or symbols. For a more decorative effect make a beaded surround.

MATERIALS

1.5mm galvanized wire
Glass bead

TOOLS AND EQUIPMENT

Jewelry pliers
Snipe-nose pliers
Pencil and paper

1

Write out your name at the appropriate size and start to bend the wire following the line.

2

Bend the wire under where it travels the same line twice so that it appears as only one line.

3

For the more complicated shapes with tighter bends, use the pliers to make the curves.

4

As you go along, make sure that you are following the line as faithfully as you can.

5

End with a strong curve and use a glass bead to make the dot for the i or just as a feature.

6

You can make a simple cloud shape to frame the name to finish it off.

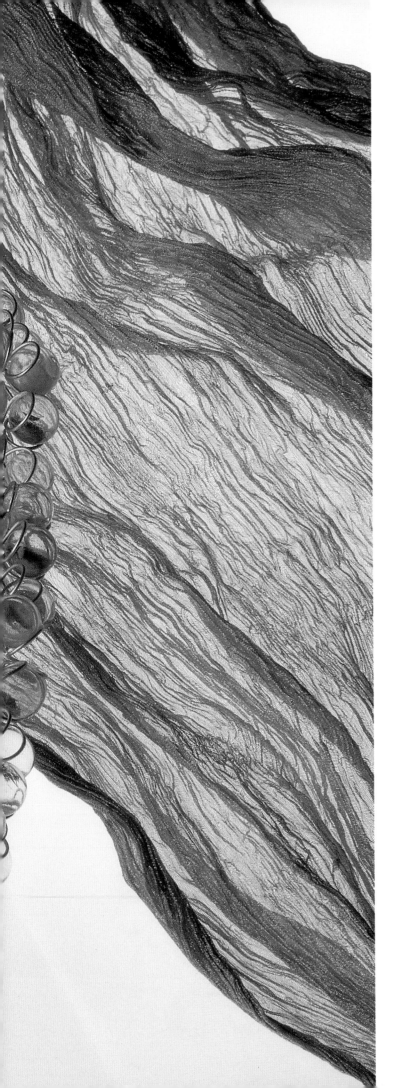

VALENTINE HEART

Simpler to make than it

looks, this valentine heart

is encrusted with glass

marbles suspended between

the loops of a coiled wire

spiral. It can be hung in front

of a window to catch the sun,

or placed around a vase of

flowers or a cluster of

candles to form a stunning

centerpiece for a romantic

meal for two. For a richer

look, try using copper wire

combined with red or blue

marbles.

MATERIALS

0.9mm and 1mm galvanized wire
Assortment of small clear marbles
Strong glue

TOOLS

¹/₂in (1cm) dowel
Wire cutters
Round-nose pliers

1

Use 0.9mm wire and the dowel to make a length of wire coil (see Basic Techniques, page 17) suitable for the required heart size. When you have finished, carefully ease the coil off the dowel.

2

Gently open out the tight loops of the wire to form a coiled spiral (see Basic Techniques, page 17).

3

Insert a length of 1mm wire through the center of the coiled spiral to give it a solid core. Try to keep the wire as central as possible through the middle of the coil for neatness. Allow the wire to project by about ½in (1cm) from both ends.

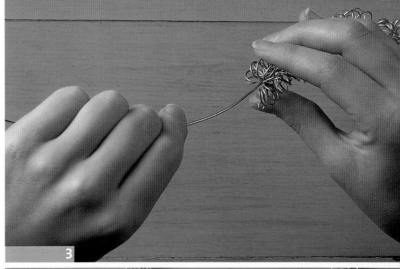

4

Bend the wire coil in half to form a V shape. This will be the bottom point of the heart.

5

Bend the two halves of the coil into curves to form the top of the heart shape. Slip the ends of the core wire through a few spirals on opposite sides of the heart, then use pliers to form the two straight ends into interlocking loops. When the heart is neatly joined, squeeze a few spirals closed to secure.

6

Press marbles between the loops of wire at regular intervals around the heart. They should stay in place if you press firmly, but if they look insecure, glue the underside of each marble at its point of contact with the wire.

ELEGANT LANTERN

It is hard to believe that this unusual and elegant lantern has been made from plain old chicken wire. Chicken wire is a fantastic medium because it can be molded and sculpted into extraordinary shapes. Here, an elongated urn has been fashioned using pliers, and then hung from scrolled wire hooks suspended from the ceiling. A glass jar is wired securely inside to hold a small candle or night light. Glass beads threaded onto wire shimmer in the light.

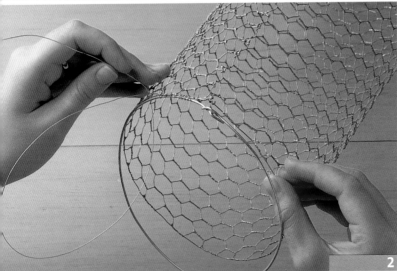

MATERIALS

Chicken wire
Galvanized wire
Florist's wire
Glass jar
Approximately 400 glass beads
Earring head wires

TOOLS

Wire cutters
Round-nose pliers
Pliers

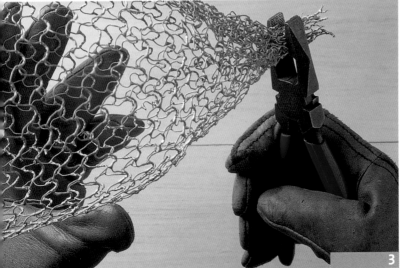

1

Cut a 16 x 20in (40 x 50cm) piece of chicken wire as shown, so that one side edge has two pronged wires projecting. Form the wire into a cylinder shape and wrap the projecting wires around the other side to secure.

2

Cut a length of galvanized wire and wrap it around the top edge of the cylinder. Join the ends together by forming interlocking loops. Bind the circle to the top of your cylinder using florist's wire.

3

Using round-nose pliers, bend each horizontal strut on your netting into a V shape. This will create heart-shaped holes and make the wire less rigid and easier to manipulate. Shape the top of the cylinder into a gradually sloping neck by squeezing in the vertical struts to narrow the cylinder.

4

Once you have formed the neck, expand the wires in the sections above and below the neck to create a fuller "belly" by opening the pliers inside the wires to make them bigger and pulling outward.

5

Approximately halfway down your lantern, start to decrease the holes again to taper the shape into a "tail" at the bottom. Once you have formed the tail, compress it into a cone shape by using the pliers as shown.

6

Attach a wire from the tip of this tail and wrap it tightly round to form a 2in (4cm) long smooth cone. Attach the end to the mesh using jewelry pliers.

7

Cut four lengths of galvanized wire and form a small spiral on the end of each one using jewelry pliers. Loop each of these wires through the top rim of your lantern at quarter intervals and bend to shape. Bind them as shown, using florist's wire.

8

Wrap two lengths of wire in opposite directions around the neck of a glass jar that fits snugly into the neck of the lantern. Twist the wires together to form a secure fixture. Thread beads onto earring head wires, allowing sufficient wire to form a hanging loop and snipping off the excess. Hang a bead in each hole of the lantern that is large enough to do so from the shoulder of the lantern downward. Insert the glass jar and use the projecting wires to attach it to the neck of the lantern. To complete, wrap a length of wire around the neck to form a collar and thread each of the four hanging wires through a bead or similar, bend over to attach. Insert a looped wire through this to hang the lantern.

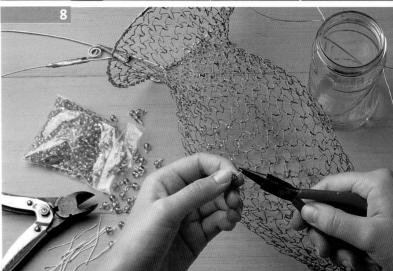

TABLE DECORATION

Create your own ornaments, table decorations, or objets d'art using wire for structure and glass for color and texture. Tripod and pyramid structures are stable and easy to create. Simply bend the wire to make feet, define their shape with beads, and join the wires together to form a top knot. Colored glass and beads can then be bound onto the structure for added decoration. Sea-washed glass (available in assorted colors and shapes from gift stores) and large recycled-glass beads have been used here for their beautiful color and texture, with small frosted seed beads woven into a mesh around the structure.

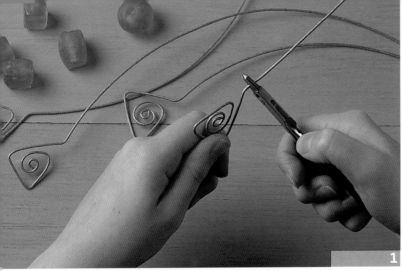

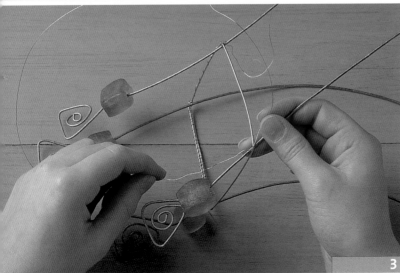

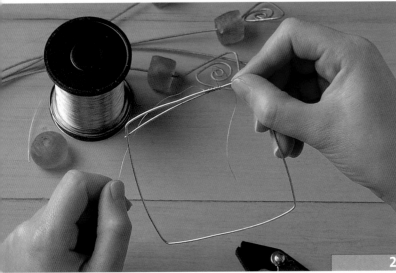

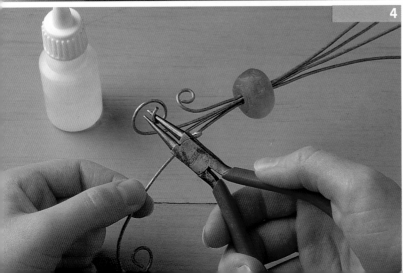

MATERIALS

1.5mm and 1.2mm galvanized wire
Fine binding wire
5 large frosted glass beads (one must have a hole big enough for all the wires to go through where they meet at the top)
Assortment of frosted glass seed beads

TOOLS

Wire cutters
Round-nose pliers
Household pliers

1

To make the frame, cut four lengths of 1.2mm wire and form a flat, open spiral at one end of each, using round-nose pliers. Use the pliers to bend the final curve of the spiral into a triangular shape to form a foot on the end of each length of wire.

2

Use household pliers to bend around 12in (30 cm) of 1.2mm wire into a square to form the base of the table decoration. Join the overlapping ends of the wire together using fine binding wire.

3

Thread a large glass bead onto each wire leg so that it sits just above the triangular foot. Secure each leg to the base frame with fine binding wire. Push the base frame down to the bottom of the four wire legs so that it sits just above the glass beads.

4

Gather together the four leg wires at the top and thread them through the remaining large bead. Using round-nose pliers, curl the end of each wire to make an attractive display.

5

Bind pieces of sea-washed glass to the frame using 1.5mm wire. Three pieces of glass were bound to each side in this example, but you can vary the number depending on the size of your structure.

6

Thread seed beads onto fine binding wire, twisting the wire at the base of each bead to hold it in position. Bind the wire randomly around the structure to form a delicately beaded mesh.

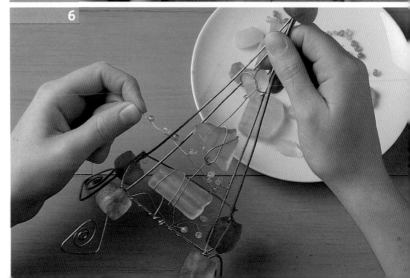

GLASS CARRIER

This pretty wire carrier allows

you to transport glasses safely to

the garden on a summer's day.

It is made from sturdy wire but

with a delicate design. The wiggly

wire is easily made with a simple

gig (see Basic Techniques, page 15,

to find out how to make one).

Moroccan tea glasses have been

chosen because of their color

and decoration, but other

glasses would work equally well.

You can adapt the design so that

it accommodates more glasses

if necessary.

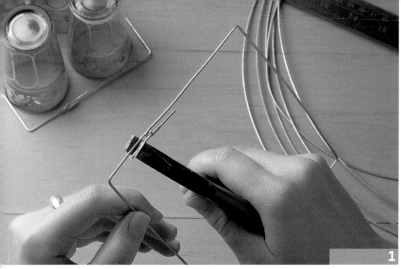

MATERIALS

Six small glasses
1.2mm galvanized garden wire
Florist's wire

TOOLS

Paper, pencil, and ruler
Wire cutters
Household pliers
Jewelry pliers
Simple gig

1

Arrange the glasses on a sheet of paper and use a pencil and ruler to draw a rectangle around them at their widest points. Cut a length of galvanized wire long enough to form into the required rectangle, allowing 3/4in (2cm) excess all around. Mark the corner points on the wire and use household pliers to bend each one to form a 90-degree angle. Form a loop at each end of the wire, interlock them, and squeeze them closed. This forms the base of the carrier. Repeat to make a rectangle for the top edge.

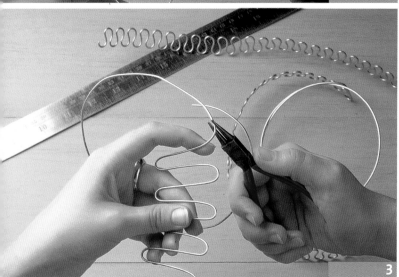

2

Cut four lengths of galvanized wire to fit lengthwise across the base rectangle and four to fit across it widthwise, allowing 3/4in (2cm) excess on each piece. Space these evenly across the rectangle and attach them by bending the ends around the base frame with pliers. Repeat for the top frame but use only one wire divider lengthwise and two widthwise. This will create six spaces for the glasses.

3

Use round-nose pliers to form a length of curved wire about 2 1/2in (6cm) wide and long enough to wrap around the frame (see Basic Techniques, page 15). Add a length of wiggly wire 3/4 (2cm) deep, made on a gig (see

Basic Techniques, page 15) long enough to wrap around
your base with enough left over to make a handle.

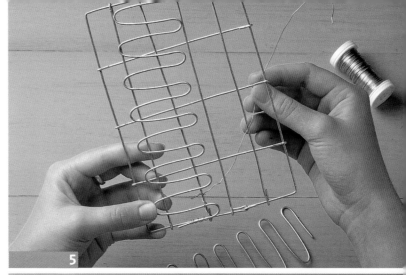

4

Using florist wire, bind the deeper wiggly wire to the top
framework, bending it to fit around the corners.

5

Make two lengths of wiggly wire to fit across your base
frame as shown and bond into place.

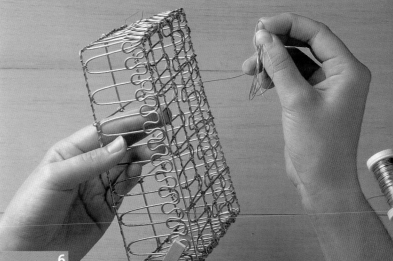

6

Position small wiggly wire around the base edge of the
sides, then place bottom frame in position and bind
these together along the edge.

7

Make the handle by binding two lengths of wire to either
side of a length of narrow wiggly wire. Bend this at a
90-degree angle at each end to form the handle shape.

8

Bind the handle onto the carrier in a central position.
make decorative spirals with the excess handle wire
using jewelry pliers.

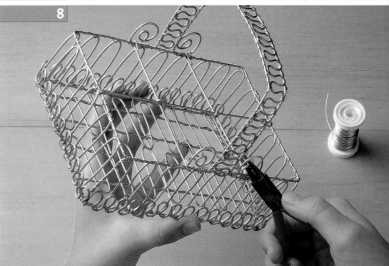

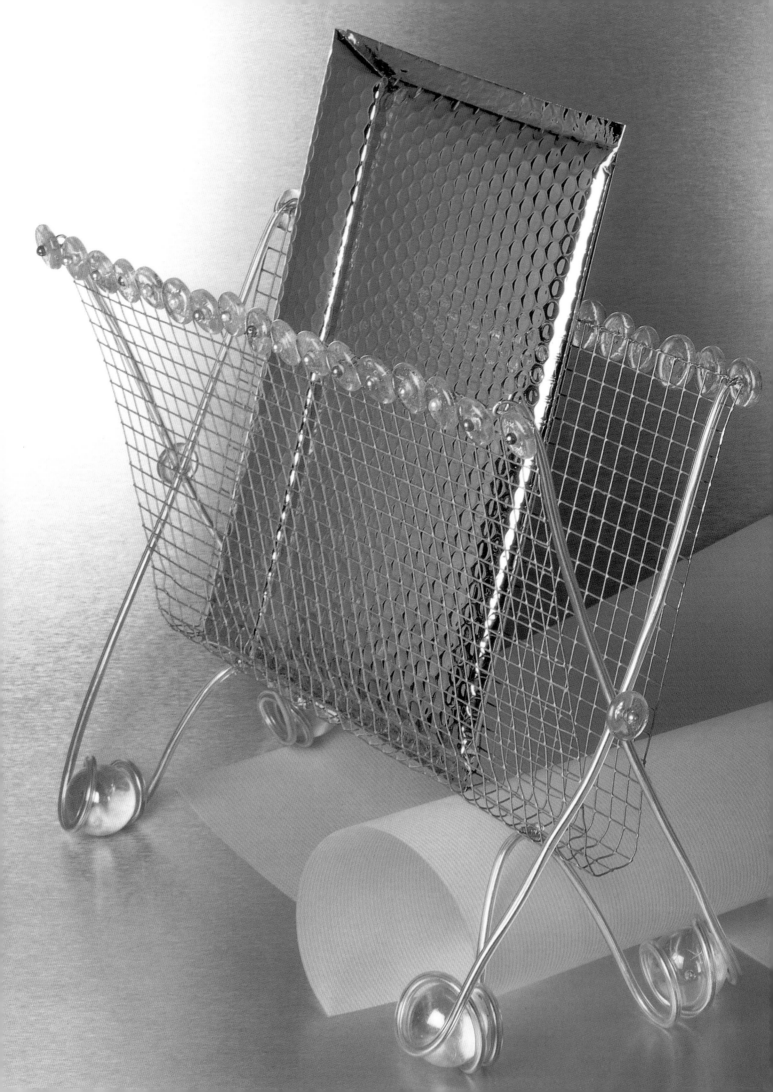

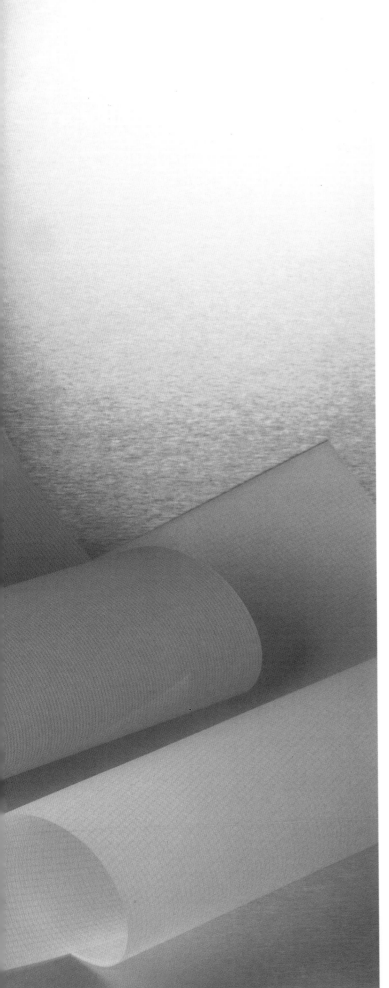

MAGAZINE RACK

Metal grating has a certain

industrial chic that, when married

with plastic tubing, iridescent

marbles, pearls, and beads,

produces a stylish design with

novelty and elegance. Metal

grating is available from garden

centers and is easy to work with,

though you may wish to wear

gloves to protect your hands from

scratches. Thick wire threaded

through plastic tubing gives

substance to the legs of the rack

and holds large marbles that are

heavy enough to anchor the

structure. The grating can be

colored using metallic spray paint.

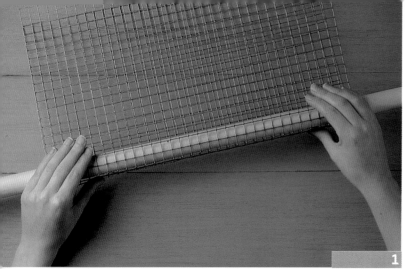

MATERIALS

Galvanized metal grating with ¹/₄in (6mm) grid

Spray paint (optional)

10ft (3m) plastic tubing (6mm diameter)

2.5mm galvanized wire

Fine binding wire

Glass disk beads

Pearl beads

4 large marbles

Strong glue

TOOLS

Wire cutters

Metal file

Long piece of thick dowel

Round-nose pliers

Chain-nose (snipe-nose) pliers

1

Cut out a rectangle of wire grating with wire cutters. The size of the grating will depend on the size of magazine rack you require. File the edges until smooth. Spray paint if desired and allow to dry. Bend the grating in half around a piece of dowel.

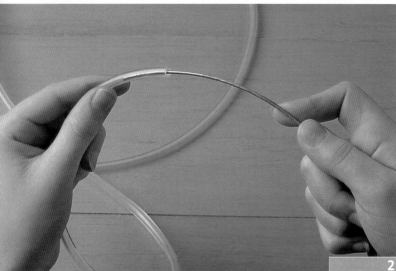

2

Cut two 5ft (1.5m) lengths of plastic tubing. Thread a length of 2.5mm wire through the center of each tube, leaving around 1in (2.5cm) excess wire at each end. Bend both wire tubes in the middle to form V shapes.

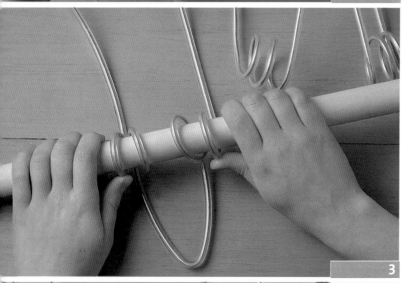

3

Measure 4¹/₂in (12cm) from the bottom of the V shape, then wrap each arm three times around the dowel to produce a pair of short coils, as shown. When you have finished, the arms of the V shape should point in the same direction as the base of the V. The coils will form the feet of the magazine rack.

4

Cross the arms of each V shape and secure in the center with binding wire. Decorate both joins by threading a disk bead and a pearl bead onto the binding wire and attach securely. Use round-nose pliers to form the four ends of the wire arms into loops.

5

Click a large marble into each coiled wire foot of the magazine rack. If necessary, put some strong glue on the underside of each marble and allow it to dry securely in place.

6

Make a decorative edging along the top of both sides of the grating by binding disk and pearl beads along them, as shown.

7

Attach the grating to the legs by slightly bending the inverted V of each leg inward and binding it to the base of the grating about five squares in from the edge, as shown.

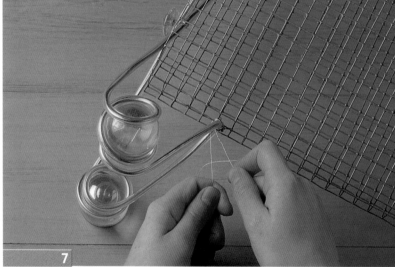

8

Hook the looped ends of the legs through the top edge of the grating and squeeze the loop firmly closed with chain-nose (snipe-nose) pliers. The structure should now be stable.

TEA GLASSES

Simple, inexpensive drinking glasses have been transformed into beautiful tea glasses by the addition of an elegant wire holder, making them ideal for fruit teas, mulled wine, or hot toddies. The flat shape of the wire surround is achieved by hammering galvanized wire. This is a simple technique that produces unusual and stylish results. For a more elaborate effect, try using colored glass and hammered copper wire.

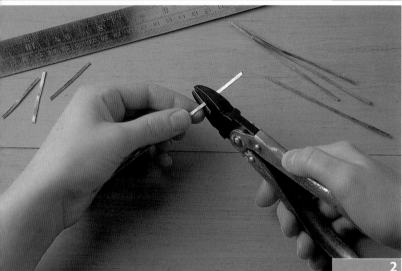

MATERIALS
2.5mm galvanized wire
Glasses

TOOLS
Sheet of metal (to hammer onto)
Hammer
Wire cutters
Chain-nose (snipe-nose) pliers

1

Place the sheet of metal on a sturdy surface and hammer along the length of the wire until it is sufficiently flat. It is best to do this in small sections because it is easier than handling long lengths of wire.

2

Cut the following lengths of wire for each glass: one to go around the top of the glass and one to go around the base (allow extra length for overlapping the ends); four lengths the same height as the glass to create two small ladders; two lengths to form a longer ladder for the handle (each one approximately twice the height of the glass); and six short lengths to create the rungs of the ladders. Add an extra 3/4in (2cm) to the length of each ladder piece to allow for creating hooks to join them together.

3

First, assemble two small ladders, using the four glass-height lengths of wire and four of the short pieces of wire. Use chain-nose pliers to bend a small hook at each end of the longer wires, then use two short pieces of wire to join pairs of the longer wires together to form a small ladder. Use pliers to bend the ends of the short wires securely around the longer wires.

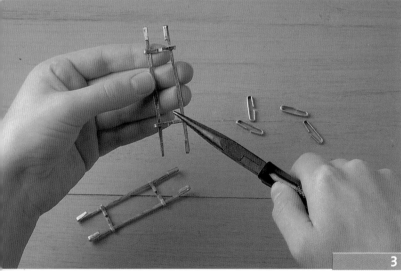

4

Make two circles of wire, one for the top of the glass and one for the bottom. Leave the excess wire in place because they will be cut later.

5

Use pliers to bend a small hook at each end of the handle wires, then use your fingers to form the wires into the desired shape. Make sure that both pieces are the same and join them together in the same way as the ladders in step 3.

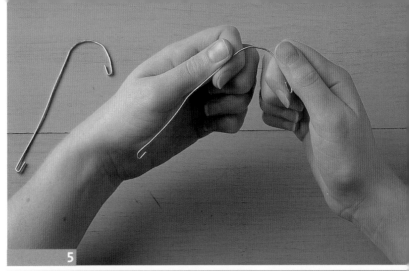

6

Using the glass to gauge the size, attach the top band of the glass holder to the handle by threading the ends of the band through the hooks of the handle. Use pliers to bend the ends back over the handle hooks to secure.

7

Attach the bottom of the handle in the same way, then cut off the excess wire from both the top and bottom bands. Use pliers to squeeze the ends firmly to create a secure fixing.

8

Attach the two ladders to the top and bottom band in the same way. Make sure the ladders are evenly spaced and that the ends are well squeezed together.

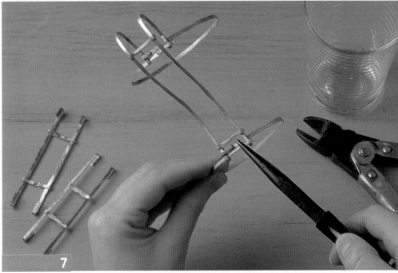

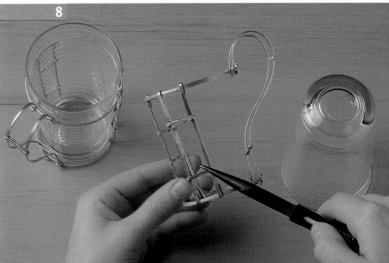

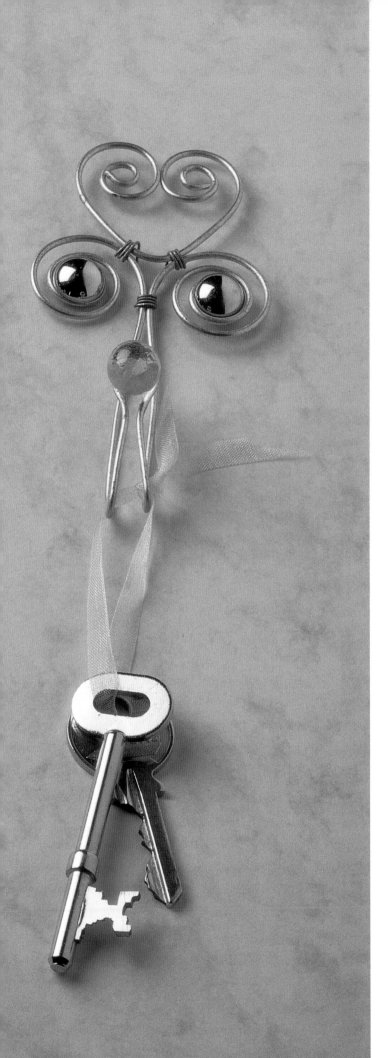

WALL HOOKS

These wall hooks are both

decorative and functional.

They are quick and easy to

make, so this project is ideal

for beginners—all you need

is some galvanized wire and

a pair of round-nose pliers to

form these attractive shapes.

When attaching the hooks to

the wall, use screws with a

decorative head for best

effect. If you wish, you can

adapt the design to make a

hanging rack by binding a row

of hooks to a wire frame.

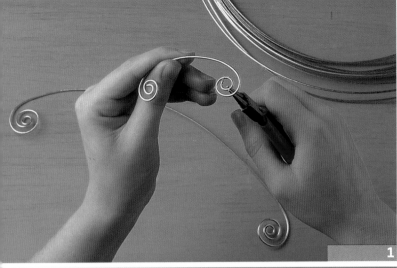

MATERIALS
1.6mm galvanized wire
Fine binding wire
Strong glue
Small glass marbles (1 per hook)
Screws (2 per hook)

TOOLS
Wire cutters
Round-nose pliers
Pencil

1

Cut a 20in (50cm) and an 8in (20cm) length of wire. Use round-nose pliers to shape both lengths of wire into double-ended spirals (see Basic Techniques, page 13).

2

Holding the shorter wire in both hands, place a finger midway along the section of wire and bend the spirals toward each other to form a heart (see Basic Techniques, page 13).

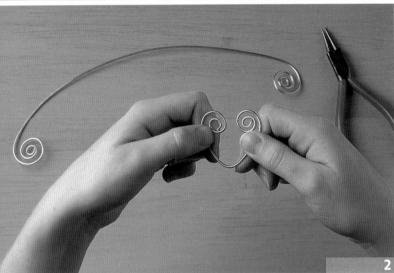

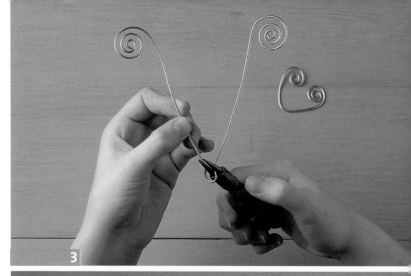

3

Find the midpoint of the longer wire, and using pliers, create a small loop by bending the wire around the pliers, as shown.

4

Place the fingers of both hands between the arms of the wire V shape and press your thumbs just beneath the spirals to bend the wire inward. The fingers will resist the bend, causing the two arms of the V shape to bow out slightly.

5

Place a pencil across the wire approximately 1in (2.5cm) up from the bottom loop and bend the wire around the pencil to form a hook.

6

Use fine binding wire to join the two long arms of the hook together just beneath the pair of spirals. Then attach the heart shape to the hook section by wrapping binding wire around the top of each spiral and the base of the heart, as shown. Use strong glue to attach a marble to the loop at the end of the hook. Attach the hook to the wall by inserting a screw through the center of each of the larger spirals.

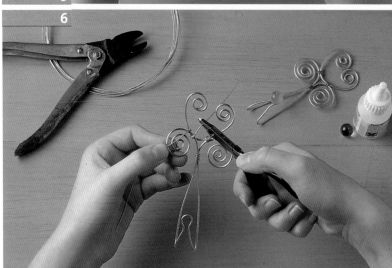

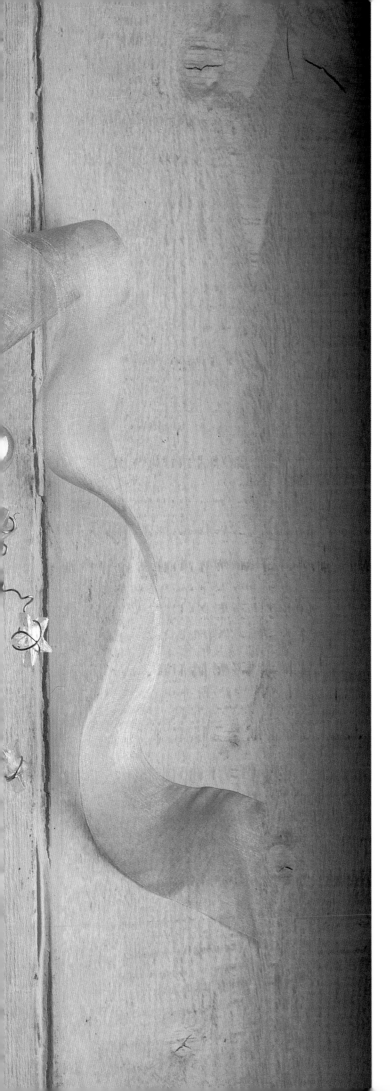

WIRE WREATH

This lovely wreath made from

interconnecting spirals of coiled

wire is the perfect winter

wonderland decoration for

Christmas festivities. Wire

springs randomly shoot out

glass stars, and iridescent

Christmas baubles are securely

suspended within the wire

loops. A sparkling ribbon adds

to the exuberance. Hang the

wreath on a door in deep

midwinter and let the snow

and frost complete the effect.

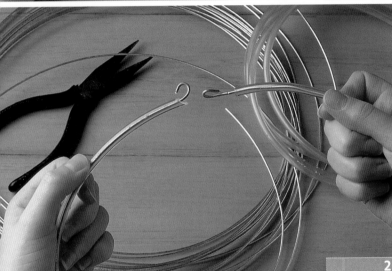

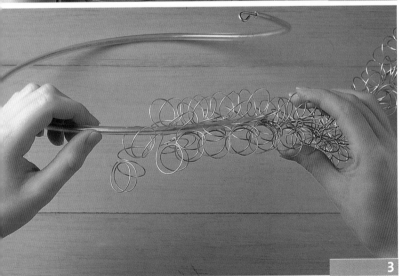

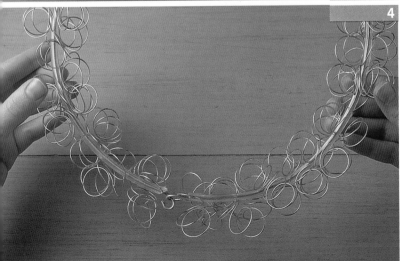

MATERIALS

0.9mm and 1mm galvanized wire

32in (80cm) plastic tubing

Glass stars

Strong glue

Small glass Christmas baubles on wires

Ribbon

TOOLS

1/8in (4mm) and 3/4in (2cm) dowel

Wire cutters

Round-nose pliers

1

Make two coiled spirals using 0.9mm galvanized wire (see Basic Techniques, page 17). Use the narrow dowel to make one of the spirals and the thick dowel to make the other. Make another coil of the same length using the narrow dowel but only go as far as step 1 in the Basic Techniques. Use your fingers to stretch the coil into an extended spring; do not bend the coils in the way that you would to form a coiled spiral.

2

Thread 1mm galvanized wire through a 32in (80cm) length of plastic tubing. Use pliers to form a loop at each end of the wire, as shown.

3

Carefully thread the wire tubing through the center of the larger coiled spiral. If this is difficult, you may prefer to wrap the coil spirally around the tubing.

4

Form the wire tubing into a circle and join the interlocking loops together. Squeeze the loops closed with round-nose pliers.

5

Cut the extended spring into 5in (13cm) lengths.
Wrap one end of each length around a small glass
star. Use pliers to make sure they are held securely,
and glue them if necessary.

6

Embed the smaller coiled spiral within the wire loops
of the wreath by pushing it in so that it is held between
the larger loops.

7

Attach the stars randomly around the wreath. Use the
ends of the wires holding the stars to secure the smaller
coiled spiral in place by binding the wires around both
the spirals and the wreath's core.

8

Bind Christmas baubles randomly around the wreath.
These should fit snugly between the larger loops, with
their wire ends entwined around the wreath's core.
Add a decorative bow to the top, binding it in place with
some wire.

WINDOW
SCREEN

When light shines through this

window screen, beautiful colored

patterns are reflected into the

room through the pretty glass

beads. Make sure you measure

the window frame accurately

before you start, and work out

the pattern of the beads using

colored pencils and graph paper.

Choose beads that look good

when held up to the light.

Make sure that all the elements

are secure before fixing the

screen to the window frame

using screws through the

corner loops.

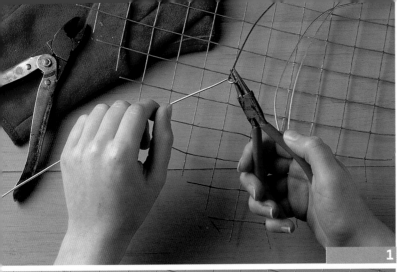

MATERIALS

Galvanized metal grating with 1in (2.5cm) grid
1.5mm galvanized wire
Frosted glass star
Fine binding wire
Frosted glass beads in various sizes
Frosted glass droplets
Strong glue
Four screws

TOOLS

Wire cutters
Round-nose pliers
Miricle pliers
Graph paper and colored pencils

1

Measure the window frame where you are going to hang the screen; double-check to make sure you have done so accurately. Carefully cut the metal grating with wire cutters to the correct size; it should fit comfortably into the space. Cut a length of galvanized wire long enough to go all the way around the edge of the grating with a little extra for the corner loops. Use round-nose pliers to form the corner loops, as shown.

72

2

Bind the wire frame around the grating by bending the cut edges of the grating around the frame using miricle pliers. Make sure it is secure and that all the wire ends are firmly bent around the frame.

3

Work out the center point of the grating and cut out a section large enough for the star to fit in.

4

Suspend the star in the hole by using fine binding wire to attach it securely to the surrounding grating.

5

Attach the beads to the grating according to the pattern you wish to create; it is best if you design it on graph paper first using colored pencils that match the different colors of the beads you are using. Here, one bead is suspended per square to form a diamond pattern on the screen. To suspend the beads, use a doubled length of fine binding wire. Wrap each strand of the wire around a corner of the square, thread both strands through the bead, then separate the strands again and wrap them around the opposite corners. This will suspend the bead in the center of the square.

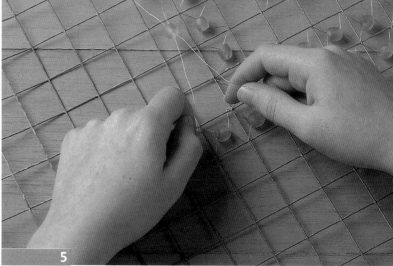

6

Position the glass droplets as desired, binding them into place and neatly tying the binding wires at the back of the screen. Secure them with some strong glue if necessary. Screw the finished screen into place on the window frame.

BEADED BOWL

Beaded wire has been fashioned into a lace-effect pattern to transform this plain glass bowl into a delicate centerpiece, perfect for a summer party. Cone-shaped beads are used for the feet because of their flat bottoms, but other shapes could be substituted if you prefer. Make sure the wires do not protrude from the base of the feet or they will scratch the surface of the table. Remember, too, that the bigger the bowl, the longer the beaded wire must be.

MATERIALS

Glass bowl
0.6mm galvanized wire
Frosted glass seed beads
Frosted glass round beads
Frosted glass cone-shaped beads
Strong glue

TOOLS

Wire cutters
Round-nose pliers

1

Choose the bowl you wish to decorate with beaded wire. Thread seed beads onto a long piece of wire, leaving some wire unbeaded at the end to allow the beads to move when you bend the wire into shape. Count the number of beads required to make a petal shape; this will depend on how large you want the petals to be. Starting with a petal, form the top edging of the bowl by looping and twisting the beaded wire to form petals at regular intervals. This example has 22 beads per petal, 8 beads between each petal, and 20 petals in total.

2

Once you have sufficient edging to surround the top of the bowl, cut off any excess wire and use pliers to form a small loop at the end. Link this around the base of the first petal and squeeze closed.

3

Thread beads onto another length of wire to form the scallops between every alternate petal. Count the number of beads required to make each scallop; this example has 24 beads per scallop. Make a space between the beads at the end of each scallop and twist the wire around the neck of the petals to connect them.

4

Form a circle of beaded wire to fit around the base of the bowl. Use pliers to form a small loop at each end of the wire and link them together, squeezing the loops closed to secure.

5

Cut lengths of wire to make the side struts of the bowl; you will need one strut per scallop. Cut the wires 1in (2.5cm) longer than the depth of the bowl measured from the center of the scallops to the bottom ring of beads. Use pliers to form a small loop at one end of each wire. Thread a round frosted bead onto each wire, then enough seed beads to reach to the bottom ring.

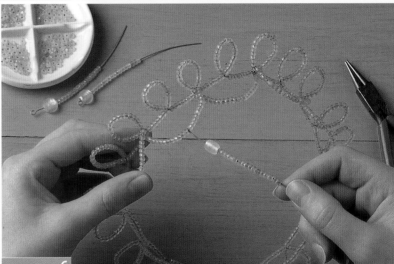

6

Hook the top loop of each wire strut onto the center of a scallop, then squeeze the loop closed with pliers. Bend the other end of each strut slightly to keep the beads from falling off.

7

Loop the end wires of each strut around the beaded base ring at regular intervals. You should be left with some unbeaded wire projecting below the base ring for each strut.

8

Thread an equal number of seed beads onto each of these wires to make the feet; two beads per foot in this example. Trim the excess wire, leaving just enough wire to slide a conical foot bead onto the end. Make sure no wire protrudes through the cone beads or they will scratch the surface where you place the bowl. When you are satisfied, use strong glue to attach the cone beads securely in place. Put the glass bowl inside the beaded frame and arrange the petals decoratively.

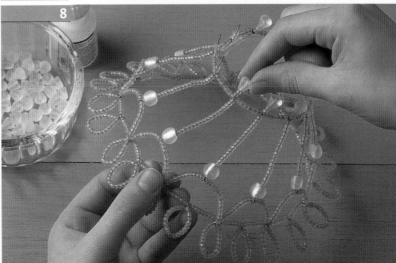

EVENING PURSE

It is hard to believe that this

glittering purse is made from

wire. Once knitted, the wire

becomes surprisingly soft and

tactile. The ends of the wires are

woven into the knitting and

used to bind the back and front

together, so there are no loose

ends. It is lined with purple

velvet, making it suitable for

carrying light items such as

lipstick. This bag has been

made in rich reds and purples,

but you could use other colors if

you prefer.

MATERIALS

Enameled copper wire in red, crimson, violet, wine,
and purple
Approximately 440 red beads
Approximately 960 beads in autumnal colors
Purple velvet
Needle and thread
Snap fastener (press stud)
Cord for handle
Tassel and bead

TOOLS

3mm and 4mm knitting needles
Scissors

1

Thread 440 beads onto the red wire, 202 beads onto the
crimson wire, 132 beads onto the violet wire, 78 beads
onto the wine-colored wire, and 108 beads onto the
purple wire. Use red beads on the red wire and
autumnal colors on the other wires. Add a few extra
beads to each wire to allow for errors.

2

Follow the pattern, keeping the stitch gauge (tension) as
even as possible. To work beads into a stitch, draw a bead
up the wire to the front of the work, knit the following
stitch, then draw up another bead to the front of the
work and knit the next stitch, and so on.

3

Join in new colors as when knitting with yarn or cut
the old color leaving a short length that you can twist
together with the new color close to the knitting.

4

To get the perfect finished shape, gently squeeze the
wire at the edges, using either your fingers or pliers.
Make sure all loose ends are worked in at this stage.

5

With wrong sides together, join the front and back pieces of the bag using matching wire. Make sure you join the lengths of wire securely, then oversew the seams.

6

Use the knitted bag as a template to cut two pieces of velvet lining. With right sides together, sew around the bottom edges using backstitch.

7

Turn the lining bag right side out and place inside the knitted bag. Turn over the top unbeaded edge of the wire bag and oversew it to the lining.

8

Sew on the snap fastener, cord handle, bead and tassel.

PATTERN

Cast on 44 stitches using 3mm needles and red wire. Knit 5 rows with no beads. Knit 10 rows with a bead between each stitch. Change to 4mm needles and crimson wire. Knit 1 row with no beads. On next and every alternate row, add a bead in each stitch; decrease 1 stitch at each end of next and every following 4th row until there are 38 stitches (10 rows). Change to violet wire. Knit 1 row with no beads. On next and every alternate row, add a bead in each stitch; decrease 1 stitch at each end of next and every alternate row until there are 30 stitches (8 rows). Change to wine-colored wire. Knit 1 row with no beads. On next and every alternate row, add a bead in each stitch; decrease 1 stitch at each end of next and every alternate row until there are 24 stitches (6 rows). Change to purple wire. Knit 1 row with no beads. On next and every alternate row, add a bead in each stitch; decrease 1 stitch at each end of next and every alternative row until 1 stitch remains. Bind off (cast off). Repeat pattern for the back of the bag without beads.

RADIANT
MIRROR

This wonderful surround,

made from galvanized wire

and a selection of pretty

beads, transforms a plain

mirror into a thing of

beauty. Frosted glass beads

have been used here to

complement the glass. Small

triangular edging beads are

perfect for the borders. You

can easily adapt the design

for mirrors of other shapes,

or to use as a picture frame.

This mirror originally had a

plastic frame, which was

removed with pliers.

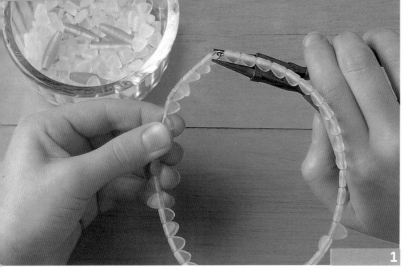

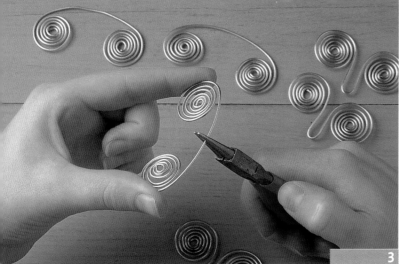

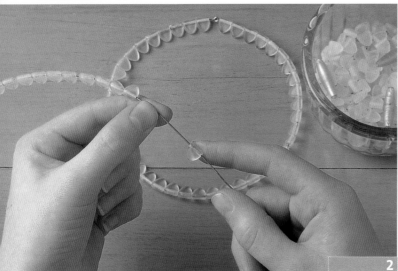

MATERIALS

Mirror without a frame

1mm galvanized wire

Frosted glass triangular edging beads

Frosted glass long beads

Fine binding wire

Enameled copper wire

TOOLS

Ruler

Wire cutters

Round-nose pliers

1

Measure the circumference of the mirror and cut a piece
of wire a little longer than this. Thread triangular edging
beads onto the wire, making sure the beads are not too
tightly packed because you will need space between
them for binding later. Use pliers to form a small loop
at each end of the wire. Link them together and squeeze
the loops closed.

2

Measure the radius of the circle you have just made,
then add to it the length of two long beads. This is the
required radius for the outer circle. Make the outer circle
in the same way as the inner circle.

3

Form 12 double-ended spirals (see Basic Techniques,
page 13). Grip the center of each wire firmly with
pliers and push the spirals backward to achieve
the shape shown.

4

Thread a piece of wire through each long bead and use pliers to form a hook at each end. Position a long bead centrally between the spirals of each wire shape. Position a bead and a spiral between the inner and outer frames. Fix them in place by squeezing the outer hook closed around both the outer frame and the wire shape. Squeeze the bead's other hook around the inner frame. Position beads and spirals evenly all around the frame.

5

Use fine binding wire to secure the spirals to the inner frame, wrapping the wire around the base of the spiral and between pairs of triangular beads.

6

Place the mirror in position on the back of the frame. Make a triangular support from enameled copper wire, binding it around the base of the long beads at appropriate intervals. Form a loop for hanging, as shown.

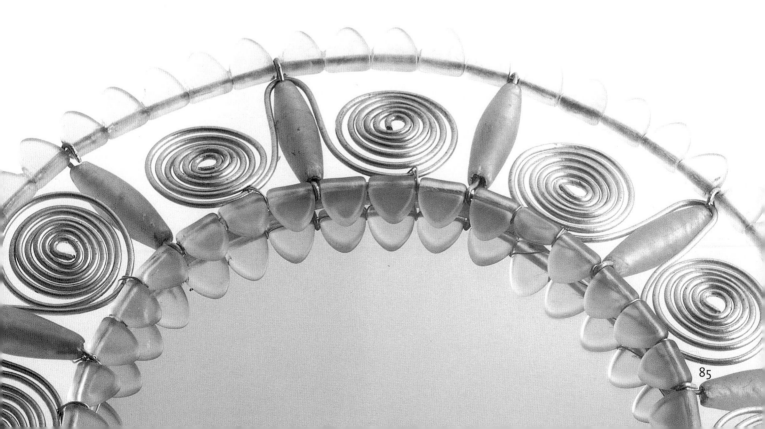

CANDLESTICKS

These distinctive candlesticks
are made from a stack of frosted
glass droplets stuck to the base
of a drinking glass and wrapped
randomly with florist's wire.
A small section of copper pipe
acts as a candleholder. You could
use sea-washed glass collected
from a beach in a similar way.
Make sure you use a strong
adhesive and allow each section
to dry before adding the next one.

MATERIALS

(for each candlestick)

Florist's wire

2 shot glasses, one slightly larger than the other

Small section of copper pipe

Round frosted glass droplets

Strong glue

TOOLS

Miricle pliers

Round-nose pliers

1

Wrap wire around the larger shot glass to create a conical coil. Do not worry about wrapping the wire around the glass neatly; the candlestick looks best with a more haphazard arrangement of coils. Remove the finished cone of wire from the glass.

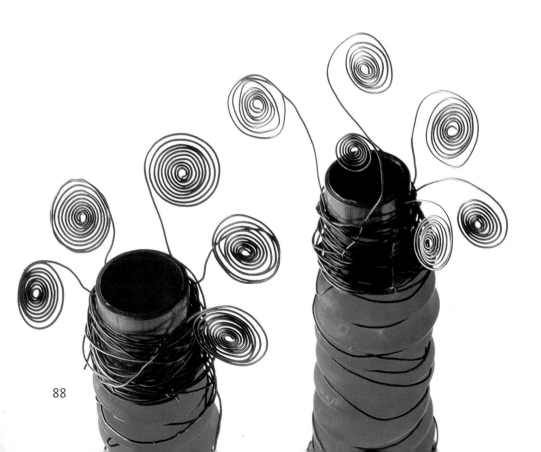

2

Coil wire around a short section of copper pipe. Make several wire spirals to decorate the top of the candlestick. Use miricle pliers to start the spiral but finish it by hand; florist's wire is soft enough to manipulate by hand and the slightly uneven spirals this produces are ideal.

3

Stick a glass droplet onto the bottom of the smaller shot glass using strong glue. Let it dry thoroughly, then glue another glass droplet on top of the first one. Again, let it dry before gluing another on top. Continue until you have reached the required height.

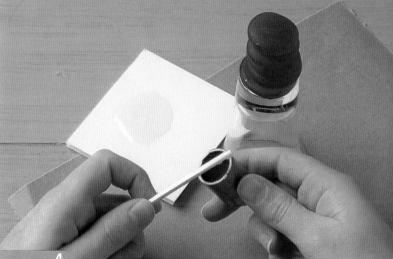

4

Glue the copper pipe on top of the tower of glass droplets and let it dry.

5

Place the large cone of wire over the candlestick. Spread out the top coils of wire so that they wind around the tiers of glass droplets at the top of the candlestick. Put the small coil of wire over the copper pipe, then bind the two coils of wire together.

6

Insert the ends of the wire spirals through the coils of wire around the copper pipe. Use round-nose pliers to bind the ends in place. Bend the spirals at attractive angles.

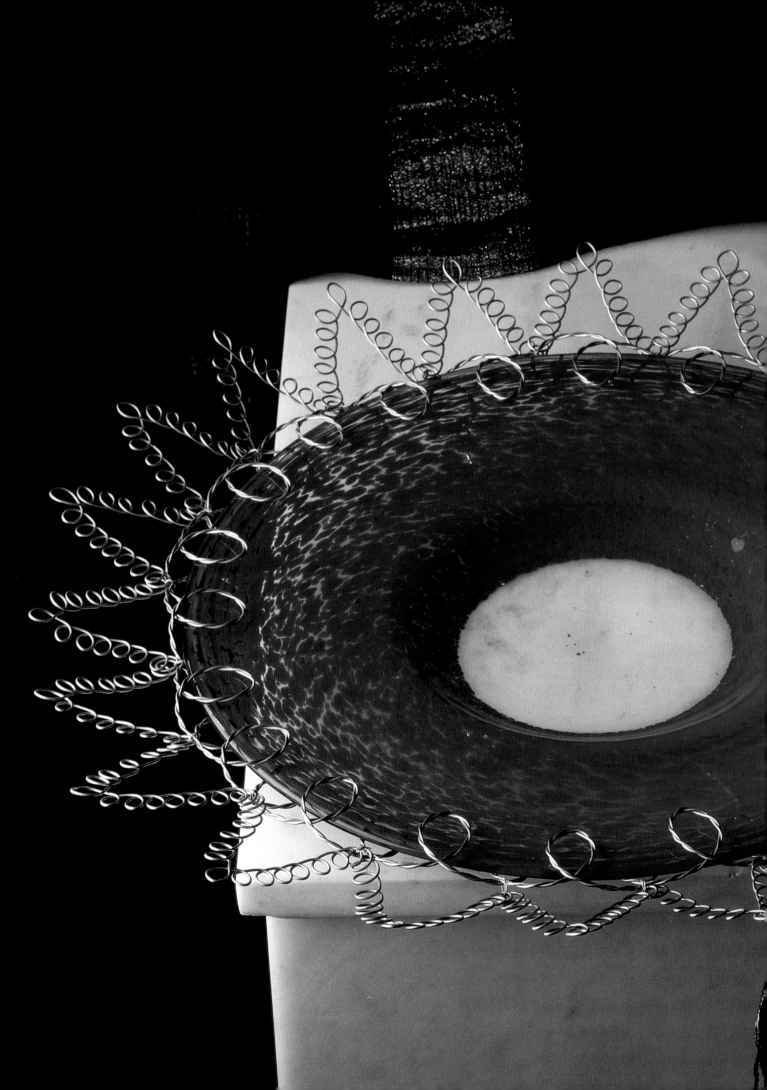

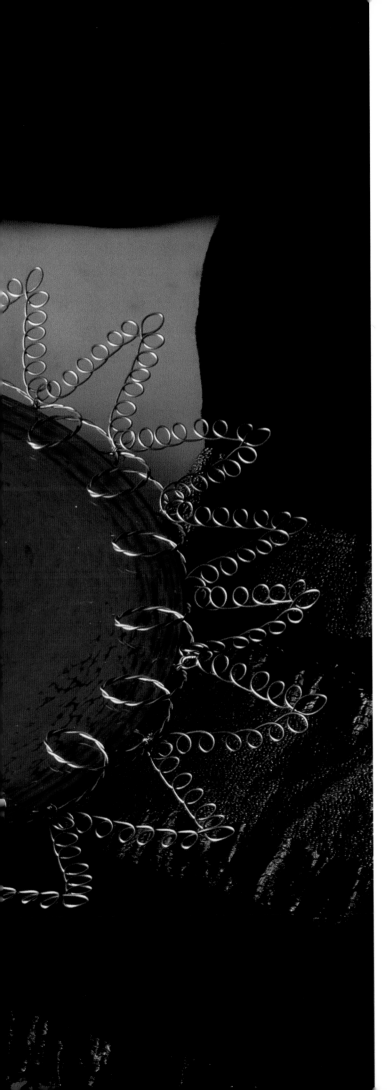

PLATE
SURROUND

This surround can be used
to add interest to a plain
plate or to frame a special
one for hanging on the wall.
The surround is made from
flattened coils of wire that
grip the edge of the plate.
Once you understand how
the surround functions, you
can try some variations,
perhaps incorporating other
decorative wire shapes or
using enameled copper wire
to add color.

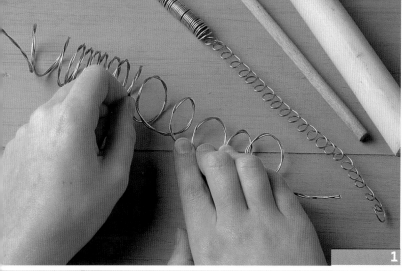

MATERIALS
0.9mm galvanized wire
Fine binding wire

TOOLS
Wire cutters
$1/2$in (1cm) and $1^1/2$in (4cm) dowel
Chain-nose (snipe-nose) pliers

1

Make two flattened coils of wire (see Basic Techniques, page 14). Use the narrow dowel to make one of the coils from a single strand of wire, and the thicker dowel to make the other coil from two strands of wire twisted together. The length of the coils will depend on the size of the plate you wish to surround. Allow a length of expanded flattened coiled wire approx $2^1/2$ times the circumference of your plate for the edging.

2

Bend the narrow flattened coil into a star-shaped edging that will fit around the circumference of the plate. Join the ends of the wire together by making small interlocking loops with pliers.

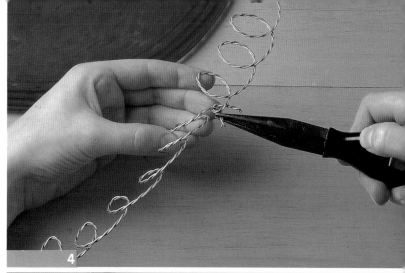

3

Insert the edge of the plate between the loops of the larger flattened coil, with the loops positioned alternately above and below the edge of the plate. When you have bent the coil into the correct size and shape, remove the plate and join the ends of the wire together by making small interlocking loops with pliers.

4

Position the star-shaped edging around the circular edging. Join the two elements together using fine binding wire at each point of contact. Tie the ends together securely and trim any excess wire.

5

If you want to hang the plate, make a hanging bracket by shaping a circle of wire to fit around the base of the plate. Use pliers to form a small loop at each end of the wire and link them together, squeezing the loops closed to secure. Cut five lengths of wire long enough to span between the base ring and the inner edge of the decorative surround. Use pliers to form a hook at both ends of each wire. Attach one end of four of the wires to the base ring, squeezing the hooks closed. Bend the remaining wire to form a U-shaped hanging hook.

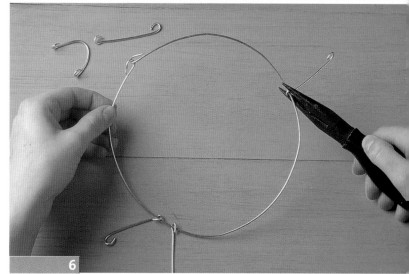

6

Insert the plate into the decorative surround, then turn it over and position the base ring on the back. Connect the loose ends of the four wires on the base ring to the decorative frame, squeezing the hooks closed. Attach the hanging loop to the base ring, as shown.

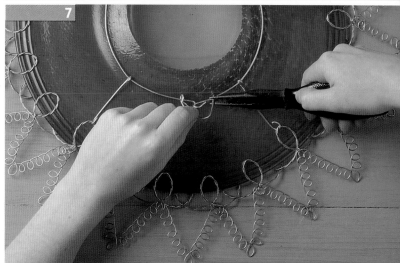

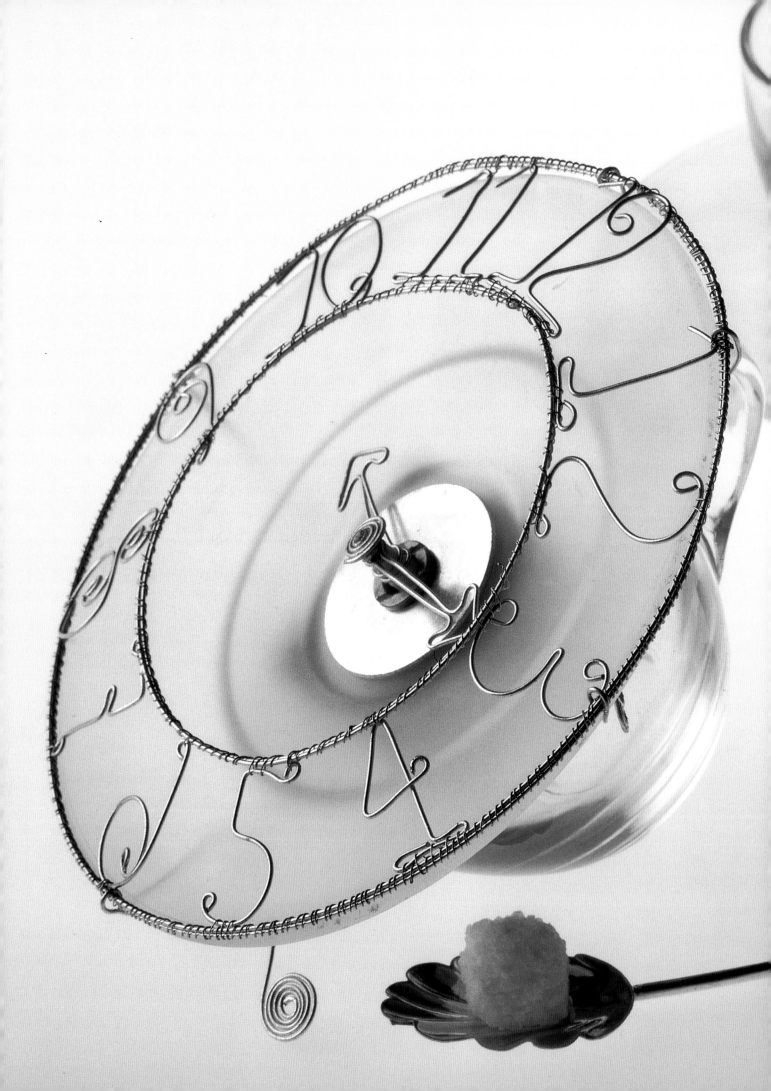

CONTEMPORARY CLOCK

This contemporary-looking clock is cunningly constructed using an ordinary glass teacup and saucer. Clock mechanisms can be purchased from craft stores or specialist suppliers, or you can dismantle a ready-made clock. Make sure the central shank is long enough to project through the saucer, taking the depth of the saucer into account. Glass drill bits are available from hardware stores.

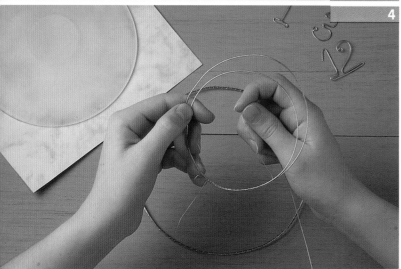

MATERIALS

Glass teacup and saucer

0.9mm galvanized wire

Clock mechanism

Fine binding wire

TOOLS

Pencil, paper, and compass

Wire cutters

Round-nose pliers

Chain-nose (snipe-nose) pliers

Glass drill bit and drill

Masking tape

1

Place the saucer face down on a piece of paper and draw around it. Find the center point of the circle and use a compass to draw an inner circle approximately 3in (7.5cm) smaller than the first one. Fold the paper into quarters through the center point. Sketch the numbers of the clock face between the two circles in such a way that each one is formed from one continuous line. Make sure the numbers are evenly positioned around the circle. Use round-nose and chain-nose (snipe-nose) pliers to shape the galvanized wire into numbers, using the numbers you have drawn as a template.

2

You may need to make several attempts before you achieve numbers you are happy with, but the end result will be worth it. Remember to make all parts of each number from a single piece of wire, including the numbers 10, 11, and 12.

3

Remove the hands from the clock mechanism, taking note of how they are assembled. Shape two arrows from galvanized wire to form new hands; remember to make the minute hand longer than the hour hand. Make sure that their bases fit into the respective hour and minute positions on the shank.

4

Form a circle of wire to fit the rim of the saucer with the ends overlapping by at least half the circumference. Join the ends with fine binding wire. Make another circle to fit the inner ring on the template, again with the ends overlapping.

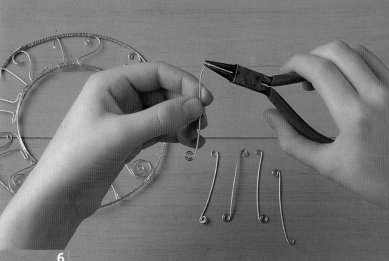

5

Wrap fine binding wire around the smaller circle, incorporating the bases of the numbers in the correct positions. Wrap binding wire evenly all around the circle, in between the numbers as well along the numbers' bases. Bind the outer circle into position around the top of the numbers in the same way.

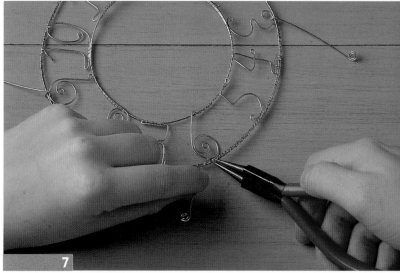

6

Make brackets to connect the wire clock face to the cup and saucer by cutting five lengths of wire long enough to reach from the rim of the saucer to the rim of the cup when the saucer is sitting on the cup. Use round-nose pliers to form a hook at one end of each wire and a spiral at the other end, as shown.

7

Attach the hooks of two of the wires to the outer circle of the clock face on either side of the number 12. Attach the hook from one of the remaining three wires to the numbers 3, 6, and 9. Squeeze the hooks closed around the rim to secure.

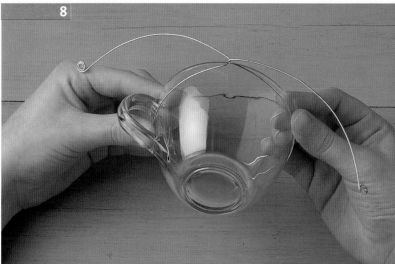

8

Cut some wire long enough to wrap twice around the rim of the cup. Use round-nose pliers to form a small spiral at each end. Thread the wire through the cup handle and around the cup. Twist the wire with a half turn to secure it tightly around the cup.

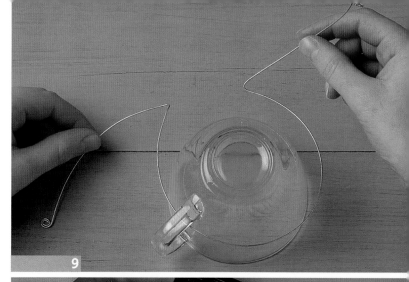

9

Unfurl the twist in the wire and remove it from the cup. This wire will act as a base for attaching the clock face and saucer to the cup later. Put the cup and the wire ring to one side.

10

Using a glass drill bit of the correct size for the clock shank, carefully drill a hole in the center of the saucer. Remember to place some masking tape over the area to be drilled and observe the safety precautions outlined (see Basic Techniques, page 17).

11

Insert the shank of the clock mechanism through the hole in the saucer from the back and attach it with the connecting screw. The mechanism should be held securely in place. Lay the saucer on top of the wire clock face, making sure they are both face down. Position the cup on top of the saucer with the handle at the 12 o'clock position. Put the wire ring around the cup once more but do not hook it in place. Pull back the wire brackets from the rim of the clock face and thread them through the wire ring around the cup, bending each one back as shown.

12

Hook the ends of the wire ring together around the cup at the 6 o'clock position. Wind the excess wire into equal-sized spirals to act as the feet for the clock to rest on.

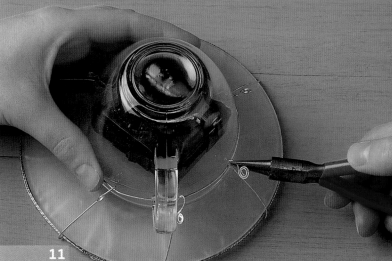

HANGING RACK

The hooks of this funky rack are
made from large glass marbles
securely bound into a sturdy wire
framework. A large version of the
classic marble is used here, but
there are many other types
available from toy stores and
specialist suppliers. When attached
to a wall, it makes an effective hat
or towel rack, but the design can
easily be adapted to make a bigger
or smaller rack as desired.

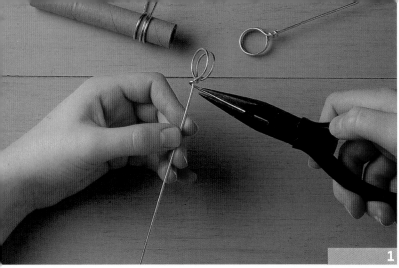

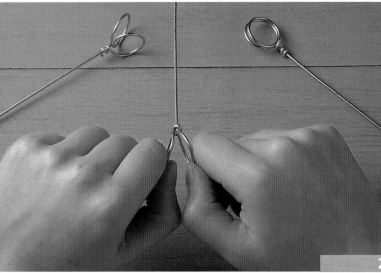

MATERIALS

0.8mm and 1.6mm galvanized wire
6 regular-sized marbles
Fine binding wire
3 large marbles
3 screws

TOOLS

Wire cutters
2 different thicknesses of dowel (to make holders
for the large and regular marbles)
Chain-nose (snipe-nose) pliers
Round-nose pliers

1

Cut three 24in (60cm) lengths of the thickest galvanized
wire. Make a holder for the regular sized marbles at both
ends of each piece of wire by winding the end of the
wire twice around a suitably sized piece of dowel or
similar item. Finish by wrapping the end of the wire
around the base of the marble holder using chain-nose
(snipe-nose) pliers. Cut off any excess wire.

2

Spread the two rings of each marble holder apart and
click a marble into place between them. Repeat this at
each end of all three pieces of wire.

3

Cut an 18in (45cm) length of thick galvanized wire and
use round-nose pliers to form three small loops evenly
spaced along its length. These will be used to house the
screws that will hold the rack to the wall. Bend the ends
of the wire upward into a neat curve, as shown.

4

Join the four lengths of wire together loosely with fine
binding wire. The binding wire will be removed later but
helps to keep four wires in place while you are working

with them. These four wires form the main horizontal rail of the hanging rack.

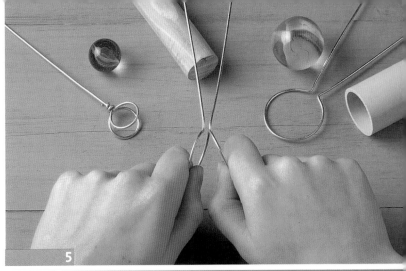

5

Make three holders for the large marbles from the thick galvanized wire. Make a split ring for each marble as described in steps 1 and 2, but do not bind around the base of them yet. Make a single wire hoop for each marble by wrapping a length of wire once around the dowel used in step 1.

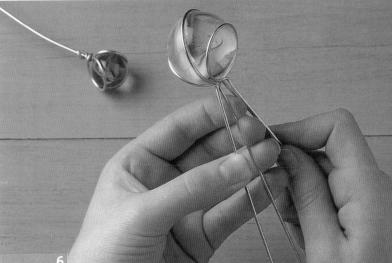

6

Click the marble into place between the each split ring and then slip a wire hoop over the middle of the marble.

7

Tightly bind the four wire ends of each large marble holder together using the finer galvanized wire. These will form the shanks of the hooks.

8

Bind each shank until it is long enough to form a hook, then bend the remaining loose wires apart at an angle of 90 degrees. Trim these wires to about 1in (2.5cm). These will later be bound into the horizontal rail of the hanging rack.

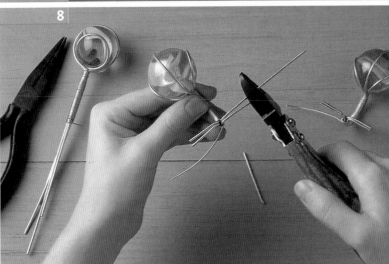

9

Form each hook into the curved shape by bending it around the thick dowel.

10

Bind the finished hooks to the horizontal rail using the narrow galvanized wire, positioning each hook directly below a screw loop.

11

Continue binding the hooks in place, removing the fine binding wire you used to hold the four wires of the horizontal rail together as you go.

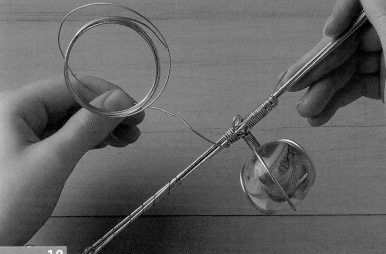

12

Spread out the decorative end wires of the horizontal rail and screw the finished rack to a wall.

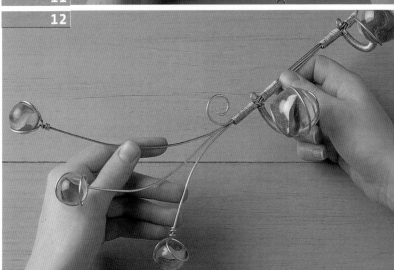

ARABESQUE LAMPSHADE

The lampshade is made from

lots of individual elements,

so this is a project for the more

experienced wire worker.

Each element is not difficult

to make, but time and

patience are needed when

binding them together to

construct the shape. Use beads

to embellish the wire and add

color. The lampshade looks

best around a soft-colored

bulb or placed over a small

ready-made shade.

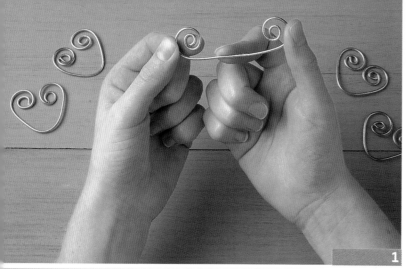

MATERIALS

1mm and 1.5mm galvanized wire
Fine binding wire
Assortment of beads
Lamp fitting
Jewelry jump rings

TOOLS

Wire cutters
Variety of pliers, including round-nose
Various thicknesses of dowel, including 2¹/₂in (6cm)
Gig (see page 15)

1

Cut eight 8in (20cm) lengths of 1.5mm wire. Use pliers to form each wire into a double-ended spiral, then bend them into heart shapes (see Basic Techniques, page 13).

2

Cut thirty 10in (25cm) lengths of 1.5mm wire and use pliers to form them into scrolls (see page 13). Make ten into short scrolls (see left-hand side of picture) and twenty into long scrolls (see right side).

3

Cut ten 13in (33cm) lengths of 1.5mm wire. Form each length into a double-ended spiral, as before. Push the ends of the spirals together to form a gentle dome shape (see left-hand side of picture). Cut eleven 11in (28cm) lengths of 1.5mm wire and form each one into an open spiral approximately 1¹/₂in (4cm) in diameter (see Basic Techniques, page 16). Bend the last bit of wire into a loop in the opposite direction (see right-hand side of picture).

4

Use fine binding wire to suspend beads inside each heart shape. Insert beads onto the center wire of the open spirals and short scrolls. You may have to cut off a little wire at the center to do this.

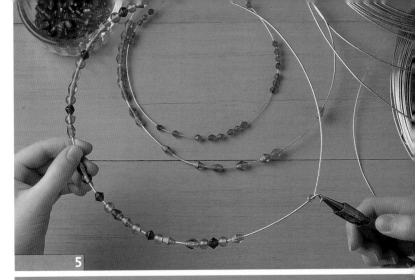

5

Cut 20in (50cm), 24in (60cm), and 27in (68cm) lengths of 1.5mm wire. Thread approximately 30 large beads onto the longest piece, 20 large beads onto the medium piece, and 25 small beads onto the shortest piece. Bend all three wires into circles, using pliers to form interlocking loops at the ends. Squeeze the loops closed.

6

Make a flattened coil 24in (60cm) long and 1in (2½cm) wide using 1.5mm wire (see Basic Techniques, page 14). Use 1.5mm wire to make a length of wiggly wire, 23in (58cm) long and in ¾in (2cm) wide (see page 15).

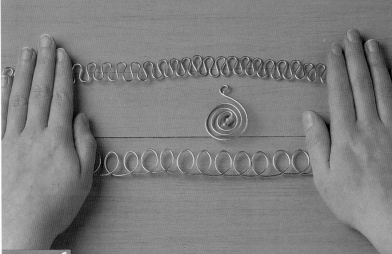

7

Coil some 1.5mm wire around 2½in (6cm) dowel or something similar. Remove the coil from the dowel and extend the loops out flat, as shown, until you have a 10-petaled flower shape. Use pliers to form a small loop at each end of the wire, link them together, and squeeze the loops closed. This will form the top of the lampshade.

8

Cut a piece of 1.5mm wire long enough to wrap around the lamp fitting two or three times. Wrap the wire around the fitting and use pliers to bend the ends of the wire out on opposite sides at a 45-degree angle. You may find this easiest with parallel pliers. Cut off the excess wire, leaving about ½in (1cm) on each side. Use pliers to form the ends into loops.

9

Position the wire lamp fitting in the center of the petaled

9

top section, connecting the links of the
wire fitting to the petals and squeezing them closed
to secure.

10

Place the largest beaded hoop around the petaled top
section. Position the domed spirals at the tip of each
petal, binding them onto the hoop and petaled top
section with short lengths of 1mm wire. Make sure you
arrange the beads evenly between the shapes around
the hoop.

11

Place a short scroll between pairs of long scrolls, binding
them together with 1mm wire at their contact points.

12

Bind the medium-size hoop to the base of the domed
spirals using 1mm wire. Space the beads evenly, as
shown. Next, bind the three-part scrolls you formed in
step 11 to the medium-sized hoop using 1mm wire.

13

Using fine binding wire, attach the smallest hoop to

10

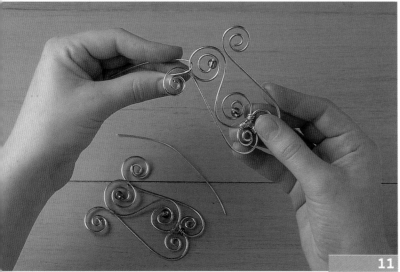

11

12

the base of the scrolls, making sure that the beads are evenly spaced.

14

Bind the flattened coil to the small hoop, then attach the open spirals to the base of the coil with fine binding wire. Attach the length of wiggly wire beneath these spirals, as shown.

15

Cut eight 2in (5cm) lengths of 1mm wire. Use pliers to form a hook at one end, suspend a droplet bead from this, and squeeze the hook closed. Thread a bead onto the remaining wire and form a loop at the top, cutting off any excess wire. Link a beaded wire to the base of each heart, then squeeze the loop closed.

16

Attach the beaded hearts to the bottom of the wiggly wire using jump rings, as shown. Make sure you squeeze the jump rings firmly closed. You can dangle more beads from the wire lampshade for additional decoration if you wish; look at the picture on page 106 for inspiration.

INDEX

RESOURCES AND ACKNOWLEDGEMENTS

THE SCIENTIFIC WIRE COMPANY
enamelled wire etc.
18 Raven Road
London E18 1HW
Tel: 020 8505 0002
wire@enterprise.net

SORO LIMITED
large range of wire
Shaire Hill Industrial Estate
Saffron Waldon
Essex CB11 3AQ
Tel: 01799 506 099

BEAD EXCLUSIVE
4 Samara Park
Cavalier Road, Heathfield
Newton Abbot
Devon TQ12 6TR
Tel: 01626 834 934
Fax: 01626 834 787
bead.exclusive@virgin.net

HS WALSH
tools: round-nose and parallel pliers etc.
21 St. Cross Street
Hatton Garden
London EC1N 8UN
Tel: 020 7242 3711

HOUSE OF MARBLES
marbles
The Old Pottery
Pottery Road
Bovey Tracey
Devon TQ13 9DF
Tel: 01626 835 358
Fax: 01626 835 315
www.houseofmarbles.com

Thank you to Andrew Gillimore for his inspired projects and general help: name plaque, glass carrier, tea glasses and hanging rack.
Thanks to 'The Holding Company', 0207 352 7495 for the silver mesh bags shown on page 42.